Proceedings in Life Sciences

Mitosis
Facts and Questions

Proceedings of a Workshop held at the
Deutsches Krebsforschungszentrum,
Heidelberg, Germany, April 25—29, 1977

Edited by
M. Little, N. Paweletz, C. Petzelt, H. Ponstingl
D. Schroeter, H.-P. Zimmermann

Administrative Editor
Vera Runnström-Reio

With 55 Figures

Springer-Verlag
Berlin Heidelberg New York 1977

M. Little, N. Paweletz, C. Petzelt, H. Ponstingl,
D. Schroeter, H.-P. Zimmermann
Institut für Zellforschung
Deutsches Krebsforschungszentrum
Postfach 10 1949
6900 Heidelberg, FRG

ISBN 3-540-08517-3 Springer-Verlag Berlin Heidelberg New York
ISBN 0-387-08517-3 Springer-Verlag New York Heidelberg Berlin

Offsetprinting and bookbinding: Brühlsche Universitätsdruckerei Lahn-Gießen.
2131/3130-543210

Preface

Two years ago, about twenty people gathered informally
in our institute to discuss mitosis. We took this
opportunity to try to separate the "hard" facts of
mitosis which are accepted by most people, from the
"soft" ones which are still open for discussion.
Surprisingly few "hard" facts survived with their
reputation still intact. This result led us to organize a similar meeting on a larger scale. The outcome
was the workshop "Mitosis: Facts and Questions", which
was held at the German Cancer Research Center in
Heidelberg from April 25-29, 1977. An introductory
lecture was given for each of nine major topics,
followed by an extensive discussion of facts, questions
and future experiments. Further details were provided
by posters.

The proceedings of the meeting are published in this
volume. We feel that many open questions and facts
described here will provide stimulating ideas and a
basis for further investigation of this fundamental
process.

The success of the workshop would not have been
possible without the help of many people. We are very
grateful to the German Cancer Research Center for its
interest and assistance, and for the support of the
Verein zur Förderung der Krebsforschung in Deutschland
represented by Prof. Dr. h.c. K.H. Bauer, the ECBO
(European Cell Biology Organization) and the Deutsche
Gesellschaft für Zellbiologie. Our sincere thanks are
also extended to our students and technicians for their
enthusiastic help, and to Mrs. Joa for typing the
manuscripts.

Heidelberg, September 1977   M. LITTLE, N. PAWELETZ,
                                C. PETZELT, H. PONSTINGL,
                                D. SCHROETER,
                                H.-P. ZIMMERMANN

Contents

Creanor, 1971a) but recent evidence from autoradiographs after pulse labelling of DNA indicates a shorter G1 of ca. 0.1 of the cycle and a longer S of 0.2 (K.A. Nasmyth, unpublished results). With these short G1 and S periods, the cell spends most of the cycle (ca. 0.7) in G2.

Although this yeast is less well known than the budding yeast *Saccharomyces cerevisiae*, it has been used extensively for genetic work (Gutz et al., 1974) as well as for cell cycle studies (Mitchison, 1970).

II. Cell Size Homeostasis

One clue to division control comes from considering the size of a cell at division. Although this size shows a good deal of variation, the mean tends to stay constant for many micro-organisms and some higher cells that are growing exponentially under any one set of culture conditions. This suggests that there is some homeostatic mechanism that regulates cell size and maintains the constant mean value. In principle, there are two ways in which this mechanism could work. The first would be to have a constant cycle time and a variable growth rate that was inversely proportional to the size of a daughter cell at the start of the cycle. Large daughter cells would grow slowly and so reduce their size to the mean value at the end of the cycle. The second would be to have a constant growth rate and a variable cycle time that was inversely proportional to daughter size. Large daughter cells would then have short cycle times.

S. pombe is a good organism in which to explore this question since it is easy to follow the growth of single cells with time-lapse microphotographs and to get an estimate of cell size from a single measurement of cell length. These advantages have been exploited by James et al. (1975) and their results have been confirmed and extended by Fantes (1977). It is clear that size homeostasis is maintained by the second of the two mechanisms above. Large daughter cells have short cycles and small daughter cells have long cycles. The quantitative relations are such that the normal size variations can be compensated within about one cycle. It is also possible to make abnormally large cells by holding a temperature-sensitive (TS) cell cycle mutant for a period at the restrictive temperature. All the TS cell cycle mutants that have been isolated in *S. pombe* continue to grow at the restrictive temperature even though division is blocked (Nurse et al., 1976). These abnormally long cells bring out another interesting aspect of the size/cycle time relationships. Above a particular size the cycle time is scarcely shortened, however long the cell. In other words there is a minimum cycle time (about 75 % of normal in these conditions) that acts as a limit to the shortening of cycle time with long cells. As a result, it takes more than one cycle for the homeostatic mechanism to reduce the size of these very long cells to normal.

Two other results from the work of Fantes are that cell size does not seem to be "inherited" (large cells do not produce daughters that are large when they in turn divide) and that there is no correlation between the size of daughters and their average growth rate during the following cycle.

The concept that emerges from this work is that cell size is regulated, within limits, by varying the cycle time. Put in another way, which is more significant for the theme of this article, the primary trigger for mitosis and cell division comes from a mechanism that measures cell size. Bearing in mind that words suchs as primary, trigger, and size are imprecise and to some extent "loaded", how could such a mechanism work? There is no shortage of models (see the review by

Fantes et al., 1975), and a simple one for a size control mechanism, colloquially a "sizer", is to have a pulse of a mitotic inhibitor produced after division. The cytoplasmic concentration of this inhibitor is reduced by growth in volume until it reaches a low enough level to allow the next mitosis.

Alternatively, an activator is produced at a rate proportional to cell mass and triggers mitosis at a critical concentration. These models are discussed later.

A "timer" is a parallel concept to a sizer, again with a degree of imprecision. In essence, it is a mechanism that ensures a constant absolute time between two biological events under certain conditions. There could be one process that determines this time or a set of sequential processes. The relevance of timers in the cell cycle is that a sizer could operate well before division and it would then be followed by a timer, which would set a fixed time interval between the moment the sizer was triggered and the act of division. This would introduce a constant lag time for the preparations for division to be completed. As we shall see, the concept of a sizer followed by a timer is an accepted view for bacteria. One measure of the imprecision of the timer concept is that it is not a "clock" of the type envisaged in the work on circadian rhythms, which, like a real clock, is independent of temperature. Cell cycle timers in bacteria are not temperature-compensated, but they do appear to keep a fixed time with different growth rates.

Returning to *S. pombe,* the results described so far suggest a sizer control operating on division but they do not show where in the cycle this control operates and whether there is a timer involved as well as a sizer. To resolve this question, we have to examine what happens with size mutants and with nutrient shifts.

III. Size Mutants and Nutrient Shifts

While isolating conditional mutants blocked in the cell cycle, Nurse also discovered a novel and very profitable group of mutants that are altered in cell size (Nurse, 1975; Nurse and Thuriaux, 1977). These are not blocked in the cell cycle and proceed through it with the same generation time as the wild type. But they are much smaller at all stages of the cycle (about half the protein and RNA content of wild type) and because of this and of their country of origin, they have been christened *wee* mutants. So far, this small phenotype has been shown in mutants of two independent genes, *wee* 1 and *wee* 2. The first mutant isolated, *wee* 1-50 (originally named *cdc* 9-50), is TS and exhibits its mutant phenotype at the restrictive temperature of 35°C. At the permissive temperature of 25°C, it is only slightly smaller than wild type.

A TS mutation of cell size at division is a powerful tool for studying the mechanisms of size control. Let us assume that the mutation affects the sizer so that it triggers division at the *"wee"* size rather than the larger normal size. If the sizer operates near nuclear division, the effect of shifting up the temperature from the normal to the restrictive should be a rapid decrease in the size of dividing cells in an asynchronous culture and an acceleration of the larger cells through G2 and into division. This will produce a semi-synchronous burst of nuclear division. If, on the other hand, the sizer operates earlier in the cycle and is followed by a timer, there should be delay equal to the timer period before size at division changes. The results of Nurse (1975) show that the first of these alternatives is what happens and support the concept of a control operating near the

time of nuclear division. It is possible, however, that the initial
assumption is incorrect and that the *wee* mutation shortens the timer
rather than affecting the sizer. This alternative can be examined by
the use of shifts in nutrients.

Fantes and Nurse (1977) have shown that cell size alters when the
growth rate is changed by using different nutrients. In general,
cell size diminishes as growth rate diminishes and cycle time increases. This also occurs in bacteria and here there is a neat explanation in terms of a sizer followed by a timer (Donachie et al.,
1973). The sizer initiates DNA replication at a constant size irres-
pective of growth rate. During the subsequent timer period, however,
cells in a poor medium grow less than cells in a rich medium and
therefore divide at a smaller size. Some of the evidence comes from
shift-up experiments when cells are transferred from poor to rich
medium, and from the reverse shift-down situation. When these experiments are done with *S. pombe*, the results are different from those
with bacteria and they are not consistent with the bacterial model.
The arguments are analogous to those used for the temperature shift
experiments with *wee* 1-50. After a shift-down, cells are accelerated
through G2 and into nuclear division, and size at division starts to
fall abruptly shortly after the change. The pattern is broadly similar
to what occurs in the *wee* 1-50 shift, and in this case there is no
genetic lesion. After a shift-up, there is a rapid inhibition of nuclear division followed a little later by a plateau in cell number and
then subsequently by a rapid rise in number and a sharp increase in
division size. This is to be expected from a sizer at nuclear division
that is reset by the nutrient change to operate at a larger size.

The combined evidence from the size mutants and the nutrient shifts
argues strongly for a sizer operating at the time of nuclear division
in the wild type and against the existence of a timer. A necessary
corollary, however, is that the size has to be modulated by the nutrient
conditions, and it has to be admitted that this lacks the elegant
simplicity of the bacterial model. It is also unclear what control is
exercised over division in the size mutants at the restrictive temperature.

The next stage is to try to identify the components of the sizer and
to understand how it triggers division. We have not got very far in
this but there are some clues in the genetic analysis of the *wee* genes
(Nurse and Thuriaux, unpublished results). A further search for small
mutant cells has produced 37 mutants alleles (independently isolated)
of the *wee* 1 gene. Only a few of these are TS, but all of them produce
small cells of about the same size. This suggests that the *wee* 1 gene
product is inactivated in the mutants and that its normal function
in the wild type is inhibitory and restrains the sizer from operating
until the normal division size has been reached. In marked contrast,
only one mutant of *wee* 2 has been found. This mutant also has an interesting and significant relation to one of the cell cycle genes *cdc* 2
whose gene product is required for nuclear division (Nurse et al.,
1976). *Wee* 2 maps very close to or within *cdc* 2. This suggests that
the *wee* 2/*cdc* 2 gene has the complex function of both controlling division and generating a product needed for division. One simple model
would be to have the *wee* 1 gene product binding reversibly to the
wee 2/*cdc* 2 gene product and inactivating it. The *wee* 1 product would
be diluted out by growth and would eventually release sufficient of
the *wee* 2/*cdc* 2 product to initiate nuclear division. *Wee* 2 would be
a rare mutation that would decrease the binding but still allow the
product to initiate division, whereas the other mutants of *cdc* 2
(eight TS mutants have been isolated) would stop initiation. Nutrient

modulation would come in through an alteration of the binding. I must stress that this is only a very provisional model and it may have become outmoded before this article is published. It does not fit all the facts and it assumes a connection between *wee* 1 and *wee* 2 that is not yet clarified; but it is an illustration of the kinds of models that we are considering.

IV. Size and DNA Synthesis

Sizers have been considered so far in their relation to nuclear division but one can ask whether the same type of control also operates in initiating DNA synthesis. The question arose originally because of the situation in *wee* 1-50 (Nurse, 1975). Although the cycle time of the small cells of this mutant at the restrictive temperature is the same as wild type, the position of the S period is not. The S period is centered at 0.3 of the cycle, as compared to 0.0 in wild type. G1 is therefore longer than in wild type and G2 is shorter. Two alternative explanations are either that the *wee* 1 gene has a pleiotropic effect and alters the time of DNA initiation as well as cell size at division or that another control becomes operative. The evidence of Nurse and Thuriaux (1977) suggests that the latter explanation is more likely and that a size control is involved. In the small cells of *wee* 1-50 at the restrictive temperature, DNA synthesis starts when the cell size reaches a value of 6-7 pg protein/cell. If a sizer controls DNA initiation and is set to operate at this size (i.e., this is the minimum size for initiation) then it should also operate in other types of small cells that are generated in the wild type by methods that are different from those that result in the expression of the *wee* 1 mutant phenotype. This is in fact what happens. Small cells can be made by three different procedures: (1) germinating spores; (2) reinoculation after nitrogen starvation; and (3) expression of the *wee* 2 gene. In all cases, the S period takes place at 6.0-7.5 pg protein/cell.

This suggests a sizer control on DNA synthesis, but there remains the problem of the S period in the normal wild-type situation that occurs at a much larger size of about 13 pg protein/cell. Here we can invoke a second type of control. The signal from the sizer has already been given but DNA synthesis does not occur until nuclear division has been completed. This argues a dependency relation that is borne out from the study of the TS conditional cycle mutants. The presence of a G1 shows that there may be an irreducible minimum time for the preparation of the S period, but recent evidence mentioned in Section I indicates that the G1 may be very short indeed.

We are left then with two modes of control of DNA initiation. There is a sizer that operates in small cells, but this becomes cryptic in the normal wild type and the S period then takes place as soon as is possible after nuclear division.

V. Imprecision

If there were a natural variation of growth rate between individual cells and an accurate sizer triggering division, one would expect all cells to divide at the same size but after varying cycle times. There is some evidence that size is less variable than cycle time. Fantes (1977) found a coefficient of variation of 6.6-7.8 % for length at division and of 13.7-14.0 % for cycle time. Earlier measurements with different strains and growth conditions bear this out [coefficients of division length of 7.9-9.1 (Mitchison and Creanor, 1971b) and of cycle time of 9.7-17.8 (Mitchison, 1975; Faed, 1959; Gill, 1965)]. Even so, there is a marked variation of size at division and this persists during the growth of a culture, so the sizer appears to be imprecise or sloppy. However, this statement must be a guarded one, since "size" is not itself a precise term. It should not have escaped the reader that I have loosely equated it with protein content, mass, volume, and length. For models that depend on achieving a critical concentration of an inhibitor or an activator, cytoplasmic volume is probably the best definition of size. Although cell length gives a rough measure of cytoplasmic volume in *S. pombe*, it does not give a precise one. A strict transformation of length into cytoplasmic volume would have to take into account the wall thickness, the rounded ends, the nuclear volume, and, most important, cell diameter, which does show changes during the cycle (Johnson and Lu, 1975). It is therefore conceivable, though not probable, that cytoplasmic volume at division is much more accurately controlled than length at division. Since the mechanism of the sizer is not clear, it is not worth discussing the reasons for its imprecision, though it is obvious that an inhibitor-dilution model would not be precise if only a small number of molecules were produced at each burst of synthesis (Sompayrac and Malløe, 1973).

It is not inappropriate to consider here the variation in the phases of the DNA cycle (G1, S, and G2). I suggested in Section VII that there may be a minimum length for G1, and there may also be a minimum length for G2. These two, together with S, would give the minimum cycle time described in Section II. It is, however, clear that G1 can be extended and that G2 can be both extended and shortened from its normal time, and proportion of the cycle (0.70). There is a lengthened G1 in *wee* 1-50 at the restrictive temperature (Nurse, 1975) and after temporary inhibition of DNA synthesis by deoxyadenosine (Mitchison and Creanor, 1971b). G2 is shortened in these two situations, and it can be lengthened after spore germination or after recovery from nitrogen starvation. What is not clear is how the normal variation in cycle time is distributed among the three phases, though it is unlikely to be concentrated in the short G1.

VI. Growth and Division

DNA synthesis, nuclear division, and cell division appear to be a dependent sequence of events. Each event does not occur unless the preceding event has been completed. Growth (the synthesis of most macromolecules) does not lie in this sequence since it has been known for many years that it will continue when DNA synthesis is blocked (e.g., Swann, 1957). I have formalised this in terms of two sequences, the growth cycle and the DNA-division or DD cycle, which are normally coupled but can be dissociated (Mitchison, 1971; 1975). Growth continues in *S. pombe* after the DD cycle has been blocked either by chemical inhibitors (Mitchison and Creanor, 1971b) or in TS cycle mutants (Nurse et al., 1976).

The two types of sizers that have been outlined for nuclear division and for DNA synthesis are controls that are exerted on the DD cycle by the growth cycle. Growth could be a smooth exponential increase limited by nutrients and the sizers would be sufficient to ensure that the periodic events of the DD cycle occurred at the right time. There would be no need to have any control working in the reverse direction from the DD cycle onto growth.

Growth in *S. pombe*, however, is not a smooth exponential process in many parameters, and this raises the question of whether there are periodic controls analogous to the sizers. Growth control is not strictly relevant to the theme of this Workshop, so I will only sketch the outlines of a picture that is far from clear at the moment. We thought some years ago that a number of enzymes were synthesised

periodically, as is DNA. We were misled by perturbations induced in
synchronous cultures and it now seems that 18 out of 19 enzymes examined are synthesised continuously (Mitchison, 1977b). Other parameters
of growth also increase continuously but careful examination shows that
this increase is not exponential. Instead there is an increase at a
constant rate (linear growth) until a point once per cycle where the
rate doubles. This pattern is shown by total dry mass (Mitchison,
1957), three enzymes (Mitchison and Creanor, 1969), ribosomal protein
and total RNA (Wain and Staatz, 1973), messenger and ribosomal RNA
(Fraser and Moreno, 1976), and CO_2 evolution in minimal medium (Creanor,
manuscript in preparation). One possible control mechanism is a gene-
dosage control. When the genes double during the short S period, the
rate of production of messenger RNA doubles. If the amount of messenger
RNA is rate-limiting for protein synthesis, the rate of protein synthesis should also double after a time lag (Fraser and Moreno, 1976).
This would be a control exerted by the DD cycle on the growth cycle.

Gene dosage, however, is not an adequate explanation for the results
on CO_2 evolution. The linear pattern, with a rate change once a generation time, continues after the DD cycle has been blocked by inhibitors of DNA synthesis and of nuclear division. The cells do not divide but they do continue to grow and so become abnormally large. The
control cannot therefore come directly from the DD cycle. Instead,
and tentatively, I would suggest that there is another sizer that
operates on growth or on some components of growth and causes rate
doublings. This is a situation where the growth cycle would be self-
regulating, though it is not impossible that one of the DD cycle
sizers could also operate on growth. A single mechanism, for example,
could trigger both nuclear division and a rate change in growth,
and would still be effective on growth when its effect on nuclear
division had been blocked by an inhibitor. This would be an "independent single timer (IST) sequence" (Mitchison, 1974).

VII. Principles

It may be helpful to illustrate the controls that have been discussed in Figure 1 and also set down a list of principles about
the control of the cell cycle in *S. pombe*. I must emphasise that these
are not laws or Euclidean axioms but rather a set of working hypotheses, and very much subject to change.

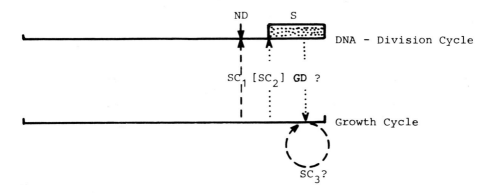

Fig. 1. Cell cycle controls in *Schizosaccharomyces pombe*. ND, nuclear division;
S, period of DNA synthesis; SC_1, size control on nuclear division; SC_2, size control
on DNA initiation (cryptic in normal wild-type cells); SC_3, size control on rate
changes in growth; GD, gene dosage control on rate changes in growth

1. Nuclear division is initiated in normal cells when they reach a critical size.
2. The mechanism that measures cell size and controls division: (a) operates at or shortly before nuclear division; (b) is altered by nutrients; (c) is altered by mutations in two genes, *wee* 1 and 2; (d) probably involves both an inhibitor and an activator.
3. This mechanism ensures a homeostatic control of cell sizer by a control on division. Cell size is not inherited.
4. Cycle time is controlled by this mechanism, but there is a minimum cycle time that cannot be shortened.
5. A similar mechanism for measuring cell size also operates on the initiation of DNA synthesis in small cells. But it is cryptic in normal cells.
6. These mechanisms are imprecise, so cell size at division varies.
7. Neither G1 or G2 is invariant. Both can be extended and G2 can be shortened.
8. Growth may be controlled either by gene dosage or by a size-measuring mechanism.

C. Discussion

The suggestion that nuclear division is triggered by attaining a critical cell size is by no means novel. It was put forward by Hertwig (1908) at the beginning of this century and it has been discussed many times since then (e.g., Swann, 1957; Mazia, 1961). Direct experimental evidence, however, is meagre. Prescott (1956), confirming earlier work of Hartmann (1928), showed that division in *Amoeba* could be stopped for many days by periodic amputations of cytoplasm. The implication is that the cells did not divide because they were never allowed to reach the critical size. Growth continued in the amputated cells but at a slower rate than normal. Prescott also showed that small and large daughter cells (generated by experimental treatment) divided at the same final size after cycle times that were inversely proportional to size. The *Amoeba* experiments are important and would be worth repeating in greater detail. Although the amputation results suggest a sizer operating during the long G2, they do not define exactly when it operates. A timer could be running during the period of several hours before mitosis when there is no growth in size in this organism.

The regulation of size by altering cycle time has been shown in a general way in several cell systems other than *S. pombe*. Mammalian cells continue to grow after division has been blocked by DNA synthesis inhibitors ("unbalanced growth"). When released from the block, the subsequent cycle is short (Galavazi and Bootsma, 1966) and cell size reverts to normal. The same happens in *Tetrahymena* after repetitive heat shocks (Zeuthen and Rasmussen, 1972). There is an equivalent phenomenon in *Physarum* and this has been analysed in detail in two interesting recent papers by Sudbery and Grant (1975; 1976), which follow the effects in subsequent cycles after ultraviolet irradiation and inhibitor treatments. The analysis is in terms of mechanisms regulating the ratio of DNA/total protein and there is no distinction between the regulation of mitosis and of the initiation of DNA synthesis since these two events are nearly coincident in *Physarum*, in which, like *Amoeba*, there is no G1. Sudbery and Grant conclude that their data do not fit many of the models considered in the earlier paper by Fantes et al. (1975) but that they are consistent with two of them. One is an unstable inhibitor of mitosis produced at a rate proportional to the amount of DNA. There is rapid turnover, so an equilibrium amount is reached rapidly. The inhibitor concentration is reduced by growth and dilution until a level is reached that triggers mitosis. This principle was first suggested by Yčas et al. (1965). The second model is a structural one that "counts" molecules. Sub-units are produced at a rate proportional to mass and are bound (reversibly) to sites to produce a "structure". When this structure is completed, mitosis is triggered. The units are not destroyed after mitosis but new sites are formed proportional to the amount of new DNA. This model is almost identical to one suggested earlier for *Physarum* by Sachsenmaier et al. (1972). It also has some similarities to the earlier model of "division proteins" suggested by Zeuthen and his colleagues for *Tetrahymena* (Zeuthen and Rasmussen, 1972). These two models fit the data from the experiments of Sudbery and Grant and also the important fusion experiments in *Physarum*, which show that a mitotic activator (or inhibitor) must be present in the cytoplasm (e.g., Rusch et al., 1966, Chin et al., 1972). However, as Sudbery and Grant (1975) point out, the *Amoeba* amputation experiment is much more easily explained by the unstable inhibitor model. This is also true of the experiments of Frazier (1973) in which DNA synthesis in *Stentor* is initiated prematurely by a decrease in the nuclear/cytoplasmic ratio through adding cytoplasm or removing parts of the nucleus.

The results of the *S. pombe* experiments do not provide definitive evidence for or against these models, but there are two points from our results that are relevant to the present dialogue on models: 1) The genetic evidence on the *wee* 1 locus suggests that there is at least one important inhibitory or negative control, and 2) the acceleration of cells through G2 and into nuclear division, which occurs after a temperature shift in *wee* 1-50 and after a shift-down in nutrients, argues against a structural model (Fantes and Nurse, 1977). If a mitotic structure has to be completed, not only would the rate of synthesis of sub-units have to increase but also this increase would have to be a transient one that occurred only for a short time after the shift. This is possible but unlikely.

The principle of a minimum cycle time that emerges from the results with abnormally large cells of *S. pombe* also applies to other cells in similar situations. Fantes et al. (1975) list in their Table 2 minimum cycle times for *Physarum, Saccharomyces, Amoeba,* and *Tetrahymena*. These times are between 50 and 67 % of the normal cycle time, but the authors suggest that with different growth rates the minimum cycle time is likely to be constant in time rather than being a constant proportion of the normal cycle. This is borne out in *Physarum* where Sudbery and Grant (1975) found a minimum cycle time of 6 h in a medium where the normal time was 8-9 h, and of 7 h in another poorer medium where the normal time was 16-17 h. Sudbery and Grant (1976) also found a cycle time of 7 h when there was no growth (in plasmodia that were irradiated and starved). Whatever therefore happens to growth, there is a "parallel pathway" involving the "preparations for division" that the cell has to complete before the trigger for mitosis can be pulled. These phrases, incidentally, all come from the classic work on mitosis by Mazia (1961). The minimum cycle time must include the S period and those parts of G2 (and G1 where it exists) that are incompressible. This perhaps does not say very much since we do not know what events are incompressible and, as we shall see below, the S period can be drastically shortened in early embryos. I have discussed the concept of parallel pathways elsewhere (Mitchison, 1974) and have mentioned one of the places where it applies in bacterial division (Donachie et al., 1973).

It would be cowardly to finish a discussion on size control at division without mentioning the embarrassing subject of eggs and early embryos. The cells here divide without growing, though some protein synthesis is necessary. As a result, they halve their size at each division, and size controls of the kind that have been discussed cannot be in operation. The easiest way out is to assume that they are running on the minimum cycle time, but this raises the problem that this time is much shorter than any found in adult cells. The S period in amphibian embryos is about 100 times shorter than that in adults (Mitchison, 1971). If the S period is short in embryos, why cannot it be compressed in adult cells? There are, as yet, no solutions to these problems, and it will need careful examination of embryos to determine when, if at all, a size control starts to operate. On present evidence, it does not seem as though there is an abrupt change from one type of control to another since Graham and Morgan (1966) have shown a steady increase in G1, S, and G2 from the 4th to at least the 18th hour of development in *Xenopus* endoderm cells. There is, however, an interesting transition in the axolotl blastula, which is worth further study. Signoret and Lefresne (1971) have shown that the early cell cycles are synchronous, short and relatively constant (coefficient of variation of about 4 %). After the tenth cycle, however, the cycles become asynchronous, longer, and more variable (coefficient of variation of 12-20 %). The transition is quite sharp and it is tempting to feel that a new pattern of cycle control is appearing at that point.

Size control for DNA initiation is well established in *Escherichia coli*, though it is still in dispute whether the mechanism involves an activator or an inhibitor (Donachie, 1968; Pritchard et al., 1969). What is clear is that new rounds of replication begin when the "initiation mass" reaches a critical value. This mass is independent of growth rate. Thereafter, there is a constant time (about 1 h) until division. This single size control regulates both DNA synthesis and division size. Initiation mass is not simply total cell mass but cell mass divided by the number of chromosome origins. Some analogous size control mechanism must presumably operate in eukaryotes since cell size usually increases with the degree of ploidy (Yčas et al., 1965). Diploid cells of *S. pombe*, for example, are nearly twice the size (protein content) of the normal haploid cells.

The situation in eukaryotes is less clear. A size control for DNA synthesis in *Tetrahymena* has been suggested by Worthington et al. (1976), and, as mentioned above, for *Stentor* by Frazier (1973). *Physarum* and *Amoeba* do not normally have a G1 and it has not been possible to generate one artificially. A size control for DNA initiation cannot therefore be separated from a size control for division. Mammalian cells usually have a G1 that is more variable in length than S + G2 (Prescott, 1976). It is attractive therefore to suggest that the main control point is at DNA initiation and that there is a constant timer running thereafter, as in *E. coli*. Could this control be a sizer? The evidence in favour is that the cell mass at DNA initiation is less variable than at mitosis (Killander and Zetterberg, 1965a, b). The evidence against is that the length of G1 is not correlated with cell size in certain experimental situations (Fox and Pardee, 1970; Fournier and Pardee, 1975). This evidence has been discussed by Nurse and Thuriaux (1977), and they suggest that the conflict could be resolved if the situation in mammalian cells resembled that in *S. pombe*. If the cells were small, there would be a size control. If they were large, the size control would be cryptic and the length of G1 would be reduced to its incompressible minimum.

S. pombe appears to have two size controls that can operate on the DD cycle whereas most of the models only involve one size control. Since, however, cytoplasmic inducers are components of the models, it is significant that there is evidence from mammalian cell fusion experiments that there are two inducers, one for DNA synthesis (Johnson and Rao, 1971) and another for mitosis (Rao et al., 1975).

The controls of both cycle time and division size are imprecise in *S. pombe*. The same is true in other cell systems. The coefficient of variation of cycle time in cultured mammalian cells ranges from 9 to 26 % with no obvious relation to the mean cycle time (data from Dawson et al., 1965; also Killander and Zetterberg, 1965a; Miyamoto et al., 1973). This range is similar to that in *S. pombe*. It is sometimes believed that the cycles of cultured mammalian cells are more variable than those of bacteria, which are thought to have a deterministic control mechanism. This is not so. Schaechter et al., (1962) found coefficients of variation of cycle time from 15 to 21 % in three bacteria. They also found, as in *S. pombe*, that the coefficients of variation of length at division were smaller - 8.5 to 13 %. The reason for these variations is not known, though those for cycle time are an essential component of one controversial theory for the control of mammalian cell cycles (Smith and Martin, 1974).

Acknowledgments. It is a pleasure to record my thanks to Dr. Paul Nurse and Dr. Peter Fantes for their comments and criticisms. Much of the work on *S. pombe* was supported by grants from the Science Research Council.

References

Chin, B., Friedrich, P.D., Bernstein, I.A.: Stimulation of mitosis following fusion of plasmodia in the myxomycete *Physarum polycephalum*. J. Gen. Microbiol. 71, 93-101 (1972)

Creanor, J.: Carbon dioxide evolution during the cell cycle of the fission yeast *Schizosaccharomyces pombe*. In preparation (1978)

Dawson, K.K., Madoc-Jones, H., Field, E.O.: Variations in the generation times of a strain of rat sarcoma cells in culture. Exptl. Cell Res. 38, 75-84 (1965)

Donachie, W.D.: Relationship between cell size and the time of initiation of DNA replication. Nature (London) 219, 1077-1079 (1968)

Donachie, W.D., Jones, N.C., Teather, R.: The bacterial cell cycle. Symp. Soc. Gen. Microbiol. 23, 9-44 (1973)

Faed, M.J.W.: Division and growth relationships in single cells. Ph. D. thesis, University of Edinburgh, 1959

Fantes, P.A.: Control of cell size and cycle time in *Schizosaccharomyces pombe*. J. Cell Sci. In press (1977)

Fantes, P.A., Grant, W.D., Pritchard, R.H., Sudbery, P.E., Wheals, A.E.: The regulation of cell size and the control of mitosis. J. Theo. Biol. 50, 213-244 (1975)

Fantes, P.A., Nurse, P.: Control of cell size at division in fission yeast by a growth-modulated size control over nuclear division. Exptl. Cell. Res. In press (1977)

Fournier, R.E., Pardee, A.B.: Cell cycle studies of mononucleate and cytochalasin-B-induced binucleate fibroblasts. Proc. Natl. Acad. Sci. U.S. 72, 869-873 (1975)

Fox, T.O., Pardee, A.B.: Animal cells: noncorrelation of length of G1 phase with size after mitosis. Science 167, 80-82 (1970)

Fraser, R.S.S., Moreno, F.: Rates of synthesis of polyadenylated messenger RNA and ribosomal RNA during the cell cycle of *Schizosaccharomyces pombe*. J. Cell. Sci. 21, 497-521 (1976)

Frazier, E.A.J.: DNA synthesis following gross alterations of the nucleocytoplasmic ratio in the ciliate *Stentor coeruleus*. Develop. Biol. 34, 77-92 (1973)

Galavazi, G., Bootsma, D.: Synchronization of mammalian cells *in vitro* by inhibition of DNA synthesis. II. Population dynamics. Exptl. Cell Res. 41, 438-451 (1966)

Gill, B.F.: The effects of ultraviolet radiation during the cell cycle. Ph. D. thesis, University of Edinburgh, 1965

Graham, C.F., Morgan, R.W.: Changes in the cell cycle during early amphibian development. Develop. Biol. 14, 349-381 (1966)

Gutz, H., Heslot, H., Leupold, U., Loprieno, N.: *Schizosaccharomyces pombe*. In: Handbook of Genetics. King, R.C. (ed.). New York: Plenum Press, 1974, vol. 1 pp. 395-446

Hartmann, M.: Über experimentelle Unsterblichkeit von Protozoen-Individuen. Ersatz der Fortpflanzung von *Amoeba proteus* durch fortgesetzte Regenerationen. Zool. Jahrb. 45, 973-987 (1928)

Hartwell, L.H.: *Saccharomyces cerevisiae* cell cycle. Bacteriol. Rev. 38, 164-198 (1974)

Hertwig, R.: Über neue Probleme der Zellenlehre. Arch. Zellforsch. 1, 1-32 (1908)

James, T.W., Hemond, P., Czer, G., Bohman, R.: Parametric analysis of volume distributions of *Schizosaccharomyces pombe* and other cells. Exptl. Cell Res. 94, 267-276 (1975)

Johnson, B.F., Lu, C.: Morphometric analysis of yeast cells. IV. Increase of the cylindrical diameter of *Schizosaccharomyces pombe* during the cell cycle. Exptl. Cell Res. 95, 154-158 (1975)

Johnson, R.T., Rao, P.N.: Nucleo-cytoplasmic interactions in the achievement of nuclear synchrony in DNA synthesis and mitosis in multinucleate cells. Biol. Rev. 46, 97-155 (1971)

Killander, D., Zetterberg, A.: Quantitative cytochemical studies on interphase growth. I. Determination of DNA, RNA and mass content of age determined mouse fibroblasts *in vitro* and of intercellular variation in generation time. Exptl. Cell Res. 38, 272-284 (1965a)

Killander, D., Zetterberg, A.: A quantitative cytochemical investigation of the relationship between cell mass and the initiation of DNA synthesis in mouse fibroblasts *in vitro*. Exptl. Cell Res. 40, 12-20 (1965b)

Mazia, D.: Mitosis and the physiology of cell division. In: The Cell. Brachet, J., Mirsky, A.E. (eds.). New York and London: Academic Press 1961, vol. 3, pp. 77-412

Mitchison, J.M.: The growth of single cells. I. *Schizosaccharomyces pombe*. Exptl. Cell Res. 13, 244-262 (1957)

Mitchison, J.M.: Physiological and cytological methods for *Schizosaccharomyces pombe*. In: Methods in Cell Physiology. Prescott, D.M. (ed.). New York and London: Academic Press 1970, vol. 4, pp. 131-165

Mitchison, J.M.: The Biology of the Cell Cycle. London: Cambridge Univ. Press 1971

Mitchison, J.M.: Sequences, pathways and timers in the cell cycle. In: Cell Cycle Controls. Padilla, G.M., Cameron, I.L., Zimmerman, A. (eds.). New York and London: Academic Press 1974, pp. 125-142

Mitchison, J.M.: Cell cycle control models. In: Growth Kinetics and Biochemical Regulation of Normal and Malignant Cells. In press, 1977a

Mitchison, J.M.: Enzyme synthesis during the cell cycle. In: Cell Differentiation in Microorganisms, Higher Plants and Animals. In press, 1977b

Mitchison, J.M., Creanor, J.: Linear synthesis of sucrase and phosphatases during the cell cycle of *Schizosaccharomyces pombe*. J. Cell Sci. 5, 373-391 (1969)

Mitchison, J.M., Creanor, J.: Further measurements of DNA synthesis and enzyme potential during the cell cycle of the fission yeast *Schizosaccharomyces pombe*. Exptl. Cell Res. 69, 244-247 (1971a)

Mitchison, J.M., Creanor, J.: Induction synchrony in the fission yeast *Schizosaccharomyces pombe*. Exptl. Cell Res. 67, 368-374 (1971b)

Miyamoto, H., Rasmussen, L., Zeuthen, E.: Studies of the effect of temperature shocks on preparations for cell division in mouse fibroblast cells (L cells). J. Cell Sci. 13, 889-900 (1973)

Nurse, P.: Genetic control of cell size at cell division in yeast. Nature (London) 256, 547-551 (1975)

Nurse, P., Thuriaux, P., Nasmyth, K.: Genetic control of the cell division cycle in the fission yeast *Schizosaccharomyces pombe*. Mol. Gen. Genet. 146, 167-178 (1976)

Nurse, P., Thuriaux, P.: Controls over the timing of DNA replication during the cell cycle of fission yeast. Exptl. Cell Res. In press (1977)

Prescott, D.M.: Relation between cell growth and cell division. II. The effect of cell size on growth rate and generation time in *Amoeba proteus*. III. Changes in nuclear volume and growth rate and prevention of cell division in *Amoeba proteus* resulting from cytoplasmic amputations. Exptl. Cell Res. $\underline{11}$, 86-98 (1956)

Prescott, D.M.: The cell cycle and the control of cellular reproduction. Advan. Genet. $\underline{18}$, 99-177 (1976)

Pritchard, R.H., Barth, P.T., Collins, J.: Control of DNA synthesis in bacteria. Symp. Soc. Gen. Microbiol. $\underline{19}$, 263-297 (1969)

Rao, P.N., Hittelman, W.N., Wilson, B.A.: Mammalian cell fusion. VI. Regulation of mitosis in binucleate HeLa cells. Exptl. Cell Res. $\underline{90}$, 40-46 (1975)

Rusch, H.P., Sachsenmaier, W., Behrens, K., Gruter, V.: Synchronization of mitosis by the fusion of the plasmodia of *Physarum polycephalum*. J. Cell Biol. $\underline{31}$, 204-209 (1966)

Sachsenmaier, W., Remy, U., Plattner-Schobel, R.: Initiation of synchronous mitosis in *Physarum polycephalum*. A model for the control of cell division in eukaryotes. Exptl. Cell Res. $\underline{73}$, 41-48 (1972)

Schaechter, M., Williamson, J.P., Hood, J.R., Koch, A.L.: Growth, cell and nuclear divisions in some bacteria. J. Gen. Microbiol. $\underline{29}$, 421-434 (1962)

Signoret, J., Lefresne, J.: Contribution a l'étude de la segmentation de l'oeuf d'axolotl: I - Définition de la transition blastuléenne. Ann. Embryol. Morphogenèse. $\underline{4}$, 113-123 (1971)

Smith, J.A., Martin, L.: Regulation of cell proliferation. In: Cell Cycle Controls. Padilla, G.M., Cameron, I.L., Zimmerman, A. (eds.). New York and London: Academic Press 1974, pp. 43-60

Sompayrac, L., Maaløe, O.: Autorepressor model for control of DNA replication. Nature (New Biol.) $\underline{241}$, 133-135 (1973)

Sudbery, P.E., Grant, W.D.: The control of mitosis in *Physarum polycephalum*. The effect of lowering the DNA: mass ratio by UV irradiation. Exptl. Cell Res. $\underline{95}$, 405-415 (1975)

Sudbery, P.E., Grant, W.D.: The control of mitosis in *Physarum polycephalum*: the effect of delaying mitosis and evidence for the operation of the control mechanism in the absence of growth. J. Cell Sci. $\underline{22}$, 59-65 (1976)

Swann, M.M.: The control of cell division: a review. I. General mechanisms. Cancer Res. $\underline{17}$, 727-758 (1957)

Wain, W.H., Staatz, W.D.: Rates of synthesis of ribosomal protein and total ribonucleic acid through the cell cycle of the fission yeast *Schizosaccharomyces pombe*. Exptl. Cell Res. $\underline{81}$, 269-278 (1973)

Worthington, D.H., Salamone, M., Nachtwey, D.S.: Nucleocytoplasmic ratio requirements for the initiation of DNA replication and fission in *Tetrahymena*. Cell Tissue Kinet. $\underline{9}$, 119-130 (1976)

Yčas, M., Sugita, M., Bensam, A.: A model of cell size regulation. J. Theo. Biol. $\underline{9}$, 444-470 (1965)

Zeuthen, E., Rasmussen, L.: Synchronized cell division in Protozoa. In: Research in Protozoology. Chen, T.T. (ed.). Oxford: Pergamon Press 1972, vol. 4, pp. 11-145

Discussion Session I: Cell Cycle, Timing of Events, Chromosome Cycle
Chairman: D. WERNER, Heidelberg, F.R.G.

D. Mazia, Berkeley, California, USA
An objective of this workshop is to list the *undisputed* facts about mitosis. Can we identify such undisputed facts about the relation between cell size and cell division?

J.M. Mitchison, Edinburgh, U.K.
There is an inverse relation between cell size and cycle time. That is one undisputed fact. There is also the principle of the minimum generation time.

J.R. McIntosh, Boulder, Colorado, USA
Hartwell has identified a cell division cycle mutant in *Saccharomyces* that fails to synthesize DNA but initiates a new bud every 90 min or so. Can this apparent timing be explained by growth control, or do you have to admit the existence of a timer?

J.M. Mitchison, Edinburgh, U.K.
No, you can explain it by growth control perfectly well.

P.F. Baker, London, U.K.
In considering the role of cell size, is it possible to distinguish between volume effects and surface area effects? Both presumably alter when a cell grows and the surface area/volume ratio may be a critical factor in determining the onset of division.

J.M. Mitchison, Edinburgh, U.K.
It is not certain what the critical size factors are which determine division. In the case of *Schizosaccharomyces pombe* one can only say that a correlation exists between cell length and division. In other organisms correlations have been found to other parameters such as DNA: mass ratios.

H. Ponstingl, Heidelberg, F.R.G.
I wonder whether we could agree on a sort of timetable. Starting with G1 we could go through the cell cycle phases and discuss the events which take place.

J.G. Carlson, Knoxville, Tennessee, USA
Morphologically a portion of G1 is occupied by decondensation of the telophase chromosomes. In the chinese hamster ovary cell this may represent up to half the duration of this phase of the mitotic cycle.

J.H. Frenster, Atherton, California, USA
In interphase lymphocytes, the administration of mitogens such as phytohemaglutinin and concanavalin convert the Go cells to the G1 phase. These cells then begin the synthesis of ribosomal RNA which was *not* synthesized in the Go cells. Messenger RNA, on the other hand, is synthesized throughout both Go and G1 phases. The transition between G1 and S phase seems to require clearing the monoribonucleotides found on the DNA template during the RNA synthesis of G1 phase and substituting monodeoxyribonucleotides characteristic of the DNA synthesis of S phase. This finding is based on the data that ribothymidine can mainly inhibit mitosis by blocking the transition from G1 to S phase, but that this inhibition can be reversed by much lower concentrations of thymidine.

C.A. Pasternak, London, U.K.
But I thought RNA synthesis continues during S phase.

J.H. Frenster, Atherton, California, USA
Not on the same locus that is being replicated. After replication you have a resumption of synthesis.

Chairman:
The G1/S transition is observed as the initiation of the first replication sites. Do any other events characterize this transition?

E. Jost, Heidelberg, F.R.G.
In HeLa red cell heterokaryons, it seems to be the case that the F2c histone has to be removed before the onset of the S phase. That is an example that the removal of one protein is obviously important for the onset of DNA synthesis in the red cell.

J.F. Lopez-Saez, Madrid, Spain
In my opinion, there is general agreement in considering the initiation of the S period as a positive, inducible event. In this sense, the studies with multinucleate cells have demonstrated, as a general rule, the synchronous initiation of the nuclei sharing the same cytoplasm. There are two important exceptions, namely: 1) the heterophasic binucleate cells obtained by fusion of cells in G1 with others in G2, and 2) the heteroploid multinucleate cells induced by multipolar anaphase and cytokinesis inhibition – about 15 % of their nuclei are unable to replicate and are also unable to carry out mitosis.

J.M. Mitchison, Edinburgh, U.K.
In budding yeast and fission yeast there must be a minimum cell size and a minimum protein content before the G1/S transition can take place. This might also be true for mammalian cells.

M. DeBrabander, Beerse, Belgium
Regarding the role of minimal cell size in DNA synthesis initiation: Do minicells produced by cytochalasin B initiate DNA synthesis?

J.H. Peters, Köln, F.R.G.
The cytoplasmic content of karyoplasts (or "minicells") produced from A9 mouse fibroplasts by cytochalasin B is heterogeneous. Those surrounded by a minimum of cytoplasm and cell membrane are only able to continue an S phase that has already started. They are unable to initiate a new cell cycle; in parallel, they are unable to attach. Karyoplasts containing larger amounts of cytoplasm and cell membrane are able to attach, grow, and divide.

P. Malpoix, Rhode-St-Genese, Belgium
Cytochalasin B may be useful as a tool to explain the regulatory importance for cell division of cytoplasmic and surface factors versus nuclear volume and activity. The fate of nucleate fragments with *variable* amounts of cytoplasm can be followed.
Moreover, starting with embryonic cells having limited possibilities of survival in vitro, the repeated inhibition of cytokinesis leading finally to passage through a multinucleate state can result in the rapid establishment of permanent cell lines capable of unlimited proliferation and aberrant differentiation (poster 33). This implies experimental modification of control mechanisms for both and offers a model system for their analysis.

J.F. Lopez-Saez, Madrid, Spain
For the transition G1/S a certain protein synthesis is required at the end of the G1 period. After this transition, protein synthesis appears to be unnecessary for completion of DNA replication in yeast and onion meristem cells.

H. Sauer, Würzburg, F.R.G
It has been shown (by Muldoon and others), that, in *Physarum*, protein synthesis is required throughout the S phase. Other experiments with cycloheximide (by Haugli and others) indicate that protein synthesis is needed for initiation as well as elongation and ligation of DNA replication. Control experiments with cycloheximide-resistant mutants make it very likely that a blockade of protein synthesis, not the side effects, causes inhibition of DNA synthesis.

Chairman:
There is also agreement that in all mammalian cells protein synthesis is necessary throughout the S phase, because the S phase in these cells is blocked immediately by cycloheximide, at any time during S phase.

C.A. Pasternak, London, U.K.
I would like to make a general point concerning protein synthesis and a particular one concerning enzyme synthesis. Dr. Mitchison and others often distinguish between

"step events" and continuous synthesis of enzymes. I would like to propose that enzymes that are synthesized in a stepwise manner are related to step function whereas enzymes that are synthesized continuously carry out some other role. For example: Enzymes concerned with DNA synthesis will be synthesized in a stepwise manner prior to DNA synthesis. Histones are nonenzymic proteins and are synthesized in a stepwise manner.

R. Braun, Bern, Switzerland
In synchronized HeLa cells, m RNA for histones is present in the cytoplasm only during the S phase. Recently, Melli et al. (University of Zürich) presented evidence that histone m RNA is made at a similar rate throughout the cell cycle. They suggest that it is preferentially degraded outside of the S phase.

J.R. McIntosh, Boulder, Colorado, USA
Can anyone tell me whether the other chromosomal proteins were also synthesized during S phase?

H. Sauer, Würzburg, F.R.G.
There is no evidence, in *Physarum*, that synthesis of acidic nuclear protein is correlated to or restricted to S phase. These results are obtained by gel analysis (by Lestourgeon and Magun). Furthermore, there is no good evidence that cytoplasmic proteins are synthesized in a manner other than linear, as shown by analyses of radioactive gel patterns and activities of several enzymes by Ernst and Wegener in our laboratory.
Thymidine kinase (TK) is an established peak enzyme in *Physarum*, which has long been suspected to correlate with the DNA division cycle. However, TK-less mutants (developed by Haugli) grow more or less normally. Among other possibly relevant nuclear proteins in *Physarum*, RNA polymerases A and B are synthesized linearly. Their level does not correlate with the transcription in the cell cycle. However, a small fraction of RNA pol. B - defined as "engaged" enzyme - does correlate with a high level of poly A tRNA synthesis in S phase but only if DNA replication takes place (Replication - Transcription - Coupling, Poster 9).

J.H. Peters, Köln, F.R.G.
In lymphocyte mitogen stimulation, DNA and RNA polymerase, as well as two different types of ribonuclease H, are synthesized at different times of the cycle in parallel with their proposed function.

Chairman:
Is the S phase complete when no more thymidine is incorporated?

H. Sauer, Würzburg, F.R.G.
As for the duration of S phase, it can be stated that in *Physarum*, thymidine incorporation defines the "S phase", but ligation of replicated DNA to full-length molecules takes up the rest of the cell cycle, except for mitosis.

R. Braun, Bern, Switzerland
In addition, actual replication of a minority of genes can take place in the G2 phase. Data from various laboratories including my own show that the genes for ribosomal RNA (rDNA) in *Physarum* replicate throughout the cell cycle. This rDNA is organized in a peculiar manner: it is not integrated into chromosomes, but is present as episome-like molecules. Each molecule (38×10^6d) is a palindrome with two head-to-head coding sequences. Every nucleolus contains about 150 such palindromes.

N. Paweletz, Heidelberg, F.R.G.
Must DNA synthesis take place before mitosis can start?

R.B. Nicklas, Durham, North Carolina, USA
In at least in some cases, such as the second meiotic division, it is unnecessary.

D. Mazia, Berkeley, California, USA
One can make the case of meiotic division without DNA replication more general by including polyploid cells that reduce their chromosome number.

R. Dietz, Tübingen, F.R.G.

In *Paris quadrifolia* there are supernumerary second divisions in which the number of chromatids per cell is reduced beyond the haploid level. Moreover, A. Bajer found divisions with undivided chromosomes in endosperm of *Haemanthus*. Thus, DNA synthesis is not an absolute requirement for spindle formation and chromosome distribution.

D. Wheatley, Aberdeen, Scotland, U.K.

Operationally we can ask what else must a cell do (after DNA has been completed, as delineated by high resistance to many different inhibitors of replication)? In HeLa S-3 there is a small requirement for protein synthesis; in *Tetrahymena* there is definitely a much longer period during which protein synthesis is required. In both cases the cells complete these tasks, which we might construe as deterministic events, in G2. Synthesis alone is not the only consideration. We must also think of how the proteins in G2 are utilized in, e.g., the hypothetic division structure of Zeuthen's model. When all these tasks have been accomplished, it appears (see Poster 12) that a cell may be *free* to enter the process of division. The rate at which it takes up this option is determined by many factors, both intracellular and extracellular. In effect any phase of the cell cycle may show the deterministic and probabilistic elements of the Smith and Martin theory.

Ch. Petzelt, Heidelberg, F.R.G.

Is the synthesis of a specific protein necessary in G2? Or is just a critical mass necessary for division?

J.M. Mitchison, Edinburgh, Scotland, U.K.

There is no conclusive evidence from cell cycle mutants, but there is a clear transition point. That does not rule out absolutely nonspecific proteins, but it is a good indication that specific proteins are necessary.

J.R. McIntosh, Boulder, Colorado, USA

As far as I can recall from a number of reports in the literature, if you inhibit protein synthesis before cell division, it will go on, and one can inhibit it for as long as about an hour. But when you inhibit protein synthesis for a longer period then mammalian cells will not go into division. Is that an acceptable statement of fact?

J.M. Mitchison, Edinburgh, Scotland, U.K.

That seems to be an acceptable fact for many cases of mammalian cells.

J.F. Lopez-Saez, Madrid, Spain

In onion cells, the development of prophase appears to be dependent on the synthesis of a certain protein. In the presence of puromycin, cycloheximide, or anisomycin, the early-middle prophases cannot reach metaphase, and moreover they return morphologically to interphase, as demonstrated with synchronous cells.

A. Braun, Bern, Switzerland

A requirement for protein synthesis late in the G2 phase has been inferred in many systems. The relevant studies depend nearly exclusively on the use of inhibitors and have therefore to be interpreted with caution. Side effects of the inhibitors on uptake of small molecules and on their metabolism may occur.

J.F. Lopez-Saez, Madrid, Spain

We realize that the protein synthesis inhibitors can cause certain side effects which can be mistaken for specific ones. To minimize this problem we selected well-known inhibitors and we determined the minimum dose required for 90-95 % inhibition of protein synthesis rate. After that we tested several; for example, puromycin, cycloheximide, and anisomycin. If the results are essentially similar, it seems logical to assume that the effect must be induced by their action on protein synthesis.

J.R. McIntosh, Boulder, Colorado, USA

In the case of *Physarum*, where protein synthesis is necessary in late G2, if you block protein synthesis, do you block chromosome condensation or do you block the whole division process?

R. Braun, Basel, Switzerland
In *Physarum*, proteins required for mitosis are made late in the G2 phase, some
10 min before mitosis. These proteins are not only required for mitosis as a whole
process, but also for one of the earliest parts, namely, chromosome condensation.

J.R. McIntosh, Boulder, Colorado, USA
Other authors have fused HeLa cells at different stages of the cell cycle, and as
I recall, a metaphase cell can induce chromosome condensation in a cell earlier in
the cell cycle. So you can observe chromosomes that are condensed at various stages
in S or even at G1. So, experimentally, chromosomes which are not replicated can
begin to condense.

S. Ghosh, Calcutta, India
In some organisms, the cells must pass through G2 phase to reach prophase, whereas
in other organisms, passing though G2 phase may not be an absolute necessity for
cells entering prophase. Because some protein is synthesized in G2 phase that is
necessary for cells entering prophase (as in HeLa, *Physarum*, or *Allium*), it is
likely that this protein is synthesized in mid or late S phase in other organisms,
where the G2 phase may overlap the S phase.

P. Malpoix, Rhode-St-Genese, Belgium
When are contractile proteins synthesized? Lestourgeon has described their presence in the nucleus and in chromatin where they could be detected as acidic
proteins. Could we have further information on this?

B.M. Jockusch, Basel, Switzerland
Actin and myosin are synthesized throughout the cell cycle in *Physarum*; myosin at
the same rate throughout, actin at a higher rate in G2 phase. Both are transported into the nucleus at the end of the G2 phase, prior to the onset of mitosis.

We disagree with Dr. Lestourgeon on the point of actin being a chromatin constituent in *Physarum*. In our chromatin preparations, neither actin nor myosin is
present among chromosomal proteins. Nuclear actomyosin in cells with "intranuclear"
mitosis (*Physarum*) is probably *not* a contaminant since: (1) the concentration of
actin in late G2 is higher in the nucleus than in the cytoplasm; and (2) actin-like
filaments and myosin-like filaments can be seen in mitotic nuclei in thin sections,
in situ.
Is tubulin present in interphase nuclei of cells with an open mitosis?

M. DeBrabander, Beerse, Belgium
New immunocytochemical techniques (see poster 25) show that tubulin is probably
not present in the interphase nucleus of mammalian cells in cultures that have an
open mitosis. Due to the limitations of immunocytochemical techniques, however,
this cannot be regarded as absolute proof.

M. Osborn, Göttingen, F.R.G.
We do not see intranuclear reaction with tubulin antibody. The question is, which
kinds of structures are recognized with what efficiency? All we know from tubulin
antibodies is that they do recognize microtubules, but the relative efficiencies
in recognition of microtubules and soluble tubulin in the nuclei remains to be
determined.

A.S. Bayer, Eugene, Oregon, USA
In cells of higher plants and animals, a certain number of microtubules are incorporated into telophase nucleus when the new nuclear envelope forms. These
microtubules disassemble and are no longer detectable during interphase. Tubulin
molecules are also present in the cytoplasm. Therefore, removal of tubulin from
the nucleus would require existence of the specific "active transport" acting
across the nuclear envelope. It is an open question, however, in what form or
state tubulin molecules persist in the interphase nucleus.

U.P. Roos, Zürich, Switzerland
I should like to add that in lower eukaryotes with a more or less closed spindle
a substantial number of spindle microtubules persist in daughter nuclei until late
telophase, but subsequently disappear. It would be interesting to know whether
their breakdown products contribute to or constitute an intranuclear pool of
tubulin.

D. Wheatley, Aberdeen, U.K.
Regarding the synthesis of tubulin, like Forrest and Klevecz, we have shown that
there is an increase in the rate of tubulin synthesis in the late S and G2 phases
of the cell cycle in mammalian cells, although the increases we reported were not
as dramatic as Lawrence and Wheatley suggested in G2. More importantly, however,
in our cells (HeLa in suspension culture) it was firmly established that the synthesis of further amounts of tubulin at this time in the cell cycle was superfluous
to the needs of the cell to enter and successfully complete division.

J.R. McIntosh, Boulder, Colorado, USA
The behavior of tubulin in mammalian cells appears to be somewhat similar to the
burst of actin synthesis in G2 mentioned by Dr. Jockusch. Is there, in fact, a
signal for the onset of G2, or does G2 simply begin when DNA synthesis stops.

J.M. Mitchison, Edinburgh, U.K.
If the question implies that G2 has a constant time, then this is clearly not the
case in a number of organisms. In yeast, G2 is compressible, but to what extent,
I am not certain.

J.F. Lopez-Saez, Madrid, Spain
Our group has demonstrated the different durations of this period for nuclei
sharing the same cytoplasm in trinucleate and tetranucleate cells induced by
cytokinesis inhibition. The mononucleate, normal cells have a G2 period of about
4 h, while under similar conditions in the faster replicating nuclei of multinucleate cells, this period lasts about 6 h, and in the slower replicating nuclei,
4 h. It appears that the G2 period involves processes that require the completion
of DNA replication, and there might be a certain compensation among the G2 duration
of the different nuclei in multinucleate cells.

J.G. Carlson, Knoxville, Tennessee, USA
A morphologically static G1 phase is missing in the neuroblast of the grasshopper
embryo. DNA replication begins in middle telophase at about the time the nucleoli
are reappearing and the nuclear membrane is reforming, and extends through all of
interphase into very early prophase. This raises the question of the existence
in these cells of a true G2 phase, since this was originally defined by Howard
and Pele as part of interphase. Further, in mammalian cells the chromosome condensation characterizing prophase can be seen as early as 15 min to 1 h in different cells before breakdown of the nuclear membrane; therefore the latter part
of the so-called G2 phase is actually part of mitosis.

D. Mazia, Berkeley, California, USA
When the study of the cell cycle was in its beginnings, the aim was to account
for the history of definite events. The now-conventional phases came in as a way
of reporting the results of a certain type of experiment, using DNA synthesis to
provide landmarks. Does not the present discussion suggest that the phases G1
etc., have acquired a life of their own, and have they not outlived their usefulness? Cases where there is no G1 or G2 are not so rare; what is the point of
saying that such cases are exceptions? The events we have in mind do take place.
If a cell has no G1 by convention, surely the required *events* do take place,
but they happened during the previous cycle. The same would be true for other
"variant" histories, even those as extreme as we see in eggs. It is the events
that interest us, not the calendar. It may be more useful to think of the cell
cycle as a bicycle. There is a growth wheel and a reproduction wheel. They can
be geared together in many ways, with the same outcome. In the alleged "normal
cycle" they are tightly geared. In eggs the growth wheel may turn several times;
then the reproduction wheel turns several times. These differences do not matter
to cells, as long as the necessary events take place and the long-term outcome
of the reproduction of cells is achieved. Should they matter to cell biologists
other than as descriptions of the life histories of different cells?

Session II

Surface Signals and Cellular Regulation of Growth

G. M. EDELMAN, P. D'EUSTACHIO, D. A. McCLAIN, and S. M. JAZWINSKI
The Rockefeller University, New York, New York 10021, USA

A. Introduction

Several lines of investigation are now converging toward the goal of understanding the organization and specificity of the cell surface and the effect of these factors on the behavior of the cell as a whole. These studies are aimed in part at understanding how signals are transmitted from the cell membrane to the cell interior, and conversely, how structures in the cell interior can influence the properties of cell surface molecules. The first part of this paper will review evidence for the existence of cellular structures that mediate such transmembrane control events.

Given the existence of structures involved in membrane-cytoplasmic signalling and composed in part of cell motility and structural elements, we have hypothesized that these structures may provide a mechanism for the coordination of cell division, cell movement, and cell-cell interactions (Edelman, 1976). In the second part of this paper, we will review evidence that perturbation of the assembly responsible for transmembrane control can critically alter growth control in normal cells.

A satisfactory understanding of the biochemistry of the cell cycle requires not only the analysis of these initial events, however, but also analysis of the subsequent sequence of events resulting in DNA synthesis and cell division. This requires both that the order of biochemical events leading to the initiation of DNA synthesis be established and that assays allowing the isolation of the molecules responsible be devised. In the last part of this paper, we will describe such an assay, and its use to define cytoplasmic activators of DNA synthesis that appear to be subject to growth control.

B. Anchorage Modulation

The current picture of the cell surface differs radically from that of a decade ago. It is now clear that the lipid bilayer is a fluid structure (Hubbell and McConnell, 1969) and at least some of the surface proteins embedded in it can diffuse randomly and independently of one another (Frye and Edidin, 1970; Poo and Cone, 1974). If these receptors are cross-linked by the addition of a multivalent ligand, however, their mobility is sharply altered. Thus, when divalent antibodies bind to cell surface receptors, the receptors form small clusters, or patches, which can then gather at one pole of the cell to form a cap. Cap formation, but not patch formation, requires active cell metabolism, and is inhibited, for example, in the presence of sodium azide (Taylor et al., 1971; Yahara and Edelman, 1972).

When the cells are treated with nonsaturating amounts of the plant lectin concanavalin A (Con A), which binds to the carbohydrate portion of various cell surface glycoproteins, subsequent addition of specific antibodies against various receptors fails to induce either patches or caps (Yahara and Edelman, 1972, 1973). This restriction, or anchorage modulation, can be reversed by removing the lectin, and therefore does not result from permanent interference with metabolism or from cell death. Analysis in the electron microscope of lymphocytes treated with ferritin-labeled Con A indicates that this surface modulation operates at the level of individual receptors (Yahara and Edelman, 1975a). Anchorage modulation has been observed for a variety of receptors on lymphocytes, and for the H-2 antigen on several other cell types (Yahara and Edelman, 1975b).

An important property of anchorage modulation is that it is propagated; i.e., binding of Con A to a well-defined small region of the cell surface can induce global modulation. To show this, we have prepared Con A bound to platelets and latex beads and have shown that local attachment of these particles to a small fraction of Con A receptor sites on the cell surface inhibits the mobility of Ig and other receptors (Yahara and Edelman, 1975c). Similarly, cells bound at one region of their surface to nylon fibers derivatized with Con A show inhibition of receptor mobility (Rutishauser et al., 1974).

Perhaps the most direct evidence for the existence of a propagated effect of Con A on the mobility of membrane receptors all over the cell comes from experiments using the fluorescence photobleaching recovery method (Schlessinger et al., 1977a). Using this technique, the lateral diffusion of surface membrane components labeled with fluorescent

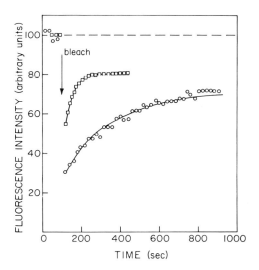

Fig. 1. Fluorescence photobleaching recovery curves of two cells labeled with RaP388 Fab. One cell had no Con A platelets (□); the other had approximately 60 platelets covering 13 % of its area (o). Both recoveries fit the theory for a single diffusion coefficient within experimental error (superimposed curves). $D = 1.25 \times 10^{-10}$ cm^2/s and $D = 1.8 \times 10^{-11}$ cm^2/s, respectively

antibodies can be measured quantitatively over distances of a few microns (Fig. 1). Under these conditions, all of the surface receptors on 3T3 cells were in one of two major states, anchored, A_1, or free, F. When Con A-coated platelets were used to cross-link glycoproteins in small localized regions of the cell surface, however, the diffusion coefficient of F state receptors decreased, i.e., they were converted to a less mobile state, A_2. The same result was obtained with soluble Con A. The size of the immobile (A_1) fraction of the receptor population was not changed by treatment with Con A platelets or with low doses (20 µg/ml) of soluble Con A.

The modulation effect showed a threshold and a plateau; i.e., occupany of greater than 4 % of the cell surface induced the effect but larger occupancies did not increase it. The effect was seen in all regions of the cell surface despite the fact that the platelets were localized on only a small region of the surface. The propagated nature of the modulation of receptor mobility was confirmed by the position independence of the effect. The propagation time was faster than the relatively long periods required for the measurements and thus cannot yet be estimated. Modulation by Con A platelets was partially reversed by microtubule-disrupting agents; the extent of the reversal was independent of the area occupied by Con A platelets. At the levels tested, these drugs did not affect modulation by soluble Con A.

These experiments eliminate two possible mechanism of anchorage modulation: (1) that Con A modulates by trapping receptors in a meshwork of the cell surface glycoprotein receptors; and (2) that Con A modulates only by cross-linking mobile glycoproteins directly to anchored or frozen membrane receptors.

What properties of the lectin are responsible for these effects, and what structure in the cell mediates them? Experiments on the structure of Con A and on the effects of drugs have yielded provisional answers to some parts of these questions.

The valence of Con A may be changed by suitable chemical treatment, and this change in valence alters its modulating properties (Gunther et al., 1973). When Con A is succinylated it dissociates from a tetramer to a dimer, but its binding specificities for carbohydrates are unchanged. Cells treated with succinyl-Con A show unaltered receptor mobility and can undergo patch and cap formation normally. Treatment with anti-ConA, however, restores the ability of bound succinyl-Con A to restrict the movement of immunoglobulin receptors. Fab fragments of antibodies to Con A did not restore these phenomena (Yahara, I., Edelman, G.M., Wang, J., unpublished observations), and anti-Con A alone had no effect. Thus, cross-linking of cell surface glycoproteins appears to be required for initiating anchorage modulation.

The effects of low temperature (4°C) and various antimitotic drugs indicate that integrity of cytoplasmic structure, particularly the cell's microtubular system, is necessary for Con A-induced anchorage modulation. If colchicine or various *Vinca* alkaloids are added to lymphocytes in concentrations ranging from 10^{-6} to 10^{-4}M, anchorage modulation is reversed in many of the cells (Yahara and Edelman, 1973b), and this effect is itself reversed by removal of the drugs. Inactive derivatives of colchicine such as lumicolchicine, which has no effect on microtubules, do not reverse anchorage modulation. Experiments with colchicine-resistant cell lines indicate that colchicine must enter the cytoplasm to affect anchorage modulation (Aubin et al., 1975). As indicated above, colchicine also partially reverses the Con A-induced inhibition of receptor mobility assayed by the fluorescence photobleaching recovery method (Schlessinger et al., 1977a). Treatment of modulated cells with a brief cold shock sufficient to depolymerize cellular microtubules (Behnke and Forer, 1967; Roth, 1967; Tilney and Porter, 1967) also releases the anchorage modulation (Yahara and Edelman, 1973a).

What arrangement of submembranous components could account for these phenomena? Capping induced by Con A can also induce the redistribution of cytoplasmic microtubules (Albertini and Clark, 1975). Nevertheless, electron microscopic studies (Yahara and Edelman, 1975a) suggest that microtubules are not present at the inner lamella of the lipid bilayer and therefore, even if cell surface glycoproteins penetrate the bilayer into the cytoplasm, some form of linkage between these receptors and microtubules would be required. Possible candidates for

this role include certain microfilamentous structures found just
under the membrane. It is likely that such actin-like molecules are
involved in capping of patched receptors, for several investigators
have found that cytochalasin B, a drug that affects certain microfilaments, also inhibits capping (de Petris, 1975; Edelman et al.,
1973). Recently, it has also been shown (Sundqvist and Ehrnst, 1976)
that in the presence of cytochalasin B microfilaments can be induced
to co-migrate with surface receptors. The further observation that
cytochalasin B and colchicine can act synergistically to inhibit
capping argues that both microfilaments and microtubules are linked
to the control of membrane receptor mobility (Sundqvist and Ehrnst,
1976).

Nevertheless, cytochalasin B does not reverse anchorage modulation,
in accord with the deduction that some other linkage besides cyto-
chalasin-sensitive microfilaments may be required for the proposed
interaction between receptors and microtubules. Aside from cytocha-
lasin-resistant microfilaments, the linkage might consist of various
assembly states of tubulin subunits (Edelman et al., 1973) or of additional proteins, possibly α-actinin or myosin (Ash and Singer,
1976).

On the basis of these observations, various models (Edelman et al.,
1973; Yahara and Edelman, 1975a; Berlin et al., 1975) have been proposed to account for anchorage modulation. One of the simplest models
(Fig. 2) suggests that the appropriate surface-modulating assembly

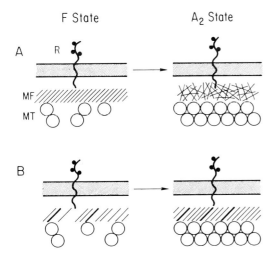

Fig. 2A and B. Schematic representation of different modes by which a surface modulating assembly (SMA) might act to cause a change in receptor state from F to
A_2. The various elements of the SMA have not been drawn to scale. (A) Modulation
event induced by local cross-linkage of glycoprotein receptors (R) results in
alteration of the submembranous components of the SMA with gelation of fibrils
(MF) and restricted diffusion of the receptor. (B) Modulation results in enhanced
binding of the cytoplasmic base of the receptor to submembranous structures. In
either case, it is assumed that intact microtubules (MT) are essential for modulation to occur

(SMA) has a tripartite structure: 1) a subset of glycoprotein receptors that penetrate the membrane and confer specificity on the system; 2) various actin-like microfilaments and their associated proteins, such as myosin, conferring the properties of coordinated movement necessary for capping; and 3) dynamically assembling microtubules, both to provide anchorage of the receptors and to allow propagation of signals to and from the cell surface.

This model assumes that the receptors normally exist in two states, anchored and free. Cross-linking certain glycoprotein receptors alters the various equilibria between the microfilaments and microtubules and their subunits, inducing a propagated assembly of microtubules and fixation of microfilaments, and increasing the proportion of anchored receptors. Conversely, changing the state of the cytoplasmic microfilaments and microtubules can alter the mobility and distribution of surface receptors. Disruption of microtubules by drugs would still leave the microfilaments and their associated proteins free to induce capping.

The disorderly gelation of subunits of actin or tubulin in a submembranous location (Fig. 2A) must be considered as an alternative to specific interactions of receptors with cytoskeletal components (Fig. 2B) as a mechanism for restriction of receptor diffusion. This mechanism can be reconciled with the observation that microtubule-disrupting agents partially reverse modulation, particularly if the postulated gel is also dissociated by these agents or if intact microtubules are necessary for formation of the gel. The fact that cytochalasin B retards receptor diffusion but not lipid diffusion (Schlessinger et al., 1976a, b, 1977b) raises the possibility that it promotes disorderly gelation. As yet, a clear-cut choice between the two models in Figure 2 cannot be made, nor can the possible involvement of additional submembranous proteins be assessed directly.

C. Relation of the State of SMA Components to Normal Growth Control

All of the data discussed so far indicate that there is indeed communication from the cell surface to the cytoplasm and back to the surface, and thus they provide strong evidence for transmembrane control. Although a complete molecular definition of the structures responsible for this effect is lacking, it appears to involve the cytoskeletal and motility-related proteins. We have hypothesized further that the interactions among these molecules are of importance to growth control. A number of experiments on the effects of inducing anchorage modulation, of treating cells with antimitotic drugs, and of transforming cells with viruses, are consistent with the hypothesis that the state of components of the SMA can affect the regulation of cellular growth at crucial decision points in the G_1 phase of the cell cycle.

One useful system in this analysis has been the mitogenic stimulation of lmyphocytes by Con A and succinyl-Con A. The dose-response curves for mitogenesis of these two lectins is shown in Figure 3. Whereas tetravalent Con A shows both a rising stimulatory and a falling inhibitory limb, divalent succinyl-Con A shows only a stimulatory effect on mitogenesis. This fact indicates that the effect of Con A is probably the composite of two separable events, one stimulatory and the other inhibitory. We will briefly review evidence that both limbs of the dose-response curves are related to different states of the SMA.

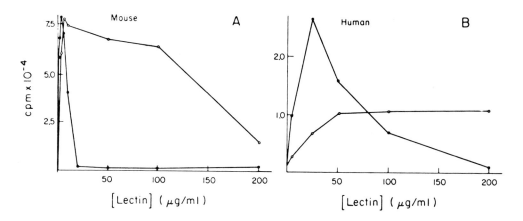

Fig. 3A and B. Stimulation of lymphocytes by Con A and succinyl-Con A. (A) Dose-response curves showing the incorporation of [3H]-thymidine after stimulation of mouse spleen cells by Con A (●———●) and succinyl-Con A (o———o). (B) Dose-response curves showing the incorporation of [3H]-thymidine after stimulation of human peripheral blood lymphocytes by Con A (●———●) and succinyl-Con A (o———o)

The relation of the falling inhibitory portion of the Con A dose-response curve to anchorage modulation is based on two lines of evidence: 1) Con A, but not succinyl-Con A, is capable of inhibiting both mitogenesis and cell surface receptor mobility (Gunther et al., 1973); and 2) Con A inhibits receptor mobility with the same dose-response curve as it inhibits mitogenesis (Yahara and Edelman, 1972). Further analysis of the inhibitory effects of high doses of Con A has shown that it causes these lymphocytes to become committed to mitogenesis while also generating a dominant but reversible negative growth signal (McClain and Edelman, 1976). Cells that have become committed but are also simultaneously blocked from entering S phase by the high doses of Con A can begin synthesizing DNA if the lectin is released by adding a competitive inhibitor of binding (Fig. 4). Experiments done in agarose cultures in which lymphocytes are kept from contact with each other suggest that the reversible inhibitory signal is mediated by structures in the individual cells and is not a result of agglutination. The inhibitory effects of ConA therefore appear not to be trivial; i.e., they are not due to nonspecific cell death, but rather represent a clamping of otherwise stimulated cells.

While these results are only correlative, more direct evidence supports the hypothesis that colchicine-sensitive structures are involved in the regulation of the crucial commitment event of lymphocyte mitogenesis. In accord with this hypothesis, we observed that colchicine inhibits the incorporation of [3H]-thymidine in mouse splenic lymphocytes stimulated by Con A (Wang et al., 1975a). Similar effects have been seen in PHA-stimulated lymphocytes (Medrano et al., 1974). Further analysis of this phenomenon revealed that in human lymphocytes, colchicine inhibited both lectin-induced blast transformation and entry into the S phase (Fig. 5).

Mitogenesis was also blocked by vinblastine and vincristine but not by lumicolchicine, a photo-inactivated derivative of colchicine (Wilson and Friedkin, 1966). The effect of colchicine on mitogenesis was not due to inhibition of nucleoside transport, inasmuch as transport is also blocked by lumicolchicine, but is not affected by vin-

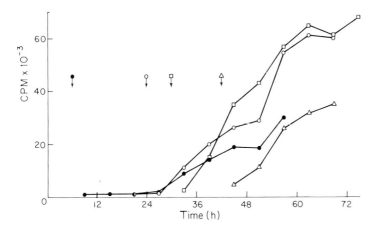

Fig. 4. Kinetics of DNA synthesis in cultures treated with high doses of Con A and released with αMM at various times. Parallel sets of human lymphocyte cultures with 150 μg/ml of Con A were treated with αMM at 6 h (●————●), 18 h (o————o), 30 h (□————□), or 42 h (△————△). The arrows indicate the point at which αMM was added for a given culture. The kinetics of DNA synthesis were monitored with 6-h pulses of 6 μCi of [3H]-thymidine at 6-h intervals. Data are plotted as the mean of duplicate tubes at the midpoint of the pulse

blastine (Wilson et al., 1974). The inhibition of mitogenesis could not be attributed to effects on the mitotic spindle, because it occurred prior to the earliest evidence of cell division. In addition, colchicine did not appear to alter cell viabilities at the doses used and the effects of the drug were reversible. Colchicine also did not affect DNA synthesis in cells synchronized by treatment with hydroxyurea after Con A stimulation. All of these observations prompted the inference that colchicine must inhibit stimulation prior to entry into the S phase.

To determine more precisely the stage of the cell cycle at which colchicine acts to inhibit stimulation and to determine whether all responding cells are susceptible to inhibition simultaneously, it is first necessary to consider the heterogeneity of the response of the cell population to Con A.

The kinetics of cellular commitment in the stimulation of mouse splenocytes (Gunther et al., 1974) and human lymphocytes (Gunther et al., 1976) by ConA have recently been analyzed. As used in these studies, the term commitment refers to the point at which a cell becomes stimulated and no longer depends upon the presence of the mitogenic agent. The kinetics of commitment for a cell population then can be assayed by exposing the cells to Con A, removing the Con A at various times thereafter by treatment with the competitive binding sugar α-methyl-D-mannoside (αMM), and assaying the number of cells capable of synthesizing DNA 48 h after the initial addition of the lectin. Under these conditions, cells become committed in all-or-none fashion. For mouse lymphocyte populations, all cells do not become committed simultaneously, but rather at various times up to 20 h after the addition of the mitogen (Fig. 6) (Gunther et al., 1974). Similar results have been obtained by others for lymphocytes stimulated with PHA (Younkin, 1972) or Con A (Lindahl-Kiessling, 1972; Powell and Leon, 1970). Due to complications introduced by daughter

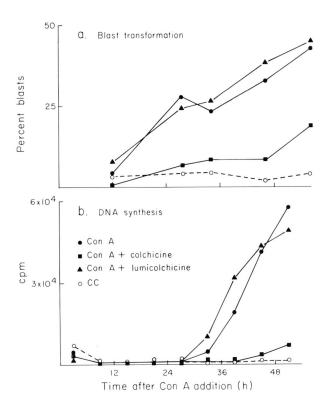

Fig. 5a and b. A comparison of the effects of colchicine and lumicolchicine on the kinetics of appearance of blast cells and on the kinetics of [3H]-thymidine incorporation in human lymphocytes stimulated by Con A. Colchicine and lumicolchicine were present from the beginning of the experiment. At the indicated times, aliquots of cultures were removed and the total number of viable lymphocytes and blast cells were counted in a hemocytometer. The data are averages of determinations on duplicate cultures and are expressed as the percentage of blast cells present at various times after the addition of Con A (a). For measurements of DNA synthesis (b), parallel cultures were pulsed at various times with [3H]-thymidine for 6 h. Data are plotted at the midpoint of this pulse. ●, Cultures containing Con A (20 µg/ml); ■, cultures containing Con A (20 µg/ml) + colchicine (10-6M); ▲, cultures containing Con A (20 µg/ml) + lumicolchicine (10-6M); o, cell control (CC)

cell proliferation these conclusions concerning cellular commitment apply only to those cells that have entered their initial S phase by 48 h.

Once the kinetics of commitment were determined for human lymphocytes, colchicine was added at various times to human lymphocyte cultures containing Con A and the response was again measured in terms of total DNA synthesis, the percentage of blasts, and the percentage of blast cells that synthesized DNA (Fig. 7) (Gunther et al., 1976). As before, the later the colchicine was added, the greater the response observed in the culture. Furthermore, the shape of the curve was similar to that observed for inhibition by αMM (Fig. 6).

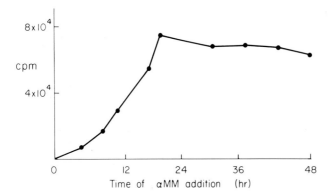

Fig. 6. The effect of αMM, added at different times after the start of the culture, on the incorporation of [3H]-thymidine by Con A-stimulated lymphocytes. Cultures (0.3 ml) containing 3 µg/ml Con A were made 0.1 M in α MM at various time points. The cultures were continued until 48 h at which time 1 µCi of [3H]-thymidine was introduced in 0.05 ml of medium. The cells were harvested for analysis at 72 h. Data points shown are the averages of measurements on duplicate cultures

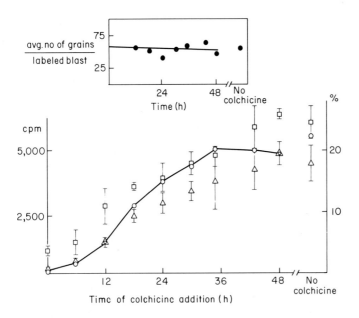

Fig. 7. The effect of colchicine added at various times on the stimulation of human lymphocytes by Con A. Cultures initially contained 20 µg/ml Con A and were made 1 µM in colchicine at the times indicated. The error bars represent ± 1 S.D. for triplicate counts of the same cell smear. *Abscissa*: times after initiation of culture (h); *ordinate*: (left) [3H]-thymidine incorporation expressed as cpm (o); (right) % of cells present as blasts (□) or labeled blasts (Δ)

Blast staining and autoradiographic experiments showed that an increasing number of cells respond to stimulation with later additions of colchicine. In addition, the average number of grains per labeled blast cell was constant. This suggests, therefore, that with longer exposure to the lectin an increasing number of cells becomes refractory to inhibition by colchicine. Colchicine does not bind to Con A, nor does it inhibit the cell binding activity of the lectin (Edelman et al., 1973). Thus, two independent reagents, αMM and colchicine, acting by different mechanisms, can block mitogenic stimulation with similar kinetics analyzed at the level of individual cells. This similarity between the kinetics of inhibition by colchicine and by αMM is consistent with the hypothesis that colchicine blocks stimulation near the time that a cell becomes committed to DNA synthesis and cell division.

If only uncommitted cells can be inhibited by colchicine, then the simultaneous addition of αMM and colchicine to cultures containing Con A should produce an effect equivalent to that produced by the addition of either reagent alone. To test this hypothesis, three series of parallel cultures received Con A at the same time. αMM was added to one series at various times thereafter; a second set of cultures received colchicine; and the third set received both αMM and colchicine simultaneously. All three series of cultures produced similar rising curves for later times of addition of the reagents (Fig. 8). Most important, simultaneous addition of αMM and colchicine produced the same degree of inhibition as observed when either reagent was added separately, consistent with the hypothesis that both treatments are affecting the same cell population.

Our findings on the heterogeneity of lymphocytes, manifested in the variable induction periods required for stimulation by Con A, suggest an interpretation analogous to that proposed by Smith and Martin (1973) for cell cycle kinetics in general. We suppose that lymphocytes may be in two states, resting and activated. Cells in the resting state may be challenged by exogenous stimulants such as antigens or lectins and move into the activated state. In the activated state, the cellular activities are deterministic and pass through the classical cell cycle phases toward division. If the transition from the resting to the activated state on stimulation occurs with a constant probability but in a random fashion, then a wide intrapopulation variability would be observed in the kinetics of activation. It is this transition from the resting to the committed activated state that is apparently blocked by colchicine.

To confirm and extend these results, it was of interest to test whether the phenomena described above were also true for a different and more homogeneous cell type. We chose normal second passage chicken embryo fibroblasts (CEF) for this analysis (McClain et al., 1977). CEF were arrested in the G_1 phase of the cell cycle by incubation in serum-free medium (Temin 1972). Addition of serum to these starved fibroblasts resulted in a wave of DNA synthesis beginning at about 8 h (Fig. 9). Addition of colchicine to a final concentration of 1 μM at the same time as the serum delayed the entry of cells into the S phase and reduced the number of cells centering the S phase. Similar results were obtained with mouse 3T3 fibroblasts. The amount of label incorporated per cell in S phase (estimated from grain counts of autoradiograms) was not affected by colchicine, and thus the effect of the drug did not appear to be due to inhibition of $[^3H]$-thymidine transport. Lumicolchicine, which does not dissociate microtubules, did not inhibit the entry of the CEF into the S phase. Other agents that disrupt microtubules - podophyllotoxin and vinblastine - had the

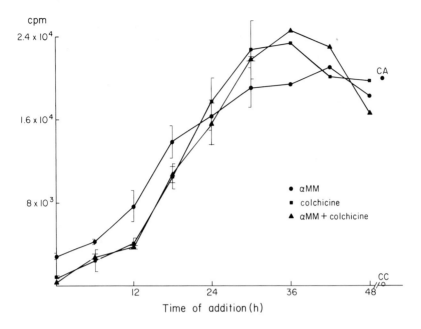

cpm

Fig. 8. Effect of the simultaneous addition of αMM and colchicine on the stimulation of human lymphocytes by Con A. Cultures initially contained 20 μg/ml Con A and at the indicated times parallel cultures received either αMM (o) (0.1 M final concentration); colchicine (■) (1 μM final concentration); or both reagents (▲). All cultures were assayed for [3H]-thymidine incorporation 48 h after the initiation of the experiment. Three independent experiments were performed, two with triplicate cultures for each condition and one with duplicate cultures. Within each experiment, the cpm for each condition were averaged and these values were normalized to the value for αMM added at 48 h (100 %). The normalized values from the three experiments were then averaged to give the points shown in the Figure. The error bars represent ± 1 S.D. of this final average. *Abscissa*: time after initiation of culture (h); *ordinate*: [3H]-thymidine incorporation expressed as % by the normalization procedure just described

same inhibitory effect at 1 μM as colchicine. The effect of colchicine was reversible: CEF exposed to the drug but then washed free of it responded to serum as well as untreated cells.

Like lymphocytes stimulated by Con A, fibroblasts stimulated by serum did not become committed synchronously, and the kinetics of inhibition by colchicine closely paralleled the kinetics of commitment to serum stimulation (McClain et al., 1977). To study the kinetics of commitment, CEF that had been arrested by serum starvation were exposed to serum. At various times thereafter, the serum was removed. Twelve hours after the initial exposure to serum, the number of cells in S phase was determined. In accord with previous observations (Temin, 1972), committed CEF began appearing soon after serum was added and the number of these cells increased for a period of at least 12 h (Fig. 10). If after adding serum, colchicine was added at various times, reduced numbers of cells in S phase were observed 12 h after the serum addition (Fig. 10). Less inhibition was seen the later the colchicine was added. When serum was removed at the same time that colchicine was added, the same amount of inhibition was observed as a function of time as with serum removal or colchicine addition alone.

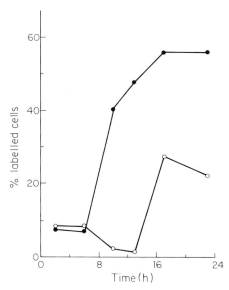

Fig. 9. Inhibition of DNA synthesis by colchicine after release of serum starvation in CEF. 2×10^5 cells were seeded onto coverslips in 2 ml of medium with serum. After the cells settled, the coverslips were transferred to serum-free medium. After 60 h, 0.22 ml of fetal calf serum was added (time 0 on abscissa) (o). Parallel cultures were also made 1 μM in colchicine (●). At various times, 6 μCi $[^3H]$-thymidine were added to the cultures; after 2 h the cultures were processed for autoradiography

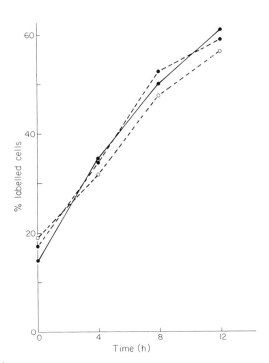

Fig. 10. Effect of colchicine on commitment to serum stimulation. Serum-starved cells on coverslips were placed in medium containing 10 % fetal calf serum at time 0. At the times indicated, coverslips were washed with buffered saline and put in serum-free medium (●——●); colchicine was added to 1 μM (o---o); or both (●---●). At 12 h the cultures received 6 μCi $[^3H]$-thymidine, and at 13 h they were processed for autoradiography

This suggests that both of these treatments of serum-stimulated CEF were affecting the same subpopulations of cells at any given time.

ConA also inhibits the serum stimulation of resting fibroblasts (McClain et al., 1977). Autoradiographic analysis confirmed that the inhibition of growth reflected a restriction of the number of cells entering S phase. In contrast, dimeric succinyl-Con A did not inhibit growth. Since the lectin and its derivative have similar binding specificities, the lack of inhibition by succinyl-Con A suggests that the inhibition by Con A is not due to competition for or interference with the binding of serum growth factors to the fibroblast surface. As in the lymphocyte, the doses of native Con A that prevent entry of the serum-stimulated cells into the S phase also cause anchorage modulation of the chick embryo fibroblast surface.

D. Alteration of the SMA in Transformed Cells

If indeed the state of components of the SMA is crucial to growth control, one might predict that they would be altered in transformed cells. To analyze this issue further, we have studied the effects of infection by Rous sarcoma virus (RSV) (Edelman and Yahara, 1976). Several coordinated changes in growth control, cell shape, and cell metabolism occur when chick fibroblasts are transformed by avian sarcoma virus (Kawai and Hanafusa, 1971). Inasmuch as a single viral gene (*src*) now appears to be responsible for transformation (Stehelin et al., 1976), the proposal that the SMA may be a central regulator of cell growth, shape, and interaction (Edelman, 1976) raised the possibility that one or more of the components of the SMA may be targets of the *src* gene product.

To test this hypothesis, we stained normal and transformed mouse and chick fibroblasts with fluorescent antibodies to visualize their microtubular and microfilamentous structures. In contrast to the orderly arrangements in untransformed cells (Lazarides, 1976; Weber et al., 1975), the transformed cells of mouse and chick showed highly disordered patterns. There was a loss of the regular actin bundles making up stress filaments, with the concomitant appearance of a more diffuse pattern of staining. In addition, the normal tubulin pattern consisting of a complex array of microtubules radiating from one or two central positions, was replaced by a fluffy pattern with multiple spots (SV 3T3 cells) or by a more diffuse pattern with a central concentration of stain (chick fibroblasts transformed with SR-A strain RSV). This is in accord with results from other laboratories using stains for tubulin, actin, and myosin (Brinkley et al., 1976; Pollack et al., 1975; Ash et al., 1976).

The striking new observation was that cells infected with a temperature-sensitive, transformation defective mutant of RSV (*ts*NY68) showed the normal pattern at the restrictive temperature and the disordered pattern at the permissive temperature. This was a reversible shift, and in a shift-down experiment, the alteration took only 1-2 h to appear in the majority of cells. Normal chick fibroblasts showed ordered patterns at both temperatures, and cells transformed with wild-type RSV showed disordered patterns at both temperatures, indicating that this early effect of expression of the *src* gene is not due to alteration of tubulin itself by the change in temperature.

In addition, the capacity of ConA to induce cell surface modulation was markedly diminished in cells at the permissive temperature (Edelman and Yahara, 1976). Inasmuch as microtubules and associated structures are essential for anchorage modulation, it is reasonable to attribute this effect to the altered structure of the microbubules at the permissive temperature. At present, however, definitive proof

must await a detailed analysis of the effect of different microtubule assembly states on surface modulation.

Perhaps the most challenging problem posed by these findings is
whether the observed alterations of the SMA are only one of many consequences of transformation and are therefore unlinked to alterations
of growth control, or whether a protein product resulting from the
action of the *src* gene alters the SMA and therefore alters growth
control. If the site of action is nuclear and the gene is present in
a single copy, then it is difficult to understand how so many coordinated changes (growth, shape, and metabolism) could occur in
transformation unless regulatory genes are affected (Fig. 11, I). On
the other hand, a *src* gene product, or a structural gene product induced by the presence of the *src* gene, could bind to one or more
sites in the cytoplasm and alter a variety of regulatory and metabolic events (Fig. 11, II and III).

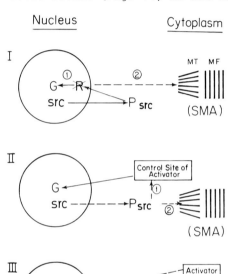

Fig. 11. Three alternative models for action of the *src* gene on both the SMA and
on growth and replication control (G).
(I) P_{src}, the *src* gene product (1), affects a regulator gene (R) with independent effects on G and SMA (2); (II) P_{src}
alters G by affecting (1) a control site
for a cytoplasmic activator and (2) independently altering shape and surface properties by alteration of the SMA; (III)
P_{src} alters SMA (1) by releasing an activator (2) (or removing a control protein) which in turn alters G to favor continuous growth

An analysis of the relative effects of temperature shifts and inhibition of protein synthesis on the kinetics of alteration in the microtubule staining pattern has led us to reject model I of this scheme
(McClain et al., 1977). If the *src* product itself were not acting on
the SMA, but rather at a stage *prior to translation* to induce the appearance of a *separate* SMA effector (model I), a shift to 41°C or the
inhibition of protein synthesis both would prevent new synthesis of
that separate effector. In both cases, the return of the SMA to the
normal phenotype would result from the natural decay of the effector
substance already synthesized, and the cells should revert to the normal phenotype after the temperature shift no more rapidly than after
blocking protein synthesis. In contrast, if the *src* gene product
acted directly on the SMA (models II and III), a shift to 41°C should

Table 1. Effects of ConA and colchicine on growth of synchronized *ts*NY68-infected cells

Condition	Peak [3H]-thymidine incorporation (cpm)	% inhibition by Con A (30 µg/ml)	% inhibition by colchicine (1 µM)
Cells grown at 41°C, serum added	7640	53	47
Cells shifted to 37°C, serum added	1900	59	68
Cells shifted to 37°C, no serum added	840	100	54

3×10^5 cells infected with *ts*NY68 plated in 60-mm dishes were grown for 24 h at 41°C in complete medium, then for 48 h at 41°C in serum-free medium. Then 10 % calf serum was added, or the cells were shifted to 37°C, or both. Con A or colchicine was added at the time of the shift. At 4 1/2-h intervals, incorporation of [3H]-thymidine into acid-precipitable material was measured. Cells at 41°C without serum incorporated 430 cpm. Peak DNA synthesis was observed at 12-13 h at 41°C and 21-22 h at 37°C

Further analysis is clearly required to pinpoint the target of the *src* gene product. Preliminary studies (Edelman and Yahara, 1976) indicate that the capability of tubulin from transformed cells to polymerize in vitro is not impaired, suggesting that the *src* gene acts at a higher regulatory level than the tubulin itself to impair SMA assembly. This has recently been confirmed and extended in another system (Wicke et al., 1977). Similarly, the causal relationships of the effects of colchicine and Con A on both the SMA and growth control of normal and transformed cells remain to be established.

All of these observations provide evidence that the state of the SMA is well correlated with the growth regulatory state of the cell for a variety of cell types, both resting and proliferating, normal and transformed. Our present conception of the events leading to cell division is given in Figure 14. Actions of any of a number of diverse

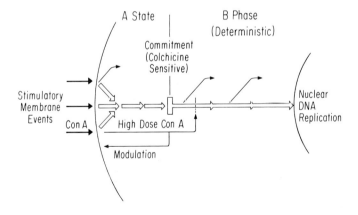

Fig. 14. Schematic of events in mitogenic stimulation

stimuli at the cell surface probably initiate several biochemical
events. Although the basis of stimulation remains unknown, a number
of recent findings have suggested that calcium ion transport and increased levels of cyclic GMP both play prominent roles. It has been
shown, for example, that intracellular levels of both calcium ions
(Allwood et al., 1971; Whitney and Sutherland, 1972) and cyclic GMP
(Hadden et al., 1972; Watson, 1975) increase after mitogen binding.
In addition, the ionophore A-23187, which mediates calcium ion transport, and TPA, which elevates cyclic GMP levels (Goldberg et al.,
1974), are independently mitogenic for lymphocytes and are synergistic
with lectins (Maino et al., 1974; Mastro and Mueller, 1974; Wang et
al., 1975b). Another of the early effects seen after mitogen binding
is the increased metabolism of the phospholipids, particularly phosphatidylinositol (Fisher and Mueller, 1968), which may be controlled
by changes in the intracellular concentration of free Ca^{2+} (Allan
and Michell, 1974). It is not clear which of these events are on the
direct required pathway for initiation of DNA synthesis; some may reflect parallel effects of lectin binding that are not necessary for
eventual replication.

It appears that high doses of Con A inhibit stimulation at some point
after the commitment event for a cell (McClain and Edelman, 1976).
It is not known, however, whether the observed effect of similar
doses of Con A in inhibiting surface receptor mobility is causally
related to the inhibition of mitogenesis.

In contrast, for both fibroblasts and lymphocytes, there is a colchicine-sensitive event in mitogenic stimulation that is so far kine-
tically indistinguishable from the commitment event for each cell.
This commitment event represents the final integration of various
stimulatory signals and is the point at which a cell becomes irre-
vocably fixed in a determinate pathway leading to the S phase. Colchicine is also known to block entry of neuroblastoma cells into S
phase (Baker, 1976) and to block the serum-induced increase of ornithine decarboxylase activity seen in the G1 phase of fibroblasts
(Chen et al., 1976). All of these results suggest that the effect of
colchicine blockade is a general one acting in a variety of cells.

According to this scheme, only those events that occur during or
after commitment should be inhibited by colchicine. Thus, the fact
that colchicine inhibits phosphatidylinositol turnover (Schellenberg and Gillespie, 1977), but not Ca^{2+} influx (Green et al., 1976)
in stimulated lymphocytes places these events respectively near and
before the commitment point. There is some dispute whether colchicine
inhibits mitogen-induced increases in RNA synthesis, protein synthesis, lipid metabolism, and amino acid transport (Betel and Martijnse,
1976; Hauser et al., 1976; Resch et al., 1977). The resolution of these
points of difference in either direction does not directly affect
the by now well-established phenomenon of inhibition by colchicine
of an early event in cell stimulation. The results only put restraints
on the definition of precommitment and postcommitment events, and of
sequential vs. parallel events in the stimulation pathway.

In any case, the body of evidence now at hand suggests that cell surface interactions and microtubular states can all affect the regulation of commitment to DNA synthesis. The nature of the commitment
signal itself remains to be established. In the light of evidence
suggesting stochastic entry into S phase, it may be that colchicine
perturbs the fluctuation or state of a key initiator of growth present in one or a few copies. One possible candidate is the centriole,
another the microtubule organizing centers. Even after commitment

mediated by such initiators of growth, however, events leading to surface modulation prevent further expression of mitogenic signalling. Thus, the idea arises that growth regulation from within and without the cell may involve two factors: (1) a stochastically fluctuating initiating element present in small amounts; and (2) a regulated set of thresholds for cellular response after a critical committing fluctuation.

E. Initiation of DNA Replication

While the interactions of these components clearly play a key role in the early events leading to the generation of a mitogenic signal and to commitment, they clearly are not sufficient by themselves to cause DNA synthesis and cell division. In the last portion of this paper we will discuss our preliminary attempts to analyze more directly the necessary dependent events late in the pathway leading to DNA synthesis in the committed cell. In particular, are there factors that appear in committed cells and that are capable of *inducing* DNA replication in the nucleochromatin of a resting cell? It now appears to be possible to define a cell-free assay for these factors (Jazwinski et al., 1976). In particular, cytoplasmic extracts prepared from rapidly growing cells of the mouse lymphoma cell line P388 induce DNA synthesis in nucleochromatin obtained from frog liver or spleen cells. The amount of synthesis observed was proportional to both the amount of cytoplasmic extract added and to the amount of nucleochromatin added. The activity was heat-labile, was nondialyzable, and was abrogated by tryptic digestion. The reaction required Mg^{2+} as well as the four deoxynucleoside triphosphates. Three lines of evidence indicate that the reaction represents authentic DNA replication and not repair synthesis: (1) the reaction was ATP-dependent; (2) DNA synthesis in the assay mixture was discontinuous; and (3) replication "eyes" (Fig. 15) appeared in the DNA during the reaction.

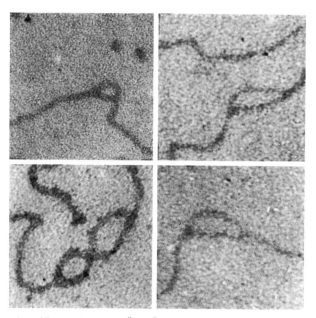

Fig. 15. Replication "eyes". Frog liver nuclei (3.6 x 10^6) were incubated with or without 50 μl of P388 cell extract

Further support for the notion that initiation of DNA replication occurs in the system is provided by the fact that nuclei derived from resting cells that are not in the process of DNA replication responded to stimulation by the cell extract. Moreover, the "initiation" activity was present only in extracts prepared from growing cells, whereas resting cells did not possess stimulatory activity (Jazwinski et al., 1976).

Perhaps the most striking evidence for a direct relationship between this initiation activity and the control of cell growth has been provided by the extension of these studies to yeast cells. Extracts of the cytoplasm of disrupted spheroplasts of *Saccharomyces cerevisiae* stimulated DNA synthesis in frog spleen nucleochromatin when assayed as described above. Temperature-sensitive mutants of the cell division cycle (*cdc* mutants 4, 7, 8, and 28 (Hartwell et al., 1974)) grown at the permissive temperature (23°C) likewise yielded extracts that were capable of stimulating DNA replication. When the cells were incubated for one generation at the nonpermissive temperature (36°C), however, their extracts showed little or no activity. All of these mutants are deficient in events of the dependent pathway leading to initiation of DNA synthesis in the yeast cell cycle. A *ts* mutant, *cdc* 10, deficient in the separate pathway for cytokinesis, showed little or no loss of activity at the nonpermissive temperature. These data indicate that the "initiation" activity, as assayed in vitro, is subject to control in the yeast cell cycle, and its appearance may be one of the terminal events in the pathway leading to DNA synthesis (Jazwinski and Edelman, 1976).

Fractionation of the active components in the yeast extract, coupled with further utilization of the *cdc* mutants in this and in other assays, should clarify the connection between commitment and initiation events. Additional studies relating the particular control events discussed earlier to the appearance of factors capable of inducing DNA replication open up the prospect of reducing certain features of cell cycle control to molecular terms.

Acknowledgments. We are grateful to Dr. Hidesaburo Hanafusa for his generous gift of the viruses used in this study and for his advice on their culture. This work was carried out under the support of U.S. Public Health Service grants AI-11378, AI-09273, and AM-04256 form the National Institutes of Health. SMJ is a postdoctoral fellow of the Helen Hay Whitney Foundation.

References

Albertini, D.F., Clark, J.I.: Membrane-microtubule interactions: concanavalin A capping induced redistribution of cytoplasmic microtubules and colchicine binding proteins. Proc. Natl. Acad. Sci. U.S. 72, 4976-4980 (1975)

Allan, D., Michell, R.H.: Phosphatidylinositol cleavage in lymphocytes: requirement for calcium ions at a low concentration and effects of other cations. Biochem. J. 142, 599-604 (1974)

Allwood, G., Asherson, G.L., Davey, M.J., Goodford, P.J.: The early uptake of radioactive calcium by human lymphocytes treated with phytohemagglutinin.Immunology 21, 509-516 (1971)

Ash, J.F., Singer, S.J.: Concanavalin A-induced transmembrane linkage of concanavalin A surface receptors to intracellular myosin-containing filaments. Proc. Natl. Acad. Sci. U.S. 73, 4575-4579 (1976)

Ash, J.F., Vogt, P.K., Singer, S.J.: Reversion from transformed to normal phenotype by inhibition of protein synthesis in rat kidney cells infected with a temperature-sensitive mutant of Rous sarcoma virus. Proc. Natl. Acad. Sci. U.S. 73, 3603-3607 (1976)

Aubin, J.E., Carlsen, S.A., Ling, V.: Colchicine permeation is required for inhibition of concanavalin A capping in Chinese hamster ovary cells. Proc. Natl. Acad. Sci. U.S. 72, 4516-4520 (1975)

Baker, M.E.: Colchicine inhibits mitogenesis in C1300 neuroblastoma cells that have been arrested in G_0. Nature (London) 262, 785-786 (1976)

Behnke, O., Forer, A.: Evidence for four classes of microtubules in individual cells. J. Cell Sci. 2, 169-192 (1967)

Bell , J.G., Wyke, J.A., Macpherson, I.A.: Transformation by a temperature sensitive mutant of Rous sarcoma virus in the absence of serum. J. Gen. Virol. 27, 127-134 (1975)

Berlin, R.D., Oliver, J.M., Ukena, T.E., Yin, H.H.: Control of cell surface topography. Nature (London) 247, 45-46 (1975)

Betel, I., Martijnse, J.: Drugs that disrupt microtubuli do not inhibit lymphocyte activation. Nature (London) 261, 318-319 (1976)

Brinkley, R.R., Fuller, G.M., Highfield, D.P.: Cytoplasmic microtubules in normal and transformed cells in culture: analysis by tubulin antibody immunofluorescence. Proc. Natl. Acad. Sci. U.S. 72, 4981-4985 (1976)

Chen, K., Heller, J., Canellakis, E.S.: Studies on the regulation of ornithine decarboxylase activity by the microtubules: the effect of colchicine and vinblastine. Biochem. Biophys. Res. Commun. 68, 401-408 (1976)

dePetris, S.: Concanavalin A receptors, immunoglobulins, and θ antigen of the lymphocyte surface: interactions with concanavalin A and with cytoplasmic structures. J. Cell Biol. 65, 123-146 (1975)

Edelman, G.M.: Surface modulation in cell recognition and cell growth. Science 192, 218-226 (1976)

Edelman, G.M., Yahara, I.: Temperature-sensitive changes in surface modulating assemblies of fibroblasts transformed by mutants of Rous sarcoma virus. Proc. Natl. Acad. Sci. U.S. 73, 2047-2051 (1976)

Edelman, G.M., Yahara, I., Wang, J.L.: Receptor mobility and receptor-cytoplasmic interactions in lymphocytes. Proc. Natl. Acad. Sci. U.S. 70, 1442-1446 (1973)

Fisher, D.B., Mueller, G.C.: An early alteration in the phospholipid metabolism of lymphocytes by phytohemagglutinin. Proc. Natl. Acad. Sci. U.S. 60, 1396-1402 (1968)

Frye, C.D., Edidin, M.: The rapid intermixing of cell surface antigens after formation of mouse-human heterokaryons. J. Cell Sci. 7, 319-335 (1970)

Goldberg, N.D., Haddox, M.K., Estensen, R., White, J.G., Lopez, C., Hadden, J.W.: In: Cyclic AMP Cell Growth and the Immune Response. Braun, W., Lichtenstein, L., Parker, C. (eds.). New York: Academic Press, 1974, pp. 247-262

Greene, W.C., Parker, C.M., Parker, C.W.: Colchicine-sensitive structures and lymphocyte activation. J. Immunol. 117, 1015-1022 (1976)

Gunther, G.R., Wang, J.L., Edelman, G.M.: The kinetics of cellular commitment during stimulation of lymphocytes by lectins. J. Cell Biol. 62, 366-377 (1974)

Gunther, G.R., Wang, J.L., Edelman, G.M.: Kinetics of colchicine inhibition of mitogenesis in individual lymphocytes. Exptl. Cell Res. 98, 15-22 (1976)

Gunther, G.R., Wang, J.L., Yahara, I., Cunningham, B.A., Edelman, G.M.: Concanavalin A derivatives with altered biological activities. Proc. Natl. Acad. Sci. U.S. 70, 1012-1016 (1973)

Hadden, J.W., Hadden, E.M., Haddox, M.K., Goldberg, N.D.: Guanosine 3':5'-cyclic monophosphate: a possible intracellular mediator of mitogenic influences in lymphocytes. Proc. Natl. Acad. Sci. U.S. 69, 3024-3027 (1972)

Hauser, H., Knippers, R., Schäfer, K.P., Sons, W., Unsöld, H.-J.: Effects of colchicine on ribonucleic acid synthesis in concanavalin A-stimulated bovine lymphocytes. Exptl. Cell Res. 102, 79-84 (1976)

Hubbell, W.L., McConnell, H.M.: Motion of steroid spin labels in membranes. Proc. Natl. Acad. Sci. U.S. 63, 16-22 (1969)

Jazwinski, S.M., Edelman, G.M.: Activity of yeast extracts in cell-free stimulation of DNA replication. Proc. Natl. Acad. Sci. U.S. 73, 3933-3936 (1976)

Jazwinski, S.M., Wang, J.L., Edelman, G.M.: Initiation of replication in chromosomal DNA induced by extracts from proliferating cells. Proc. Natl. Acad. Sci. U.S. 73, 2231-2235 (1976)

Kawai, S., Hanafusa, H.: The effects of reciprocal changes in temperature on the transformed state of cells infected with a Rous sarcoma virus mutant. Virology 46, 470-479 (1971)

Lazarides, E.: Actin, α-actinin, and tropomyosin interaction in the structural organization of actin filaments in nonmuscle cells. J. Cell Biol. 68, 202-219 (1976)

Lindahl-Kiessling, K.: Mechanism of phytohemagglutinin (PHA) action. V. PHA compared with concanavalin A (Con A). Exptl. Cell Res. 70, 17-26 (1972)

Maino, V.C., Green, N.M., Crumpton, M.J.: The role of calcium ions in initiating transformation of lymphocytes. Nature (London) 251, 324-327 (1974)

Mastro, A.M., Mueller, G.C.: Synergistic action of phorbol esters in mitogen-activated bovine lymphocytes. Exptl. Cell Res. 88, 40-46 (1974)

McClain, D.A., D'Eustachio, P., Edelman, G.M.: The role of surface modulating assemblies in growth control of normal and transformed fibroblasts. Proc. Natl. Acad. Sci. U.S. 74, 666-670 (1977)

McClain, D.A., Edelman, G.M.: Analysis of the stimulation-inhibition paradox exhibited by lymphocytes exposed to concanavalin A. J. Exp. Med. 144, 1494-1506 (1976)

Medrano, E., Piras, R., Mordoh, J.: Effect of colchicine, vinblastine, and cytochalasin B on human lymphocyte transformation by phytohemmagglutinin. Exptl. Cell Res. 86, 295-300 (1974)

Pollack, R., Osborn, M., Weber, K.: Patterns of organization of actin and myosin in normal and transformed cultured cells. Proc. Natl. Acad. Sci. U.S. 72, 994-998 (1975)

Poo, M.M., Cone, R.A.: Lateral diffusion of rhodopsin in photoreceptor membrane. Nature (London) 247, 438-441 (1974)

Powell, A.E., Leon, M.A.: Reversible interaction of human lymphocytes with the mitogen concanavalin A. Exptl. Cell Res. 62, 315-325 (1970)

Resch, K., Bouillon, D., Gemsa, D., Averdunk, R.: Drugs which disrupt microtubules do not inhibit the initiation of lymphocyte activation. Nature (London) 265, 349-351 (1977)

Roth, L.E., Electron microscopy of mitosis in amebae. III. Cold and urea treatments: a basis for tests of direct effects of mitotic inhibitors on microtubule formation. J. Cell Biol. 34, 47-59 (1967)

Rutishauser, U., Yahara, I., Edelman, G.M.: Morphology, motility, and surface behavior of lymphocytes bound to nylon fibers. Proc. Natl. Acad. Sci. U.S. 71, 1149-1153 (1974)

Schellenberg, R.R., Gillespie, E.: Colchicine inhibits phosphatidylinositol turnover induced in lymphocytes by concanavalin A. Nature (London) 265, 741-742 (1977)

Schlessinger, J., Axelrod, D., Koppel, D.E., Webb, W.W., Elson, E.L.: Lateral transport of a lipid probe and labeled proteins on a cell membrane. Science 195, 307-309 (1977b)

Schlessinger, J., Elson, E.L., Webb, W.W., Metzger, H.: Lateral motion and valence of Fc receptors on rat peritoneal mast cells. Nature (London) 264, 550-552 (1976b)

Schlessinger, J., Elson, E.L., Webb, W.W., Yahara, I., Rutishauser, U., Edelman, G.M.: Receptor diffusion on cell surfaces modulated by locally-bound concanavalin A. Proc. Natl. Acad. Sci. U.S. 77, 1110-1114 (1977a)

Schlessinger, J., Koppel, D.E., Axelrod, D., Jacobson, K., Webb, W.W., Elson, E.L.: Lateral transport on cell membranes: mobility of concanavalin A receptors on myoblasts. Proc. Natl. Acad. Sci. U.S. 73, 2409-2413 (1976a)

Smith, J.A., Martin, K.: Do cells cycle? Proc. Natl. Acad. Sci. U.S. 70, 1263-1267 (1973)

Stehelin, D., Varmus, H.E., Bishop, J.M., Vogt, P.K.: DNA related to the transforming gene(s) of avian sarcoma viruses is present in normal avian DNA. Nature (London) 260, 171-173 (1976)

Sundqvist, K.-G., Ehrnst, A.: Cytoskeletal control of surface membrane mobility. Nature (London) 264, 226-231 (1976)

Taylor, R.B., Duffus, W.P.H., Raff, M.C., dePetris, S.: Redistribution and pinocytosis of lymphocyte surface immunoglobulin molecules induced by antiimmunoglobulin antibody. Nature (London) 233, 225-229 (1971)

Temin, H.M.: Stimulation by serum of multiplication of stationary chicken cells. J. Cell Physiol. 78, 161-170 (1972)

Tilney, L.G., Porter, K.R.: Studies on the microtubules in heliozoa. J. Cell Biol. 34, 327-343 (1967)

Watson, J.: The influence of intracellular levels of cyclic nucleotides on cell proliferation and the induction of antibody synthesis. J. Exp. Med. 141, 97-111 (1975)

Wang, J.L., Gunther, G.R., Edelman, G.M.: Inhibition by colchicine of the mitogenic stimulation of lymphocytes prior to the S phase. J. Cell Biol. 66, 128-144 (1975a)

Wang, J.L., McClain, D.A., Edelman, G.M.: Modulation of lymphocyte mitogenesis. Proc. Natl. Acad. Sci. U.S. 72, 1917-1921 (1975b)

Weber, K., Bibring, Th., Osborn, M.: Specific visualization of tubulin-containing structures in tissue culture cells by immunofluorescence. Exptl. Cell Res. 95, 111-120 (1975)

Whitney, R.B., Sutherland, R.M.: Enhanced uptake of calcium by transforming lymphocytes. Cell Immunol. 5, 137-147 (1972)

Wicke, G., Lundblad, V.J., Cole, R.D.: Competence of soluble cell extracts as microtubule assembly systems. J. Biol. Chem. 252, 794-796 (1977)

Wilson, L., Bamburg, J.R., Mizel, S.B., Grisham, L.M., Creswell, K.M.: Interaction of drugs with microtubule proteins. Federation Proc. 33, 158-166 (1974)

Wilson, L., Friedkin, M.: The biochemical events of mitosis. I. Synthesis and properties of colchicine labeled with tritium in its acetyl moiety. Biochemistry 5, 2463-2468 (1966)

Yahara, I., Edelman, G.M.: Restriction of the mobility of lymphocyte immunoglobulin receptors by concanavalin A. Proc. Natl. Acad. Sci. U.S. 69, 608-612 (1972)

Yahara, I., Edelman, G.M.: The effects of concanavalin A on the mobility of lymphocyte surface receptors. Exptl. Cell Res. 81, 143-155 (1973a)

Yahara, I., Edelman, G.M.: Modulation of lymphocyte receptor redistribution by concanavalin A, anti-mitotic agents and alterations of pH. Nature (London) 246, 152-155 (1973b)

Yahara, I., Edelman, G.M.: Electron microscopic analysis of the modulation of lymphocyte receptor mobility. Exptl. Cell Res. 91, 125-142 (1975a)

Yahara, I., Edelman, G.M.: Modulation of lymphocyte receptor mobility by concanavalin A and colchicine. Ann. N.Y. Acad. Sci. 253, 455-469 (1975b)

Yahara, I., Edelman, G.M.: Modulation of lymphocyte receptor mobility by locally bound concanavalin A. Proc. Natl. Acad. Sci. U.S. 72, 1579-1583 (1975c)

Younkin, L.H.: In vitro response of lymphocytes to phytohemagglutinin (PHA) as studied with antiserum to PHA. Initiation period, daughter-cell proliferation, and restimulation. Exptl. Cell Res. 75, 1-10 (1972)

Discussion Session II: Membranes

Chairman: W.W. FRANKE, Heidelberg, F.R.G.

Chairman:
What surface phenomena are correlated with mitosis or replication?

N. Paweletz, Heidelberg, F.R.G.
Cells in tissue culture normally start to round up when they enter mitosis and partly detach from their neighbors or the support. In the scanning electron microscope
we find an increase in the number of microvilli per square unit. During mitosis
there is also an increase in surface area during cleavage-furrow formation, probably
due to the unfolding of microvilli. Cells within the tissue also start to round up
as far as possible, since they do not completely lose contact with their neighbors,
but only become loosened in some areas. Anyhow, we can observe gross morphologic
changes of the cell surface, which may be related to changes in the composition of
the membrane itself or to changes in the subcortical layer.

C.A. Pasternak, London, U.K.
During the cell cycle of mammalian cells in suspension, the number of microvilli
increases as the cells increase in volume, so that by the time the volume has
doubled, the actual cell surface (apparent surface of a sphere + surface area due
to microvilli) has doubled also. The extra surface area due to the microvilli is
then used for the cytokinetic furrow during mitosis.
The same mechanism of cytokinesis probably occurs in nonsuspension cells, that attach
and spread on a substratum, also. But because the surface change from rounded-up
(during mitosis) to fully spread (in mid G1) is very large and is accompanied by a
loss of microvilli, the surface change during mitosis, which is relatively less
dramatic, may be obscured.
One may conclude that microvilli play a role whenever extra surface area is required: 1) during cytokinesis; 2) during cell spreading; 3) during cell swelling
(see Knutton, S. et al., Nature 1976); 4) during exocytosis and endocytosis, and
so forth.

N. Paweletz, Heidelberg, F.R.G.
Concerning surface changes in cells in suspension, we know that cells which normally are attached to some support (glass or neighboring cells) do not enter
mitosis when they remain in suspension. Cells which remain attached to the support and are treated with colchicine start with a strong blebbing during mitosis,
in particular at higher doses. So we can conclude that colchicine also acts on
the cell surface, in this case probably on skeletal elements.

B.M. Jockusch, Basel, Switzerland
A comment to Dr. Edelman's remarks on the correlation between stress fibers and
cell shape: In collaboration with Dr. Wiedmann, Basel, we have analyzed the morphology of a parent melanoma line and a number of variants obtained from the parent
line by selecting survivors in high doses of wheat germ agglutinin (WGA). The
parent line contains no stress fibers (as seen in indirect immunofluorescence with
anti-actin and anti-α-actinin), whereas the WGA-resistant variants do. This is
correlated with the in vivo behavior of these cells: While the parent melanoma
cells are highly tumorigenic, invasive, and have a high capacity to form metastases,
the variants have lost invasiveness and metastasizing capacity, although they still
form primary tumors. Both cell types have a similar cell shape and do not differ
grossly in size.
In collaboration with Dr. Wiedmann, Basel, we also obtained a concanavalin A-resistant variant from the same melanoma line: This variant is smaller, blacker than
the parent line and contains no stress fibers. It is highly malignant and readily
forms metastases. One of our WGA-resistant lines was subjected to high doses of
concanavalin A. Among the survivors was one line with the overall morphology of
the concanavalin A-resistant line described above. This "double variant" again
showed a high capacity for forming metastases, correlated with absence of stress
fibers.

G.M. Edelman, New York, USA
I think that this is a very fascinating set of observations. Perhaps one implication
is that, since these cells have stress fibers and are still tumor cells, the stress
fibers are not involved in normal control of the cell cycle. It also suggests that
the absence of stress fibers is correlated with a motile state.

M. DeBrabander, Beerse, Belgium
Another argument stresses the relationship between cell migration and metastasis or
invasion. Dr. Mareel from Ghent has recently shown that destruction of microtubules
by antimicrotubular drugs completely inhibits invasion in an in vitro organ culture
system. This is probably due to loss of the capacity for directional migration when
cells are devoid of microtubules.

M. Osborn, Göttingen, F.R.G.
Our original data showed a correlation by immunofluorescence microscopy between the
absence of bundles of actin fibers and growth in methocellulose in a series of
well-characterized SV4O-transformed cells and revertants (Pollack et al., Proc.
Natl. Acad. Sci. USA 72, 994, 1975). Growth in methocellulose has been correlated
with tumorigenicity of the same cells in the nude mouse (Shin et al., Proc. Natl.
Acad. Sci. USA 72, 4435,1975). Subsequently a similar correlation between the absence of
well-defined actin bundles and the transformed phenotype was discovered in chick
embryo cells transformed by RNA tumor viruses (Edelman, this volume; Edelman and
Yahara, Proc. Natl. Acad. Sci. USA 73, 2047, 1976; Ash et al., Proc. Natl. Acad.
Sci. USA 73, 3603, 1976). It will be interesting to see how general such a relationship is . Cells showing a noncoordinate change in these properties, such as
those described by Dr. Jockusch, may be interesting because they may provide
further insight into how tumorigenicity, shape, and actin cables are interrelated.

One further comment: I do not know of any study in which cell movement was investigated as a function of the thickness of the actin bundles. However, our general
impression is that perhaps there is an inverse correlation since cells such as
human fibroblasts, which have very thick bundles, hardly move, while other cells,
in particular also ameba, which can move very fast seem to lack these thick cables.

K. Resch, Heidelberg, F.R.G.
I want to comment on the problem of surface receptor heterogeneity. Activating
ligands such as concanavalin A bind to a variety of surface components that carry
the appropriate oligosaccharide, glycolipids, and several glycoproteins. Lymphocytes are activated to grow when only a small proportion of binding sites interact
with an activating ligand. Moreover, interaction of more binding sites with this
ligand may even inhibit cell activation. Thus the most simple assumption would be
that we have to distinguish receptors in a strict sense, i.e., which upon interaction with a ligand transmit a signal into the cell, resulting in growth or division and which are therefore part of the coupling device between ligand binding
and cell activation, from "bulk" binding sites, which are biologically inert. Indeed there are some data supporting this activation in lymphocytes.
Finally I would like to ask Dr. Edelman a question: Are there any data on the modulation of surface components, which can be defined as receptors rather than
mere binding sites?

G.M. Edelman, New York, USA
Of course, there is very great heterogeneity in the glycoproteins to which concanavalin A binds. We have used immune microprecipitation methods and membrane extraction techniques combined with gel electrophoresis to do this. The main point
is that concanavalin A, which induces surface modulation, and WGA, which does not,
gave two complex but *different* patterns. This implies that only *some* glycoproteins
are necessary for induction of modulation.

K. Resch, Heidelberg, F.R.G.
In lymphocytes I believe we have found experimental access to the problem of receptor heterogeneity. At present we can only speculate whether it will enable us
to define exactly what a receptor (in the strict biologic meaning) is. What we
found can be summarized briefly:
From homogeneous plasma membrane vesicles of T lymphocytes two fractions can be
separated by affinity chromatography on concanavalin A-sepharose. The membrane

vesicles have a mean diameter of 100-300 nm (one lymphocyte releases between 2000 to 5000 such membrane fragments) and are fairly pure according to chemical composition and marker enzymes. We have data showing that both fractions are outside out, i.e., have preserved their original orientation. More important, there is a strong suggestion that both fractions are derived from the same cell.
One of the vesicle fractions (which contain about 1/3 of the total plasma membrane) is enriched in binding sites which show a higher affinity for concanavalin A than the binding sites of the other fraction. Both fractions, in addition, are clearly distinct in their content of membrane-bound enzymes.
Thus, the plasma membrane of lymphocytes is heterogeneous with respect to the affinity of binding sites to an activating ligand (Con A) and, more intriguing, heterogeneous in the membrane areas associated with the different binding sites.

J. Robertson, München, F.R.G.
Dr. Edelman, do you, or does anyone here have information concerning a modification of the actin or myosin-like molecules of the surface-modulating apparatus - e.g., by phosphorylation or by attachment of a carbohydrate moiety - following transformation?

G.M. Edelman, New York, USA
No, there is no work yet done on phosphorylating enzymes in connection with the *src* gene product.

J.H. Peters, Köln, F.R.G.
Together with Beug and Graf we studied the shift from the malignant to the benign phenotype of enucleated chicken fibroblasts infected with a temperature-sensitive mutant of Rous sarcoma virus. It appeared that the phenotype could be changed according to the temperature shift in both directions. This argues that the expression of the malignant phenotype is independent of a nuclear feedback control mechanism.

O.G. Meijer, Amsterdam, The Netherlands
I would like to comment on the possible role of the cell coat in proliferation. It seems to be an established fact that lytic activity around the cell coat is correlated with proliferative activity. For trypsin this has been interpreted by claiming that it might digest the receptor for some inhibitory substance. But we recently found that collagenase and hyaluronidase enhance cell proliferation - of HeLa cells and human fibroblasts. Collagen and hyaluronic acid are not known to have receptor functions as far as I know. Does this not imply that the cell coat could have a mechanical role in determining the state in which proliferation might or might not occur? I suggest we include possible models for cell coat architecture in models we make for the membrane and cortex to understand the role of lytic activity in proliferation.

G.M. Edelman, New York, USA
Perhaps we ought to distinguish between necessary *specific* interactions with a particular receptor and *sufficient* conditions such as you mention for the general state of all of the cell surface receptors.

M. DeBrabander, Beerse, Belgium
The removal of cell coat material could disclose cell surface receptors for, e.g., growth factors. Could this explain the mitogenic effecs of proteases, hyaluronidases, etc.?

O.G. Meijer, Amsterdam, The Netherlands
Schnebli and others have claimed, about four years ago, that lytic activity around cells could be a causative factor for malignant behavior. This has been falsified several times by showing that trypsin inhibition did not return transformed cells to the normal state. But it did slow down the cell cycle in all phases. This indicates that lytic enzymes have a rather general role and do not function as a signal in one particular moment.

Chairman:
What changes in permeability and ion transport may be correlated with mitosis?

D. Mazia, Berkeley, California, USA
We can find the largest amount of information on membrane changes associated with the activation of cell division in the literature on fertilization of sea urchin eggs. Here is a very quiescent cell that can turn on very quickly - DNA synthesis begins 15 min after fertilization. All categories of membrane properties change at fertilization and all in the same direction - increase - whether of ionic fluxes, passive diffusion, transport processes. The situation is quite different from that in the lymphocyte, and it would be profitable to compare the two as we attempt to relate membrane changes to mitogenesis. In the egg, all systems are ready to go; it has everything the lymphocyte must make before the latter can begin processes directly related to mitosis.

C.A. Pasternak, London, U.K.
Some enzymes (like 5'-nucleotidase) gradually double in amount (like total plasma membrane protein), between G1 and G2. But others, like Na/K-ATPase, increase 10- to 20-fold (implying some sort of turnover, or change, in activity) towards the end of the cell cycle (see Graham, J.M., et al., Nature 246, 291, 1973).

Chairman:
Is there any evidence for a direct structural coupling of, e.g., microtubules or microfilaments to integral membrane proteins of the surface membrane?

M. DeBrabander, Beerse, Belgium
I fear that we have all been misguided by the old dogma that microtubules serve primarily a cytoskeletal role in mammalian cells, similarly to what has been shown in many protozoa. This comes from the shape changes that occur in cultured cells treated with antimicrotubular drugs. However, these can, in fact, not be explained by the mere disappearance of an endoskeleton. Indeed, the cells remain flattened and only assume an irregular shape in the culture plane. Moreover, cells without microtubules are perfectly able to flatten onto the substratum. Shape changes can better be explained through the loss of coordination of the cortical microfilament system. This system normally is divided into an active part of the leading front of the cell and a stable part in the rest of the cell cortex. Microtubule disappearance results in the uncontrolled activation of the microfilament system all over the cell periphery, which becomes visible as generalized undulating membrane activity. Thus, microtubules seem to play a signal-transducing rather than a skeletal role in these events. The same could operate in the transmembrane control of cell surface components. Cell surface receptors would be coupled to the cortical microfilaments; there is ample ultrastructural evidence for such a possibility. Microtubules, by serving a signal transducing role from the cell periphery to the cell center and vice versa, could control in some way the activity of the microfilament system and by doing so, the movement of the cell surface receptors.

Chairman:
Can somebody summarize the morphological facts that are known about the interaction or the proximity of cortical microtubules and surface membranes?

G.G. Borisy, Madison, Wisconsin, USA
There are certain cells in which microtubules are closely applied to the membrane, e.g., sperm. Nucleated erythrocytes of many species have a marginal band of microtubules which are applied directly underneath the cell membrane. In many protozoa there are systems of microtubules that run just under the cell membrane. However, in most tissue culture cells that I know of, there have been no observations of tubules directly contacting membranes. One sees rather a network of fine filaments, and the tubules tend to lie deeper in the cytoplasm.

D. Mazia, Berkeley, California, USA
Using colchicine, etc., we ask questions about tubulin molecules but demand answers about microtubules. We consider it a paradox when we observe, as Dr. Edelman has done, very important effects of antitubulin drugs on cell surface phenomena without finding the postulated microtubules. Are we ready to consider that the tubulin molecule plays some part in membrane structure or behavior when it is not assembled as microtubules?

M. DeBrabander, Beerse, Belgium
Using the Sternberger immunocytochemical technique we have not seen any special association of tubulin with the cell membrane.

M. Osborn, Göttingen, F.R.G.
Perhaps failure to find tubulin in the membrane or nucleus by immunocytochemical techniques should be interpreted at the moment with caution. We do not know with what efficiency any given tubulin antibody would recognize soluble tubulin as opposed to tubulin in microtubules. Nor do we know yet whether soluble tubulin might be selectively extracted by fixation.

Chairman:
May I just summarize this again? There is the existence - in membrane fractions from a variety of sources, we all are aware of the weakness of this preparative term - of the association of tubulin. In particular, I would like to mention brain membrane fractions, such as synaptosomal membranes. One substructure, which I find intriguing is the submembranous fuzzes that are closely associated with the postsynaptosomal membrane (Feit, H., et al., Proc. Natl. Acad. Sci. USA 74, 1047, 1977). When you purify it to an extreme, even after treatment with deoxycholate, you get an enrichment of α and β tubulin. The problem with all of these membrane fractions is, how can it be specifically ruled out that you did not have redistribution during homogenization? Three groups of authors, as far as I know, have checked this by adding soluble radioactive tubulin during homogenization. However, this method is useful only for soluble pools of tubulin and there might be others that do not exchange during this period of homogenization.

G.G. Borisy, Madison, Wisconsin, USA
Suppose tubulin were in the membrane. What biological function would it serve there, as a tubulin molecule, unassociated with other tubulin molecules to form a structure?

Chairman:
These fuzzes are not really *in* the membrane, they are in some dense layer *at* the membrane. There is a whole list of cytological situations where you do have microtubule organizing centers that show up as such fuzzes.
There is another line of experimental evidence that supports this claim of membrane-associated tubulin, i.e., binding experiments with colchicine, which have even more pitfalls. So the evidence for membrane-associated tubulin still must be considered weak.
Do we have evidence that there is any form of organized tubulin other than in microtubules, especially in association with membranes?

G.G. Borisy, Madison, Wisconsin, USA
There is some evidence for the association of microtubules with internal membranes of tissue culture cells. The significance of this has not yet been established. Smith has described the association of neurotubules with mitochondria and synaptosomes. Porter and Murphy and Tilney (Murphy, D.B. and Tilney, L.G., J. Cell Biol. 61, 757, 1974) have described the association of cytoplasmic microtubules with pigment granules. In tissue culture cells tubule bundles run parallel to elongated mitochondria.

H. Fuge, Kaiserslautern, F.R.G
In these cases the microtubules are connected with the membranes by filamentous units.

M. Girbardt, Jena, G.D.R.
In a fungal hypha the septum is initiated by a membrane modulation at a restricted area of the cytoplasmic membrane. This area is triggered by the mitotic nucleus during the "metaphase." Nuclei, translocated by microsurgery, induce the septum at every phase within the hypha. After completion of the division a microfilamentous ring is attached to the "activated" area of the plasma membrane, which seems to act like the contractile ring in cells of higher animals.

Chairman:
There is agreement that microfilaments are tightly associated with the inner surface of the plasma membrane. Are there specific means to bind them to the inner surface?

B.M. Jockusch, Basel, Switzerland

We would consider α-actinin for the attachment molecule of microfilaments in the membrane. It is not yet clear whether α-actinin is directly inserted in the plasma membrane, but it is clearly associated with the inner surface of the plasma membrane of tissue culture cells. In collaboration with Dr. A.C. Allison we demonstrated binding of ferritin-coupled anti-α-actinin to inside-out plasma membrane vesicles. Liposomes, when incubated with α-actinin, carry α-actinin molecules to their proper density in a sucrose gradient. Preliminary data suggest that the α-actinin molecule is indeed inserted in the lipid bilayer (fluorescent energy transfer, tryptic digestion).

G.M. Edelman, New York, New York, USA

α-Actinin is a good candidate for certain roles – but unless its density is very high it cannot explain the local effects on receptor movements for a great variety of receptors. Even if it is not dense, it would serve a very important role if it linked actin to the membrane.

Chairman:

What regulates nuclear membrane breakdown?

D. Mazia, Berkeley, California, USA

All one can now say about the relation between the chromosome cycle and the breakdown of the nuclear envelope is that the two events are dissociable. There are several cases in which the breakdown of the nuclear envelope can be inhibited, yet the chromosomes continue to condense, and fully condensed chromosomes are seen inside the nucleus.

Chairman:

How may endomenbranes be involved in regulatory aspects of replication and mitosis?

C. Petzelt, Heidelberg, F.R.G.

We have found a membrane-bound Ca^{2+}-ATPase showing cyclic fluctuations of activity during the cell cycle with an activity peak at mitosis. This has been found in sea urchin eggs as well as in mammalian cells. The enzyme differs in several aspects from the Ca^{2+}-ATPase-system of the mitochondria and can be distinguished from the Ca^{2+}-ATPase of the sarcoplasmic reticulum. Although there are several indications showing that it is located in or near the mitotic apparatus, we have not as yet been able to determine its exact location in the cell. The vesicles in or around a mitotic spindle are the obvious candidates for it. Our hypothesis is that by regulating locally the concentration of free Ca^{2+}, this system may control the establishment and/or maintenance of the spindle.

P.J. Harris, Eugene, Oregon, USA

In most animal cells one finds an accumulation of vesicles at the spindle poles during mitosis. This mass of vesicles is especially exaggerated in sea urchin eggs where it defines the clear area of the mitotic apparatus. Vesicles begin to accumulate at the forming asters long before the breakdown of the nuclear membrane. From the central mass or centrosphere they extend outward into the yolk region to form the astral rays, along with radially arranged microtubules, and by telophase the asters have invaded the entire cell. The significance of the association of vesicles with microtubules, while not proved, is suggested by Petzelt's work showing a membrane associated Ca-dependent ATPase in the sea urchin egg, and by the increasing evidence that calcium regulation may control mictrobule polymerization and depolymerization.

P.F. Baker, London, U.K.

What is the dependence on Ca of the Ca-ATPase you have isolated? Attaching any physiologic significance to a Ca-ATPase with such a low affinity for Ca worries me, as all the available evidence is consistent with an ionized Ca inside cells of less than 1 μM (see Baker, Progr. Biophys. Mol. Biol. 24, 179, 1972; Baker and Warner, J. Cell Biol. 53, 579, 1972). Although Ca might rise to high levels locally, to return it to submicromolar values, a system with high affinity for Ca is essential. This causes me to hear with caution any suggestion that the vesicular Ca-ATPase you have described is involved in Ca-sequestration in vivo.

Would it have lost a co-factor or be involved primarily with the transport of another ion?

P.J. Harris, Eugene, Oregon, USA
Whole cell measurements of cytoplasmic calcium do not tell you what concentrations may exist locally. If calcium release is followed by rapid sequestering, the concentration could be very high locally while in the remainder of the cell it could be quite low. The way in which this might work is suggested by Ridgway's observations on fertilization in Medaka eggs (Ridgway, E.B., et al., Proc. Natl. Acad. Sci. USA 74, 623, 1977). Aequorin luminescence appeared as a narrow band moving outward from the point of sperm penetration, indicating that released calcium was rapidly sequestered again.

P. Malpoix, Rhode-St. Genese, Belgium
About the possibility of local intracellular sequestering of calcium, we should remember the example of the mitochondria, which contain a much higher concentration of this ion, and of magnesium, than the cytsol.

Chairman:
Would an increase of calcium be able to control regulating events working, e.g., simply by dissociation of microtubules?

R.B. Nicklas, Durham, North Carolina, USA
It obviously does in muscle, but in intranuclear mitoses there are no vesicles inside at all.

Chairman:
Those who favor this concept would not be too reluctant to include the perinuclear cisternae as another candidate for sequestering Ca.

U.-P. Roos, Zürich, Switzerland
Cellular slime molds are lower eukaryotes considered by some authors to be protozoans, but by others to be primitive fungi. Slime mold amebae in interphase have an electron opaque body (NAB) associated with the nucleus. This body is at the center of an array of cytoplasmic microtubules, and it is typically surrounded by a number of small vesicles of unknown function. During the early stages of mitosis, the nuclear envelope is intact, except for fenestrae with which the spindle pole bodies are associated (Roos, J. Cell Biol. 64, 480, 1975; Moens, J. Cell Biol. 68, 113, 1976). On the cytoplasmic side, modest asters extend from the spindle pole bodies and some vesicles also occur in this area. Thus, there is a correlation between the occurrence of microtubules and vesicles similar to that described by Dr. Harris, but it is correct, as stated by Dr. Nicklas, that there are no intranuclear vesicles. Assuming that the vesicles sequester Ca^{2+}, one could attribute to them a function in the maintenance and/or assembly of *cytoplasmic* microtubules but not of *spindle* microtubules, unless they governed the intranuclear Ca^{2+}-concentration through the polar fenestrae.

G.G. Borisy, Madison, Wisconsin, USA
The idea for the Ca regulation of tubulin disassembly came initially from R. Weisenberg's observation on the reassembly of tubulin in vitro. He concluded that very low concentrations of Ca were required to obtain microtubule assembly in vitro. However, this was based on the use of high concentrations of Ca and high concentrations of a Ca-chelator in a rather incompletely defined system. Subsequent work on the biochemistry of tubulin assembly in vitro has not supported the idea that very low concentrations of calcium will inhibit tubule assembly. So if we wish to retain the idea that calcium may be a regulator of microtubule assembly in the spindle, it is not on the basis of information we have on in vitro assembly.

P.J. Harris, Eugene, Oregon, USA
I think there is too much evidence now for cytoplasmic calcium regulation by means of a sequestering system to rule out the role of calcium control during mitosis. In vitro studies of purified brain tubulin do not necessarily indicate what is happening in the whole cell. The control system may also involve actin and myosin as well as microtubules. Release of calcium in the centrosphere could cause a concentration of an actin-myosin network and a pulling in of spindle microtubules to move the chromosomes.

M. Schliwa, Frankfurt, F.R.G

It has been shown in in vitro experiments that millimolar concentrations of Ca ions are required to block microtubule polymerization efficiently. However, relatively little is known about in vivo effects of Ca ions on microtubules. Using the ionophore A23187, I studied the effects of Ca ions on microtubules in a cell system that allows easy and direct testing of the effects of Ca on microtubules in vivo. I observed shortening of the microtubule-supported axopodia of the heliozoan *Actinosphaerium eichhorni* in the presence of a Ca concentration as low as 10^{-5} M. Shortening clearly correlates with microtubule breakdown. This Ca level seems to be low enough to be of physiological significance.

Session III

Ultrastructure of Mitotic Cells

H. FUGE, Fachbereich Biologie, Universität Kaiserslautern, 6750 Kaiserslautern, FRG

A. Introduction

The study of chromosomal behavior during mitosis has always been one
of the most fascinating topics of cytologic research. The early light-
microscopic examination of fixed and stained cells provided evidence
for the existence of spindle fibers, structures that apparently connected the kinetochores of the condensed chromosomes with the polar centers
(chromosomal fibers), and that spanned the distance between polar
centers (continuous or interpolar fibers). Based on these observations
spindle models were formulated in which the chromosomes were "pulled"
or "pushed" by contraction or dilatation of spindle fibers toward the
poles (reviewed by Schrader, 1953). Since spindle fibers generally
could not be visualized in living cells by bright field microscopy,
several authors maintained that the fibers were artifacts caused by fixation (see Schrader, 1953). However, it could finally be demonstrated
with the polarization microscope that fiber structures are a reality
(Inoué, 1953). Zones of weak birefringence between chromosomes and
spindle poles and between the poles suggested the existence of aniso-
tropically arranged rod-like elements in these areas. Improved electron-microscopic fixation procedures, particularly the introduction
of glutaraldehyde as a fixing agent in the early sixties, resulted
in the demonstration of microtubules (MTs) as the predominant structural elements of the mitotic spindle.

Since most spindle MTs are preferentially oriented toward the poles,
they are regarded as the source of spindle birefringence by most
authors (e.g., Sato et al., 1975). This opinion is not shared by Forer
(1976) who assumes that MTs plus actin components contribute to birefringence in certain spindle regions. Present hypotheses on the possible molecular mechanism of chromosome distribution in mitosis are
focused on spindle MTs as the most probable candidates for force production, at least as integral elements of the force-producing system.
Actin-like filaments have repeatedly been demonstrated in glycerinated
spindles with (and without) heavy meromyosin binding. However, there
is still no clear evidence that they are also present in untreated
spindles in sufficient amounts and in a specific association with
chromosomes, which would suggest a participation in force production
(for references and discussion see Fuge, 1977).

It is beyond the scope of this article to discuss all the ultrastructural data and all the problems concerned with spindle architecture.
Because of the central role MTs obviously play during division, primary emphasis is placed on their different arrangement and participation in genome distribution in various eukaryotes. To discuss spindle models, differences and similarities in spindle architecture
among different eukaryotes must be understood, since a general molecular mechanism of force production, if it exists, must be compatible
with all eukaryote spindles.

Further discussions concerning the ultrastructure of mitotic cells
are found in several comprehensive reviews (Pickett-Heaps, 1969, 1974;
Luykx, 1970; Nicklas, 1971; Bajer and Molè-Bajer, 1972; Heath, 1974b;
Kubai, 1975; Fuller, 1976; Fuge, 1974b, 1977).

B. Types of Spindle Microtubules

The most complex morphologic organization is found in the mitotic
spindle of higher plants and animals. Different types of arrangements
of MTs can be distinguished in a fully developed metaphase spindle of
a higher eukaryote. The commonly used terminology is briefly recapi-
tulated. The fraction connected with chromosomes is called "chromosomal" or "kinetochore" MTs (kMTs). All other spindle MTs can be summarized under the term non-kinetochore (non-kMTs). Intensive serial
section studies have revealed that this latter group can be divided
into at least three types: (1) continuous (interpolar) MTs: MTs that
run from pole to pole; (2) interdigitating MTs: MTs that run from one
pole for a small distance into the opposite half-spindle. MTs from
both half-spindles interdigitate in the equatorial region; (3) free
MTs: relatively short MT fragments that lie somewhere between the
spindle poles. Polar centers, if they are present, are commonly surrounded by or connected with MTs radiating into the cytoplasm. If
centrioles serve as polar centers, these MTs are called centrospheral
or aster MTs (e.g., Bajer and Molè-Bajer, 1972; Fuge, 1974b; Heath
and Heath, 1976).

It must be emphasized that non-kMTs in higher plants and animals
(one exception is discussed below) do not show physical connection to
larger organelles. The only connection that is frequently observed
is the lateral association with neighboring tubules by intertubular
cross-bridges (reviewed by McIntosh, 1974).

As is shown below, these types of spindle MTs are not all present
in many lower eukaryotes, and they are involved to different degrees
in genome distribution (survey in Table 1). Although there is a certain correlation between spindle complexity and organization level of
the organism, spindle architecture can possibly be used only to a
limited extent as a guide for phylogenetic considerations, since it
is difficult to decide in each case which features are really primitive, and which are possibly a result of secondary reduction.

C. Organization of Mitotic Spindles in Lower Eukaryotes

I. Extranuclear Spindles of Dinophyceae and Hypermastigote Flagellates

For the understanding of nuclear division in some primitive eukaryotes
the distribution of DNA to the daughter cells in bacteria must be recapi-
tulated briefly. The nuclear body (nucleoid) of bacteria (e.g., of
Bacillus subtilis, Jacob et al., 1966) is connected with an invaginated
segment of the cell membrane, called the mesosome. When the cell is
about to divide the mesosome is doubled and the two daughter mesosomes move farther and farther apart, thus leading to the separation
of the genome. The molecular mechanism of mesosome separation is still
unknown. There may be localized membrane synthesis between the daughter
mesosomes, but other mechanisms can be imagined (reviewed by Heath,
1974b).

The main point in our context is that genome separation in prokaryotes
is a membrane-mediated process. The nuclear division in the phylogene-
tically primitive group of Dinophyceae can be in some way related to
prokaryote division. The free-living dinoflagellate *Crypthecodinium*
(formerly *Gyrodinium*) has been intensively studied with the serial sectioning technique by Kubai and Ris (1969), and it may serve as an
example for the probably most primitive division type found among
Dinophyceae (for review of the complete literature see Kubai, 1975).

Table 1. Types and involvement of MTs in genome separation (for references and discussion see text)

Location of MTs	Observed MT type	MT behavior during division	Genus (classification)
Extranuclear	contMT	no contMT elongation	*Crypthecodinium* (Dinophyceae)
Extranuclear	contMT	contMT elongation, no kMT shortening	*Amphidinium* (Dinophyceae)[a], *Syndinium* (Dinophyceae), *Trichonympha* (Flagellata)
Intranuclear	contMT	contMT elongation	*Mucor* (Zygomycetes), *Phycomyces* (Zygomycetes), *Pilobolus* (Zygomycetes)
Intranuclear	contMT, kMT	contMT elongation, kMT shortening (?)	*Nassula* (Ciliata), *Paramecium* (Ciliata)
Intranuclear	contMT, kMT	contMT elongation, kMT shortening	*Polysphondylium* (Acrasiomycetes), *Ascobolus* (Ascomycetes), *Saccharomyces* (Ascomycetes)
Intranuclear	contMT, kMT, idMT, 'free' MT	contMT elongation, kMT shortening, (idMT interaction ?)	*Dictyostelium* (Acrasiomycetes)[b], *Thraustotheca* (Oomycetes), *Uromyces* (Basidiomycetes)
Open division	contMT, kMT, idMT	contMT elongation, kMT shortening, (idMT interaction ?)	*Lithodesmium* (Diatomeae), *Diatoma* (Diatomeae)[c]
Open division	contMT (?), kMT, idMT, 'free' MT	kMT shortening, (contMT elongation ?), (idMT interaction ?)	Higher plants and animals

contMT, continuous (interpolar) MTs; kMT, kinetochore or chromosomal MTs; idMT, interdigitating MTs; [a] changes of MT lengths unclear; [b] 'free' MTs not detected; [c] contMT not detected

With the onset of nuclear division all chromosomes (which are prokaryote-like in organization) become associated with the nuclear envelope (n.e.) at special sites where cytoplasmic channels traverse the nucleus. These channels contain bundles of cytoplasmic MTs that are oriented parallel to the prospective division plane. During division of the nucleus the channels and the MTs persist. Although the nucleus elongates, the length of the cytoplasmic MTs remains constant. It has been suggested by Kubai and Ris (1969) that chromosome separation in *Crypthecodinium* is brought about solely by membrane growth, without active participation of the MTs. This feature, together with its peculiar chromosome structure, places *Crypthecodinium* in the neighborhood of prokaryotes. The cytoplasmic MT bundles within the channels, according to Kubai and Ris, may serve as a cytoskeleton and as guiding elements.

Several other organisms from the dinophycean group have been investigated in which cytoplasmic MTs seem to be more active during division. Two examples, the free-living *Amphidinium* (Oakley and Dodge, 1974) and the parasitic *Syndinium* (Ris and Kubai, 1974), are considered. The division of *Amphidinium* still resembles that of *Crypthecodinium*. However, a certain number of MTs within the cytoplasmic channels of *Amphidinium* contact knob-like differentiations of the n.e. On the nuclear side these knobs are connected with the chromosomes. In analogy to higher eukaryotes they can possibly be regarded as kinetochores, although the uninterrupted n.e. still separates chromosomes and MTs.

Similar morphologic features were observed in *Syndinium* (Ris and Kubai, 1974). Since division has been followed in detail, more can be said about MT dynamics in this organism. (1) Cytoplasmic MTs are continuous between separating centrioles. The centriole separation must be assumed to be caused by MT elongation. (2) Centriolar regions and kinetochore-like differentiations of the n.e. are interconnected by MTs. During centriole separation the extranuclear MTs become situated inside one single cytoplasmic channel traversing the nucleus. Since shortening of the kinetochore-centriole MTs could not be observed during division, genome distribution seems to be to some extent caused by centriole separation, i.e., growth of continuous MTs. On the other hand, the n.e. must still be involved in chromatid separation in *Syndinium*. On the basis of the ultrastructural observations it is difficult to decide which is the most important factor for separation.

Another protozoan group, the hypermastigote flagellates, must be included in the group of lower eukaryotes that are typified by the participation of the n.e. in genome separation. A well-studied example is the division of the termite symbiont *Trichonympha agilis* (Kubai, 1973). In this organism a bipolar spindle is formed in the cytoplasm, alongside the nucleus. The spindle then becomes embedded in a nuclear groove during subsequent division. Chromosomes are attached to kinetochore-like structures that are associated with the n.e. Spindle MTs insert into the kinetochores and connect them with the spindle poles. Kubai (1973) suggested that the initial separation of sister kinetochores in *Trichonympha* is achieved by a membrane mechanism. The final division of the nucleus was, however, thought to be accomplished by elongation of the spindle. The kinetochore-pole MTs do not seem to shorten during division, and thus may be regarded as (passive) orientation elements.

II. Occurrence of Intranuclear ("Closed") and "Open" Divisions

It can be summarized that in the low eukaryotes discussed previously, MTs are never in direct contact with chromosomes during division. MTs stay in the cytoplasm, and chromosomes remain within the closed n.e. In these cases it may be justified to speak of an extranuclear spindle (Kubai, 1975).

In all other organisms so far examined such delineation between MTs and chromosomes could not be observed. In terms of the behavior of the n.e., two groups of organisms can be distinguished: (1) organisms that develop a spindle within a persistent n.e., often referred to as the "closed" division type, or as intranuclear divisions; (2) organisms that are characterized by a fragmentation of the n.e. prior to spindle formation or during metaphase ("open" division type).

Closed division is common in different groups of lower eukaryotes such as animal protozoa, green algae, slime molds, and fungi. Open division is typical of higher plants and animals, but it is also found among algae and fungi. Within the first group several organisms possess polar openings (polar fenestrae), an attribute that possibly can be regarded as something like an intermediate type between closed and open division, since the fragmentation of the n.e. also starts in the polar regions in higher plants and animals (see below).

It was suggested that the closed division may represent the phylogenetic precursor to open division (e.g., Pickett-Heaps, 1969). When we look at the fungal kingdom (subdivision Eumycotina) this assumption seems reasonable. Organisms from the more primitive classes of Oomycetes (*Thraustotheca*, Heath, 1974a; *Saprolegnia*, Heath and Greenwood, 1968, 1970), Chytridiomycetes (*Phlyctochytrium*, McNitt, 1973), and Zygomycetes (*Mucor*, McCully and Robinow, 1973; *Phycomyces*, Franke and Reau, 1973; *Pilobolus*, Bland and Lunney, 1975) have a persistent n.e. during division, except for polar fenestrae in some cases. The same is still true for Ascomycetes (*Ascobolus*, *Podospora*, Zickler, 1970; *Schizosaccharomyces*, McCully and Robinow, 1971; *Saccharomyces*, Zickler and Olson, 1975; Peterson and Ris, 1976). In the most advanced class of Basidiomycetes the n.e. generally becomes fragmented about the time of metaphase (e.g., *Coprinus*, Lu, 1967; Lerbs, 1971; *Boletus*, McLaughlin, 1971). However, recent examination of the rust fungus *Uromyces* (Heath and Heath, 1976) showed that intranuclear division also occurs in Basidiomycetes. Probably rust fungi as relative primitive Basidiomycetes are more closely related to Ascomycetes, and thus have a more ascomycete-like division.

When we look at other eukaryotic groups (algae, slime molds) the value of closed versus open division as a phylogenetic marker becomes, however, questionable. Open division occurs, for example, not only in the highly developed filamentous alga *Oedogonium* (Ulotrichales) (Pickett-Heaps and Fowke, 1969), but also in the primitive flagellate *Ochromonas* (Chrysomonadales) (Bouck and Brown, 1973). If the formation of an extranuclear spindle is regarded as primitive, mitosis in diatoms has primitive as well as more advanced characteristics. In *Lithodesmium* (Manton et al., 1969) and *Diatoma* the spindle is formed in the cytoplasm, in addition to the nucleus. Later the n.e. becomes fragmented, as in higher plants, and the spindle sinks down into the nucleoplasm. In the slime mold *Physarum* (Myxomycetes) the haploid myxamebas show open division, whereas in the diploid plasmodia of the same organism, the n.e. remains intact throughout nuclear division (Aldrich, 1969).

III. Intranuclear Spindle Organization

The MTs within the nuclear channels of the dinoflagellate *Cryptheco-dinium* can only have static functions, since changes in length cannot be observed during elongation of the nucleus. In *Trichonympha* and *Syndinium* the continuous extranuclear MTs do elongate during division, and thus may induce the separation of the spindle poles. However, chromosomal MTs still seem to remain inactive and thus may only serve as static orientation elements.

When we look at different organisms with intranuclear spindles, two tendencies in spindle behavior become evident:

1. Separation of the poles by elongation of continuous MTs is accompanied by an approach of the chromosomes to the poles. This is documented by shortening of chromosomal MTs. The relative importance of chromosome-to-pole approach for genome distribution seems to increase with higher organization of the organism.

2. Interdigitating MTs seem to become more frequent at the expense of continuous MTs. This may suggest that interdigitating MTs also become responsible for spindle elongation (see discussion below).

1. Nuclear Division by Spindle Elongation

A simple type of intranuclear mitotic spindle can be observed in some organisms belonging to the fungal group of Zygomycetes (e.g., *Mucor*, McCully and Robinow, 1973; *Phycomyces*, Franke and Reau, 1973; *Pilobolus*, Bland and Lunney, 1975). A spindle of continuous MTs arises (central spindle) between two spindle pole bodies (SPB) that are apposed to the inner nuclear membrane.

Although their function is probably the same in all fungi, there are significant ultrastructural differences in polar organelles (for reference see Heath, 1974b; Kubai, 1975; Fuller, 1976; Fuge, 1977). Structural variation led to multiform terminology: nucleus-associated body, nucleus-associated organelle, SPB, kineto-chore equivalent, centrosomal plaque, centrosome. For simplification, it has been proposed to use SPB for all fungal organelles (Kubai, 1975). This proposal is also followed here.

The central spindle of Zygomycetes consists of 10 - 30 MTs in a fully developed stage (Franke and Reau, 1973; Bland and Lunney, 1975). Condensed chromosomes cannot be seen, and any MTs that are not continuous between the SPBs seem to be missing. The nucleus first becomes dumbbell shaped by elongation of the central spindle and is finally divided into the daughter nuclei. The process of ordered chromatid segregation is totally unclear.

As in the cited Zygomycetes, the n.e. remains intact throughout division in the ciliate micronucleus (e.g., *Nassula*, Tucker, 1967; *Paramecium*, Stevenson and Lloyd, 1971; Stevenson, 1972; *Blepharisma*, Jenkins, 1967). The spindle is formed inside the nucleus without participation of polar organelles comparable to SPBs or centrioles. The polar regions in *Nassula* are characterized by flat membranous vesicles opposed to the inner nuclear membrane. Chromosomes are clearly discernible and form a metaphase plate. Each chromosome is connected with a few chromosomal MTs, although true kinetochore regions cannot be observed. These chromosomal MTs obviously reach to the polar apices. Many spindle MTs exist, besides chromosomal MTs, which probably reach from pole to pole. During mitosis the biconal nucleus adopts a dumbbell shape. This is achieved by the formation of a long "separation spindle" increasing the metaphase spindle length approximately tenfold in anaphase (Stevenson and Lloyd, 1971). Since the number of MTs in the separation spindle remains fairly constant

during elongation (Tucker, 1967), it can be assumed that elongation
of the nucleus is accomplished by growth of the continuous MTs. The
movement of chromosomes could not be followed in detail, but after
spindle elongation has ceased, two portions of chromosomes are found
within the terminal knobs of the nucleus. Although a shortening of
chromosomal MTs cannot be excluded, genome distribution is obviously
mainly effected by elongation of continuous MTs.

2. More Complex Intranuclear Spindles

Several fungal kingdom organisms with closed division are considered
next. In complexity of spindle MT arrangement, they are more similar
to higher eukaryotes.

Polysphondylium (Roos, 1975) and *Dictyostelium* (Moens, 1976) belong to the
slime mold class of Acrasiomycetes. In both organisms the mitosis of
the haploid myxamebas has been followed by electron microscopy. Spindle
formation starts in early prophase in association with the SPBs. The
SPBs lie closely together within a gap of the n.e. (Moens, 1976).
They are interconnected by short continuous MTs; other MTs radiate into the cytoplasm and into the nucleoplasm. The latter point in the
direction of the chromosomes, and it can be assumed that they become
connected to the kinetochores. The further course of the division
process is characterized by a progressive separation of the SPBs that
is accompanied by elongation of the continuous MTs. Since the SPBs
move along the n.e., the continuous spindle between them traverses
the nucleus. The SPBs are finally situated at opposite poles of the
nucleus in a stage that may be homologized to metaphase. The chromosomes are connected with the SPBs by chromosomal MTs. *Dictyostelium*
(Moens, 1976) possesses layered kinetochores that are comparable to
those of animals (for review of kinetochore structure see Fuge, 1974b,
1977). The process of spindle formation suggests that sister chromatids become attached to opposite SPBs at an early stage. There is
considerable evidence that genome separation in anaphase is accomplished
in both slime molds by at least two separate mechanisms, (1) by
elongation of the central spindle between the SPBs, and (2) by poleward motion of the daughter chromosomes. The latter movement is evi-
denced by shortening of the kMTs. Zones of overlap in the interzone
during anaphase of *Dictyostelium* suggest that this organism possesses
interdigitating MTs, in addition to continuous and kMTs.

Several attempts have been made to clarify the ultrastructure of
yeast (Ascomycetes, Endomycetidae) mitosis by examination of ultra-
thin sections (McCully and Robinow, 1971; Moens and Rapport, 1971;
Zickler and Olson, 1975). These studies were impeded by the density
of the yeast nucleus and by the lack of chromosome condensation
during mitosis. The studies revealed the existence of a central
spindle between the SPBs, but not of chromosomal MTs. However, in a
recent study using thick sections of dividing *Saccharomyces* cells and
the high-voltage electron microscope (HVEM), Peterson and Ris (1976)
showed yeast mitosis to be very similar to that of the preceding
slime molds. Although the *Saccharomyces* chromosomes are indistinct,
the presence of chromosomal MTs, which shorten during anaphase, can
no longer be doubted after this study.

Unlike yeast chromosomes, the chromosomes of some organisms from the
group of Euascomycetes are very distinct during mitosis (e.g., *Ascobolus*, *Podospora*, Zickler, 1970). Their spindle architecture is similar
to that of yeasts, except for the morphology of their SPBs and the
arrangement of the continuous spindle. The SPB of *Ascobolus* is plate-
shaped (centrosomal plaque) (Zickler, 1970). This gives rise to
broadly shaped polar zones and a broad central spindle.

Interdigitating MTs are obviously present in the spindle of the slime
mold *Dictyostelium* (Moens, 1976). This type of MTs could be clearly
demonstrated in two other intranuclear spindles *Thraustotheca* (Oomycetes) (Heath, 1974a) *Uromyces* (Basidiomycetes) (Heath and Heath, 1976)
by careful analysis of serial sections. It may be suggested that
comparable analyses will detect interdigitating MTs in still more
fungi with closed division.

The appearance of interdigitating MTs can be regarded as a step in
the direction of higher complexity of the spindle. Spindle elongation
seems to be accomplished in a number of lower eukaryotes simply by
growth of MTs that are continuous between the spindle poles. Interdigitating MTs have stimulated speculation about a second possible
mechanism of spindle elongation. MTs that overlap in the equator region in metaphase could interact by sliding against each other in
anaphase, thus leading to a separation of the poles (according to
the sliding-filament hypothesis of McIntosh et al., 1969). An organism whose spindle architecture may strongly suggest elongation by
sliding is the alga *Diatoma* (Diatomeae). The fully developed metaphase
spindle seems to consist of two interdigitating half-spindles and
of chromosomal MTs. There are no indications of continuous MTs. The
chromosomes first approach the poles in anaphase, and the spindle subsequently elongates. During elongation the zone of overlap decreases
in extent. Separation of the interdigitating half-spindles is the most
reasonable explanation for this phenomenon. However, there is no experimental proof that sliding produces the driving force for separation.

D. Organization of the Mitotic Spindle in Higher Plants and Animals

The more complex and larger in size a mitotic spindle is, the more
it becomes necessary to analyze oriented serial sections of preselected mitotic stages to obtain exact data about MT arrangement. Spindles of higher plants and animals are larger and obviously more complex than those of most lower eukaryotes discussed in Section C.
Among the many plants and animal cell types whose mitosis has been
studied with the electron microscope only a limited number has so
far been examined and analyzed by means of serial sections. These
mainly are endosperm cells of the blood-lily *Haemanthus katherinae*
various cultured mammalian and marsupial cells (e.g., man: HeLa,
WI-38; rat kangaroo *Potorous tridactylis*: PtK), and spermatocytes of
the tipulid *Pales ferruginea*. Although they may not be representative
of all plants and animals the following discussion is restricted to
these cell types. The collected data may be sufficient to demonstrate
some similarities and some obvious differences in spindle ultrastructure of higher eukaryotes.

I. Microtubule Organization Centers in Spindle Formation

Higher plants and animals have in common that the n.e. becomes fragmented with the formation of the division spindle. Whereas SPBs
clearly function as microtubule organization centers (MTOC) in fungal
division the involvement of centrioles during initiation of spindle
MTs in higher eukaryotes is still a matter of controversy. The two
pairs of centrioles in animals generally become arranged at opposite
sides of the nucleus before n.e. fragmentation. During migration toward the prospective spindle poles the centriole pairs are already
surrounded by centrospheral MTs. It is one of the unsolved problems
whether this migration is initiated by growth or by interaction of the

interdigitating MTs from both centrospheres or by some other mechanism (see, e.g., discussion by Roos, 1973a).

In older electron-microscopic studies it has never been doubted that the spindle is formed by centrospheral MTs invading the nucleoplasm from both poles. The nucleus shows invaginations of the n.e. opposite to the centrospheres prior to spindle formation. This has been observed in HeLa cells (Paweletz, 1974), in rat kangaroo cells (Roos, 1973a), and in tipulid spermatocytes (Fuge, unpublished). The invaginations contain centrospheral MTs pointing in the direction of the nucleus. At the bottom of a nuclear invagination the n.e. often shows regions where the nuclear membranes have separated, possibly indicating that the n.e. is about to disintegrate at these sites (Paweletz, 1974). It seems probable that the invaginations are caused by growing centrospheral MTs. As long as the n.e. is intact, no MTs can be detected within the nucleoplasm.

Prophase chromosomes were shown to be connected, or at least closely associated with, the n.e. in mitosis (Roos, 1973a; Paweletz, 1974) and in meiosis (Rickards, 1975; Fuge, 1976). There are indications that these membrane-associated chromosomes have some relation to the centriolar regions in the cytoplasm. Paweletz (1974) assumed that HeLa chromosomes are actively transported toward the centriolar regions by the n.e. Rickards (1975) recorded saltatory movements of bivalents radially oriented to the centrospheres during diakinesis in *Achaeta* spermatocytes.

During fragmentation of the n.e. opposite to the centrospheres, centrospheral MTs invade the nuclear lumen (e.g., Roos, 1973a; Paweletz, 1974). The invasion is, of course, a dynamic process that is difficult to follow in detail with the electron microscope. This explains why there are different views concerning the further fate of the invaded centrospheral MTs. Following the interpretation of classical studies Paweletz (1974) assumed that the kinetochores become connected with the invading centrosperal MTs during the intial phase of HeLa spindle formation. By contrast, the association of chromosomes with a spindle pole in PtK2 cells has been assumed to be achieved by MTs that are newly assembled at the kinetochores, i.e., grow from the kinetochores in the direction of the centrospheres (Roos, 1976). According to Roos, the centrospheres are involved in kMT synthesis only by providing MT subunits (tubulin) diffusing into the nucleoplasm after n.e. breakdown. If PtK2 chromosomes are situated at some distance from the pole at the time of n.e. fragmentation, their orientation is accompanied by active bending and poleward movement of the kinetochore regions. Although this movement toward the pole does not unequivocally prove de novo assembly of MTs at the kinetochores, it is, in our opinion, difficult to imagine how growing centrospheral MTs, in making contact with the kinetochores (Paweletz, 1974), could induce a poleward movement.

Some observations suggest that centrioles as MTOCs probably are not as important during spindle formation as is generally assumed (general discussion by Pickett-Heaps, 1969, 1974). By flattening living spermatocytes of *Pales ferruginea* in diakinesis, the centrioles together with their associated centrospheres were prevented from occupying their normal position at the prospective spindle poles (Dietz, 1966). Nevertheless, a functioning bipolar spindle was formed. Dietz concluded that the spindle was formed under the organizing influence of the kinetochores and that centrioles are insignificant for spindle formation and orientation. It was shown in a recent electron-microscopic study of mitosis of the newt *Taricha granulosa* (Molè-Bajer, 1975)

that the orientation of the spindle (which is formed outside the nucleus before n.e. fragmentation) is not related to the position of
the centrioles. However, an organizing influence of the chromosomes
must be excluded in *Taricha*.

It has recently been proved experimentally that kinetochores have the
capability of inducing the assembly of kMTs in a tubulin medium (McGill
and Brinkley, 1975; Telzer et al., 1975; Snyder and McIntosh, 1975).
This may favor Roos's (1976) opinion that orientation of kinetochores
after breakdown of the n.e. is accompanied and induced by new assembly
of kMTs. Hence it may be argued that the invading centrospheral MTs
in animals only contribute the first non-kMTs of the spindle.

The latter argument is possibly supported by the events during
spindle formation in endosperm of *Haemanthus* (Bajer and Molè-Bajer,
1969, 1971, 1972). Spindle formation in *Haemanthus* is possibly representative of most angiosperms. Large numbers of cytoplasmic ("clear zone")
MTs invade the nucleus from two opposite sides after n.e. fragmentation.
As typical of angiosperms, polar organelles are not present (acentric
spindle). The investigators have suggested that these cytoplasmic MTs
form the first non-kMT bundles of the young spindle, and it has been
doubted (Bajer and Molè-Bajer, 1972) that invading cytoplasmic MTs
can directly become attached to kinetochores. Rather, the kMTs were
believed to grow out from the randomly oriented kinetochores and to
contact the non-kMT bundles. This contact may be responsible for the
proper orientation of the kinetochores thereafter.

The overall MT content of the spindle gradually increases up to
metaphase congress of the chromosomes (Fuge, 1971; Jensen and Bajer,
1973b; Roos, 1976). By combined in vivo observation and electron-
microscopic study, it was possible to obtain precise data on the behavior of kMTs during prometaphase movement and orientation in rat
kangaroo PtK2 cells (Roos, 1976), and in *Haemanthus* endosperm (Lambert
and Bajer, 1975; Jensen, 1976).

From the study of Roos (1976) it became evident that the direction
of prometaphase movement is a function of kMT orientation. A chromosome approaches one pole if both its kMT bundles (syntelic orientation)
or only one kMT bundle (monotelic orientation) point to that pole.
If the kMT bundles of both chromatids point to opposite poles
(amphitelic orientation) the chromosome undergoes oscillating movements.

The morphology of the kMT bundles could be correlated with chromosome
velocity in *Haemanthus* (Lambert and Bajer, 1975). Highly irregular and
diverging kMTs seem to be characteristic of fast-moving chromosomes,
whereas more parallel kMT arrangement is typical of slow or stationary
chromosomes. By counting MTs in serial sections Jensen (1976) obtained
evidence that kinetochores facing the direction of movement have approximately twice as many kMTs than those of stationary chromosomes.
In other words, application of motive forces is positively correlated
with the number of kMTs.

II. Arrangement and Distribution of Spindle Microtubules During Meta- and Anaphase

Anaphase is undoubtedly one of the most fascinating stages in mitosis.
Hence several attempts have been made to obtain exact data on the
arrangement of spindle MTs at meta- and anaphase. The methods used
were reconstruction from serial sections and examination of thick sections with the HVEM. It turned out that there are obviously great dif-

ferences in MT arrangement between mammalian and marsupial (HeLa, PtK1), plant (*Haemanthus*), and insect (*Pales*) spindles. The most outstanding points are reviewed in the following.

From their studies of mammalian as well as marsupial cells Brinkley and Cartwright (1971) concluded that 30 - 45 % of non-kMTs are continuous and the rest interdigitating MTs. According to McIntosh et al. (1975) the fraction of continuous MTs is not as high. Rather, most non-kMTs may be connected with one pole (connection means ending in one centrosphere region, but not a firm binding to the centrioles), and may extend to different degrees into the spindle. Some are short and do not reach the equator, others are longer and extend into the opposite half-spindle. This arrangement results in a "sloppy" interdigitation (overlapping) of both half-spindles. Hence, if we follow the interpretation of McIntosh et al., the bulk of non-kMTs in the mammalian spindle may be represented by interdigitating MTs.

The kinetochores of mammalian chromosomes at metaphase are connected with 20 - 40 kMTs (e.g., Brinkley and Cartwright, 1971; Roos, 1973b; McIntosh et al., 1975). According to McIntosh et al. only a fraction of these kMTs reaches into the polar region in metaphase; many of them are much shorter than the kinetochore-pole distance.

Individual non-kMTs of meta- and anaphase spindles in *Haemanthus* (Jensen and Bajer, 1973a; Bajer et al., 1975) seem to be relatively short with respect to spindle length. They are aggregated in the form of distinct bundles that may run from pole to pole, contact other bundles, or "branch". The kMTs, on the average 120 per kinetochore at metaphase (Jensen and Bajer, 1973b), do not reach the poles in metaphase. Rather, kMT bundles contact non-kMT bundles to form so-called hybrid bundles (Lambert and Bajer, 1975). kMTs of a bundle were observed to be more or less parallel at metaphase but to become divergent in anaphase (Lambert and Bajer, 1972; Bajer et al., 1975). kMTs (Jensen and Bajer, 1973b) as well as non-kMTs (Jensen and Bajer, 1973a) were shown to decrease in number during anaphase.

Tracking of tubules in serial sections of *Pales ferruginea* spermatocytes indicated that meta- and anaphase spindles consist of large numbers of relatively short and irregularly arranged non-kMT units (mean lengths: $\bar{x}$ = 2.6-3.0 μm, number of measured MTs in two spindles: n = 148) (Fuge, 1974a). According to this investigation, non-kMTs that are as long as or longer than the equator-pole distance (approx. 10 μm) do not seem to be very frequent. By contrast, LaFountain (1976) maintained that the non-kMTs of the closely related species *Pales suturalis* are relatively long. However, he concluded this from only four MTs measured in anaphase, a sample that may not be representative of the total population. Non-kMTs of tipulid spermatocytes are obviously not arranged in bundles, and there is no indication of interdigitating half-spindles (Fuge, 1973, 1974a).

Although there are many relatively long kMTs both in meta- and anaphase of *Pales* (Fuge, 1974a; LaFountain, 1976), most do not reach to the spindle apices (Fuge, 1974a). kMT bundles obviously splay out with the onset of anaphase motion and, as shown from serial section analysis (Fuge, 1974a), become intermingled with non-kMTs.

Pales chromosomes in anaphase are characterized by a peculiar association of extrakinetochore chromatin with non-kMTs (Fuge, 1972, 1973, 1974a; Fuge and Müller, 1972; Forer and Brinkley, 1977). The chromatin of the chromatid arms forms radial lamellae (syn. Leisten, flanges). A small fraction of non-kMTs becomes arranged in rows within the notches formed by the lamellae. There are indications that these MTs are connected with the chromatin by filamentous cross-bridges (Fuge and Müller, 1972; Fuge, 1973, 1975).

Parts of sex chromosome chromatids (Fuge, 1972) and free chromatid ends of mutated bivalents (dicentric fragments resulting from chiasma formation in inversion heterozygotes) (Fuge, 1975) are shifted pole-ward in anaphase, i.e., are situated in front of the kinetochores. Akinetochoric chromosomal fragments with chromatin lamellae and associated non-kMTs become displaced toward the poles in anaphase (Fuge, 1975).

All these observations may suggest that anaphase forces act not only upon the kinetochores in *Pales*, but also upon the extrakinetochore chromatin. non-kMTs may be important for force production or trans-mission (Fuge, 1973, 1974a, 1975). Laterally associated non-kMTs have been observed in other organisms [e.g., in *Spirogyra* (Conjugatophyceae), Mughal and Godward, 1973, and in *Chroomonas* (Cryptophyceae), Oakley and Dodge, 1976; for complete literature see Fuge, 1974b, 1977] and may be of general importance in mitosis.

Several investigators have tried to follow the redistribution of spindle MTs in anaphase. Preselected spindles were sectioned at right angles to the long axis of the spindle and the MTs were counted in serial sections. Plotting the values as a function of position on the spindle axis gives a distribution histogram of spindle MTs (McIntosh and Landis, 1971; Brinkley and Cartwright, 1971, 1975; Jensen and Bajer, 1973a; Fuge, 1973; McIntosh et al., 1975; Forer and Brinkley, 1977). The interpretation of the histograms with regard to MT dynamics is, however, difficult, since the length of each of the thousands of spindle MTs is unknown.

Metaphase spindles of mammalian and marsupial cells (human WI-38 cells, McIntosh and Landis, 1971; rat kangaroo PtK1 cells, Brinkley and Cartwright, 1971, 1975) have one maximum of MT concentration in each half-spindle. As most clearly shown by McIntosh et al. (1975) in a distribution diagram of a Chinese hamster CHO cell, these two maxi-ma separate from one another in early anaphase, simultaneously with the kinetochores of the sister chromatids. This strongly suggests that the half-spindle maxima in meta- and early anaphase of mammalian cells are caused by kMTs. No conclusions can be drawn from the histograms concerning possible dynamic changes of non-kMTs at this early stage of chromatid separation.

In contrast to mammalian cells, the metaphase spindle of *Haemanthus* endosperm seems to possess only one high maximum of MT concentration at the equator, with the MT numbers decreasing toward the poles (Jen-sen and Bajer, 1973a). However, MTs were counted in only a few sec-tions, which did not provide enough information to plot a definite distribution histogram. The high concentration of MTs at the equator was thought to be caused mainly by kMTs. This would be in accordance with the metaphase maxima of mammalian cells. The anaphase distribution profile of *Haemanthus* shows one maximum in each half-spindle. In con-trast to mammalian cells, these maxima are situated *behind* the kine-tochore (as viewed from the poles), at the level of the long trailing chromosome arms. This clearly indicates that the maxima in *Haemanthus* cannot be directly caused by kMTs. In the opinion of Jensen and Bajer (1973a) the maxima may be caused by large numbers of kMT fragments that have been "broken off" from the kMT bundles during anaphase move-ment, as well as by non-kMTs that have moved from the equator region in the direction of the poles.

Poleward displacement of non-kMTs during anaphase may also explain
distribution profiles obtained from *Pales ferruginea* spindles (Fuge, 1973).
MTs were counted within circular coaxial zones in the serial sections, whereby it was possible to detect those changes in MT distribution that occur *along* the spindle axis, as well as those occurring
at right angles to it. *Pales* chromosomes move at the periphery of an
axial portion of the spindle that is characterized by a relatively
high concentration of non-kMTs. Counts from a metaphase spindle revealed one broad maximum of non-kMTs at the equator region and a
progressive decrease toward the poles. This form of the histogram is
the same in both the axial portion and in the total spindle. After
the chromosomes have separated in anaphase the axial spindle portion
shows one non-kMT maximum in each half-spindle in front of the kinetochores. The histogram of the total spindle has an irregular shape,
but at least two regions of high MT concentration in the half-spindles
can be attributed to non-kMTs surrounding the trailing arms of the
moving chromosomes (Fuge, in press). The change in MT distribution
from meta- to anaphase could suggest that non-kMTs become displaced
from the equator toward the two poles in anaphase. In other words,
moving chromosomes of *Pales* may be accompanied by moving non-kMTs.

Late anaphase and telophase of mammalian cells are characterized by
a new MT maximum in the equator region. The half-spindle maxima have
disappeared at this time, obviously due to depolymerization of the
kMTs. The equatorial maximum is caused by the formation of the midbody region, a zone of preferential overlapping of interzonal non-
kMTs (Buck and Tisdale, 1962; Paweletz, 1967; Brinkley and Cartwright, 1971; McIntosh and Landis, 1971; Erlandson and de Harven,
1971; Roos, 1973a; McIntosh et al., 1975). The midbody may be the
result of interdigitation of MTs from the two half-spindles, and
thus could actively be engaged in spindle elongation. However, there
are indications that de novo polymerization and growth of non-kMTs
also occur in this region in late anaphase and contributes to midbody formation (McIntosh and Landis, 1971; Brinkley and Cartwright,
1971; Paweletz, pers. commun.).

Preferential zones of overlapping MTs in the interzone of late anaphase were observed in other organisms (e.g., in *Dictyostelium* and
Diatoma in *Haemanthus*, Lambert and Bajer, 1972; in the gall midge *Heteropeza*, Fux, 1974). They could not be detected in *Pales* spermatocytes
(Fuge, 1973).

From all cited MT counting studies it can be deduced that the overall
MT number in the spindle decreases during anaphase.

E. Different Features of Microtubules in Mitosis

In conclusion, several items concerning spindle ultrastructure are
listed which, in our opinion, may be relevant for further discussion
of spindle models.

1. A complex spindle consisting of kMTs and of different kinds on non-
kMTs, characteristic of higher plants and animals, is not the prerequisite for proper genome distribution. The distribution can likewise
be accomplished by a prokaryote-like membrane-mediated mechanism, or
by simple elongation of continuous MTs, without active involvement of
kMTs.

2. The occurrence of continuous MTs is not the prerequisite for simple
elongation. Elongation can obviously be achieved also by shorter MT

units of the half-spindles, probably by those that interdigitate in the equatorial region (*Diatoma* and mammalian spindles).

3. If kMTs occur, they can be arranged and can function in different ways:

a) They can run between kinetochores and poles, but are not actively involved in chromatid separation (*Syndinium, Trichonympha*).

b) In cases where kMTs are likely to be actively involved in separation, i.e., are observed to become shorter during anaphase, they can be arranged in two different ways: (i) They can directly connect kinetochores and polar centers throughout division (slime molds, Ascomycetes). (ii) Many of them can be considerably shorter than the kinetochore-pole distance in metaphase. The kMT bundles can splay out and intermingle with non-kMTs in anaphase (*Haemanthus, Pales*, and, with restriction, mammalian cells).

Points (i) and (ii) are more or less consistent with the two main spindle types (direct and indirect type) proposed by Schrader (1953).

4. Initiation of orientation of kinetochores in prometaphase involves de novo assembly of kMTs (PtK2, *Haemanthus*).

5. Complex spindles consisting of kMTs and different kinds of non-kMTs show several dynamic changes during anaphase (higher plants and animals).

a) kMTs and non-kMTs decrease in number. This is equivalent to a disassembly of the spindle during the course of anaphase.

b) non-kMTs undergo certain changes in arrangement that are still not clearly understood. These changes in arrangement may be responsible for a decrease in MT concentration in the interzone in mid-anaphase, and a concomitant increase at the level of the moving chromosomes (*Haemanthus, Pales*).

c) Certain non-kMTs have the tendency to associate laterally with extra-kinetochore chromatin (*Pales, Haemanthus, Spirogyra*).

The aim of this article was to give a review on observations concerning MT arrangement during mitosis. It is not intended to discuss in detail the advantages and disadvantages of present spindle models in the light of the ultrastructural facts. But it can at least be concluded that none of the spindle models seems adequate to explain chromosome movements in all eukaryotes. Sliding of kMTs against non-kMTs (McIntosh et al., 1969), as well as a zipping mechanism between kMTs and non-kMTs (Bajer, 1973) are, in our opinion, unlikely to occur in certain fungi, since lateral interaction of both MT types must be excluded on morphological grounds. On the other hand, the dynamic changes in MT arrangement observed in higher plants and animals are difficult to correlate with an assembly-disassembly mechanism proposed by Inoué and Sato (1967) and Dietz (1969).

References

Aldrich, H.C.: The ultrastructure of mitosis in myxamoebae and plasmodia of *Physarum flavicomum*. Am. J. Botany 56, 290-299 (1969)

Bajer, A.S.: Interaction of microtubules and the mechanism of chromosome movement (zipper hypothesis). I. General principle. Cytobios 8, 139-160 (1973)

Bajer, A.S., Molè-Bajer, J.: Formation of spindle fibers, kinetochore orientation, and behavior of the nuclear envelope during mitosis in endosperm. Chromosoma (Berl.) 27, 448-484 (1969)

Bajer, A.S., Molè-Bajer, J.: Architecture and function of the mitotic spindle.
  Advan. Cell Mol. Biol. 1, 213-266 (1971)
Bajer, A.S., Molè-Bajer, J.: Spindle dynamics and chromosome movements. Intern.
  Rev. Cytol., Suppl. 3, 1-271 (1972)
Bajer, A.S., Molè-Bajer, J., Lambert, A.-M.: Lateral interaction of microtubules
  and chromosome movements. In: Microtubules and Microtubule Inhibitors. Borgers,
  M., de Brabander, M. (eds.). Amsterdam: North-Holland Pub. Co. 1975, pp. 393-
  423
Bland, C.E., Lunney, C.Z.: Mitotic apparatus of *Pilobolus crystallinus*. Cytobiol.
  11, 382-391 (1975)
Bouck, G.B., Brown, D.L.: Microtubule biogenesis and cell shape in *Ochromonas*.
  I. The distribution of cytoplasmic and mitotic microtubules. J. Cell Biol. 56,
  340-359 (1973)
Brinkley, B.R., Cartwright, J.,Jr.: Ultrastructural analysis of mitotic spindle
  elongation in mammalian cells in vitro. Direct microtubule counts. J. Cell
  Biol. 50, 416-431 (1971)
Brinkley, B.R., Cartwright, J., Jr.: Cold-labile and cold-stable microtubules in
  the mitotic spindle of mammalian cells. Ann. N.Y. Acad. Sci. 253, 428-439 (1975)
Buck, R.C., Tisdale, J.M.: The fine structure of the mid-body of the rat erythro-
  blast. J. Cell Biol. 13, 109-115 (1962)
Dietz, R.: The dispensability of the centrioles in the spermatocyte divisions of
  Pales ferruginea (Nematocera). Chromosomes Today 1, 161-166 (1966)
Dietz, R.: Bau und Funktion des Spindelapparates. Naturwissenschaften 56, 237-248
  (1969)
Erlandson, R.A., de Harven, E.: The ultrastructure of synchronized HeLa cells. J.
  Cell Sci. 8, 353-397 (1971)
Forer, A.: Actin filaments and birefringent spindle fibers during chromosome move-
  ment. Cold Spring Harb. Conf. Cell Proliferation 3, 1273-1293 (1976)
Forer, A., Brinkley, B.R.: Microtubule distribution in the anaphase spindle of
  primary spermatocytes of a crane fly (*Nephrotoma suturalis*.) Can. J. Genet.
  Cytol. (in press)
Franke, W.W., Reau, P.: The mitotic apparatus of a zygomycete, *Phycomyces blakes-
  leeanus*. Arch. Mikrobiol. 90, 121-130 (1973)
Fuge, H.: Spindelbau, Mikrotubuliverteilung und Chromosomenstruktur während der
  I. meiotischen Teilung der Spermatocyten von *Pales ferruginea*. Eine elektronen-
  mikroskopische Analyse. Z. Zellforsch. 120, 579-599 (1971)
Fuge, H.: Morphological studies on the structure of univalent sex chromosomes
  during anaphase movement in spermatocytes of the crane fly *Pales ferruginea*.
  Chromosoma (Berl.) 39, 403-417 (1972)
Fuge, H.: Verteilung der Mikrotubuli in Metaphase- und Anaphase-Spindeln der
  Spermatocyten von *Pales ferruginea*. Chromosoma (Berl.) 43, 109-143 (1973)
Fuge, H.: The arrangement of microtubules and the attachment of chromosomes to
  the spindle during anaphase in tipulid spermatocytes. Chromosoma (Berl.) 45,
  245-260 (1974a)
Fuge, H.: Ultrastructure and function of the spindle apparatus. Microtubules and
  chromosomes during nuclear division. Protoplasma 82, 289-320 (1974b)
Fuge, H.: Anaphase transport of akinetochoric fragments in tipulid spermatocytes.
  Electron microscopic observations on fragment-spindle interactions. Chromosoma
  (Berl.) 52, 149-158 (1975)
Fuge, H.: Ultrastructure of cytoplasmic nucleolus-like bodies and nuclear RNP
  particles in late prophase of tipulid spermatocytes. Chromosoma (Berl.) 56,
  363-379 (1976)
Fuge, H.: Ultrastructure of the mitotic spindle. Intern. Rev. Cytol. 52, in press
Fuge, H., Müller, W.: Mikrotubuli-Kontakt an Anaphaseechromosomen in der I. meio-
  tischen Teilung. Exptl. Cell Res. 71, 242-245 (1972)
Fuller, M.S.: Mitosis in fungi. Intern. Rev. Cytol. 45, 113-153 (1976)
Fux, T.: Chromosome elimination in *Heteropeza pygmaea*. II. Ultrastructure of the
  spindle apparatus. Chromosoma (Berl.) 49, 99-112 (1974)
Heath, I.B.: Mitosis in the fungus *Thraustotheca clavata*. J. Cell Biol. 60, 204-
  220 (1974a)

Zickler, D.: Division spindle and centrosomal plaques during mitosis and meiosis in some Ascomycetes. Chromosoma (Berl.) <u>30</u>, 287-304 (1970)
Zickler, D., Olson, L.W.: The synaptonemal complex and the spindle plaque during meiosis in yeast. Chromosoma (Berl.) <u>50</u>, 1-23 (1975)

Discussion Session III: Fine Structure of Mitotic Cells

Chairman: N. PAWELETZ, Heidelberg, F.R.G.

Chairman:
The previous lecture clearly demonstrated a wide variety of structures and organelles in the cell during mitosis. This raises the question: Which organelles are essential for mitosis? Since not all structures and organelles could be discussed in detail, special emphasis was placed on chromosomes, microtubules and spindles, and vesicles. Can a spindle be formed without chromosomes?

A.S. Bajer, Eugene, Oregon, USA
Chromosome-free spindles are occasionally found in endosperm of *Haemanthus*. Their rate of occurrence can be somewhat increased by moderate pressure (100 psi) in an atmosphere of N_2O. These spindles show many properties of the normal spindle (bipolar nonkinetochore transport) and are finally transformed into phragmoplasts with cell plates. The fine structure of these spindles (J. Molè-Bajer, in preparation) closely resembles that of a normal spindle, but there is no differentiation into spindle fibres, i.e., microtubules are more envenly spaced.

D. Mazia, Berkeley, California, USA
Experiments on the formation of spindles in the absence of chromosomes are possible in cells in which we can obtain numerous poles in one way or another. The question to be asked is whether two poles can make connections to each other when they are not also connected to chromosomes. Such experiments can be done, but no one, as far as I know, has looked into this question carefully.

P. Malpoix, Rhode-St-Genèse, Belgium
The formation of cytasters can be readily induced, independently from chromosomes, by a number of different agents, e.g., heavy water. On the other hand, anastral spindles can occur, e.g., Brachet and Jencer have observed that after cytochalasin treatment of tunicate eggs, not only do numerous cytasters appear, but also anastral spindles. The condensed metaphase chromosomes are, of course, associated with the latter. So, formation of aster, and of spindle, can be initiated independently.

S.J. Counce, Durham, North Carolina, USA
Centrioles of the cleavage stage of embryos of *Drosophila* can be moved away from the nuclear membrane of prophase nuclei by ultrasonic treatment. These centrioles are capable of organizing long narrow spindles with astral structures lying adjacent to the nuclei which remain intact (Counce and Selman, 1956). Both the shape and size of such spindles differ from those of normal mitotic cleavage figures.

P. Sentein, Montpellier, France
When the asters are separated from the spindle (Poster 23), the chromosomes do not move beyond the limits of the spindle and never reach the poles. In this case, the half spindles do not behave as homogeneous structures.

L. Hens, Brussels, Belgium
With respect to the linkage between microtubules and chromosomes, it has to be mentioned that microtubules exist not only connecting chromosomes with the mitotic spindle, but also connecting chromosomes with each other. The functional interpretation of this phenomenon remains unclear.

P. Seintein, Montpellier, France
When cycloheximide (10^{-3} M) acts on segmentation mitoses of *Urodele* eggs, after 4 or 6 mitotic cycles the nuclei become separated from the asters, which associate by 2, 4 (or more) cycles to form achromatic systems. These achromatic systems without chromosomes do not form true spindles and do not show any more or less parallel organization of spindle fibers. On the contrary astral fibers from both poles cross each other in the intermediate zone.

Chairman:
Spindles, or at least parts of them, can be formed without chromosomes. For a normal course of mitosis, however, chromosomes must be incorporated into the spindle. Do chromosomes actively participate in their own dislocation?

A.M. Lambert, Strasbourgh, France
In the case of colchicine or vinblastine-treated cells, separation of condensed
chromatids occurs without visible spindle activity. After a few μm they stop, return again and, as a result, a polyploid restitution nucleus is formed. The separation does occur up to 5-6 μm, which is more than enough to get polyploidy.
During normal mitosis such chromatid repulsion may occur as a trigger of anaphase
and, then, spindle dynamics would be responsible for further separation and final
equal distribution of the genome into two daughter diploid nuclei.

S. Ghosh, Calcutta, India
In colchicine-treated cells, the sister chromatids are found to separate, although
they do not show movement toward the poles. It may be necessary to distinguish
between the movements related to chromatid separation and those related to redistribution into daughter cells.

R.B. Nicklas, Durham, North Carolina, USA
I think that if Donna Kubai or Hans Ris were here, they would not regard chromatid
separation as an exception because you are not getting genome separation in the
sense of getting the daughter chromosome units far enough apart to do any good to
the cell.

Chairman:
These statements brought up the question, to what degree are microtubules and/or
kinetochores involved in the process of chromosome movement.

M. Hauser, Bochum, F.R.G.
There is one exception, perhaps, also in macronuclear division in ciliated protozoa.
This is a clear two-step mitosis: before nuclear division, replicated chromosomes
become separated by rotational movements and during this phase remarkably few
microtubules are present (short fragments), whereas before and after the rotation
phase of chromosomal elements numerous microtubules can always be found.

P. Dustin, Brussels, Belgium
The asynchronous divisions of chromosomes indicate that they have an autonomy. Whatever the exact relation of kinetochores to microtubules is, it appears that these
organelles play an active role in microtubule assembly. My answer would thus be:
yes, chromosomes *are* active in mitosis.

R.B. Nicklas, Durham, North Carolina, USA
Independent movement of chromosomes does not suggest that movement is caused by the
chromosomes themselves.

R. Dietz, Tübingen, F.R.G.
In anaphase there are correlations between the number of kinetochore microtubules
and the chromosome size (load to be transported). In prometaphase the number of
kinetochore microtubules is correlated with chromosome velocity. These findings
indicate a rather direct role of microtubules in chromosome movement.
It may be wrong to conclude from a growing aster an increasing microtubule-organizing capacity of the centriole. In tipulid spermatocytes, potentially each chromosome becomes surrounded by its own spindle, if the asters have been displaced
far enough. During the formation of the chromosomal spindles, the astral rays grow
shorter. In regular division, however, when asters lie close to chromosomes, those
astral rays which point toward chromosomes grow rapidly after the nuclear envelope
has collapsed. This discrepancy is thought to have the following explanation: After
the breakdown of the nuclear envelope, each chromosome surrounds itself with an
assembly field. If astral rays terminate within such an assembly field, they act
as homologous seeds; subunits are added, to the ends, astral rays grow rapidly.
If no astral rays terminate within such an assembly field, there is a lag phase.
Subsequently, heterologous nucleation seems to occur at kinetochores, which initiates the formation of kinetochores microtubules. Furthermore, spontaneous
nucleation occurs next to kinetochores, which initiates the formation of free
microtubules. In this way, and without being overgrown by astral rays, each chromosome forms its own spindle.

U. Euteneuer, Frankfurt, F.R.G.
What about chromosome movements in prophase, when the nuclear membrane is still intact and no microtubules are present in the nucleus? There are some reports that describe these movements, for instance by Paweletz and Rickards.

G.K. Rickards, Wellington, New Zealand
On the question of direct activity of the chromosome, I would point out that in the prophase chromosome movements of cricket spermatocytes (Rickards, Chromosoma 49, 407, 1975), chromosome ends and kinetochores appear to be directly involved in bringing about movement. This suggests to me that perhaps in the later movements on the spindle also, kinetochores are actively involved, rather than simply being passively transported through transport of associated microtubules.

R.B. Nicklas, Durham, North Carolina, USA
To be active individually does not necessarily have to be an attribute of chromosomes, i.e., independent movement of chromosomes does not suggest that movement is caused by the chromosomes themselves.

Chairman:
There is still controversy in the answer on the activity of chromosomes. It can be stated as a fact that sister chromatid separation, at least in some organisms, is autonomous from the spindle. In prophase movement, however, there is not yet enough evidence for activity of the chromosomes. Probably microtubules and/or nuclear membranes are also involved in this process, as well as the chromosome condensation itself.
The main purpose of mitotic movement, namely, to separate the replicated genome into daughter cells, cannot be fulfilled by chromosomes with no mitotic apparatus. Can chromosomes move without kinetochores?

A.S. Bajer, Eugene, Oregon, USA
Yes. Nonkinetochore transport is the movement (elimination) of small granules and acentric chromosome fragments toward the poles during prometaphase and metaphase (plant endosperm, newt tissue culture).

D.N. Wheatley, Aberdeen, Great Britain
Can we really afford to consider the chromosome and the kinetochore separately? If the chromosome is considered on its own, then it may have no active function in mitosis, but if we consider the kinetochore as an integral component of the chromosome, then the whole chromosome complex can have some active function in mitosis.

J.H. Frenster, Atherton, California, USA
The previous discussion suggests that we are approaching the mitosis phenomena from the standpoint of whether chromosomes serve the spindle or the spindle serves the chromosomes. Actually, newer electron-microscopic techniques (Nakatsu et al., Nature 248, 334, 1974) have revealed new ultrastructural data concerning chromosomes during mitosis. While DNA helix openings are observed to decrease in number and size during normal or neoplastic cell differentiation, and during normal cell metaphase, anaphase, and early telophase, such reactions are not observed during neoplastic cell mitosis (Poster 2). This suggests that ligand molecules, such as acidic proteins and RNA capable of binding to single-stranded DNA in DNA helix openings, can be carried by such chromosomes from the maternal cell through cell division to the daughter cells, thus preserving the differentiated state through cell generations (Frenster, Cancer Res. 36, 3394, 1976) and providing for the equal partition of epigenetic systems as well as of genetic systems.

P. Malpoix, Rhode-St-Genèse, Belgium
We should also consider the importance of highly repetitive DNA in the points of attachment of chromosomes to spindle. The kinetochore regions of the chromosomes have been shown to be rich in highly repetitive "satellite" DNA.
This constitutive heterochromatin remains condensed throughout cell division. It has not been strongly conserved in evolution in that satellite DNA from different species may differ considerably in base sequence; yet deficiency or excess of it may produce severe, though variable, phenotypic effects.

Chairman:
In relation to the activity of chromosomes during mitosis another point of view
arose during the discussion. What do we know about the synthetic activity of
chromosomes during mitosis?

J.M. Mitchison, Edinburgh, Scotland
I think there is an interesting question to be asked about the function of chromosomes during mitosis. This is, how far do they serve at that time as a template
for RNA transcription, as they do in interphase. It is true that most transcription
stops in the fully condensed metaphase chromosomes of higher eukaryotes. But there
is a report of the continued synthesis of 5 S RNA in mammalian metaphase, and there
is no evidence that RNA stops during the nuclear division of yeasts whose chromosomes may not condense completely at mitosis. So partially condensed chromosomes
may produce RNA and even fully condensed chromosomes may produce a little RNA.

J.F. López-Sáez, Madrid, Spain
The low rate of transcription during mitosis is a well-known fact but it is important not to confuse a rate with a role. Thus, during prophase of meristem cells
there is a certain RNA synthesis, and our group has demonstrated that this is a
requirement of nuclear membrane breakdown and subsequent mitosis development. Also,
throughout telophase, nucleolar transcription can be observed, in in vivo and in
vitro incubations, and nucleolar reorganization is dependent on this RNA synthesis.
Therefore, 45 % (prophase) + 30 % (telophase) = 75 % of mitotic time, and for this
time transcription occurs.

H. Fuge, Kaiserslautern, F.R.G.
Additional remark with respect to transcription during prophase: There is, of
course, transcription during meiotic prophase.

H. Ponstingl, Heidelberg, F.R.G.
There is indirect evidence that histone phosphorylation plays a role in chromosome
condensation: In synchronized Chinese hamster cells, e.g., histones H1M and H3
are phosphorylated specifically from the onset of prophase to metaphase (Gurley et
al., J. Biol. Chem. 250, 3936, 1975).

Chairman:
From these contributions we can conclude that chromosomes are synthetically active
during mitosis, but obviously to a lesser degree than in interphase.
The problems about the mechanisms of chromosome condensation could not be discussed in detail. A few contributions, however, deal with the molecular structure
in relation to function during mitotic events.

D. Mazia, Berkeley, California, USA
Does the present explosion of new models of the structure of chromatin, centered on
the nucleosome, contribute anything about the chromosome cycle in mitosis? The new
models do provide for some packing of DNA in the primary thread, but do they help
our effort to understand the packing of the still-long threads into compact mitotic
chromosomes?

J.H. Frenster, Atherton, California, USA
To the degree that nucleosomes are involved in the packing conformations of DNA,
to that degree are nucleosomes important in the early mitotic passage of chromosomes in the condensed state to the daughter cells.

B.R. Oakley, Downsview, Ontario, Canada
Probably it should be brought out at this point that there are a lot of so-called
primitive organisms with a lot of different chromosome structures at mitosis. Some
of them never condense, some of them stay permanently condensed. So it would be
very difficult to make such a generalization about chromosome structure, as it involves mitosis.

Chairman:
Since only a few data are available concerning molecular structure of chromosomes
and their condensation, Dr. Mazia's question still remains open.

Leaving the field of chromosomes and coming to the structure of microtubules and spindles, the important question about the polarity of microtubules in the mitotic apparatus was raised.

J.R. McIntosh, Boulder, Colorado, USA
Microtubules formed in vitro from neurotubulin are known to be polar. From diverse cytologic observations it is likely that spindle tubules also are polar. In this context, polar means that one end of the microtubule is different from the other, no matter where you break it. This has to be distinguished from polarity of growth (a microtubule might add subunits at the end proximal or distal to a spindle pole), although polarity of growth might be a reflection of this intrinsic molecular polarity of structure.
I submit that there is an interesting, unanswered experimental question concerning the structural polarity of spindle microtubules. It would be an important contribution to our knowledge of spindle structure if someone could develop a marker for microtubule polarity that would serve in the same way as heavy meromyosin serves to display the polarity of actin filaments.
In light of the idea of microtubules possessing structural polarity, the central spindle of yeasts and other primitive eukaryotes is probably constructed from overlapping sets of microtubules that originate at one pole or the other and run some distance to the opposite pole (see diagram 1).

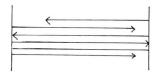

The spindles of *Diatoma* described in the lecture by Fuge possess a substantial number of microtubules that overlap with tubules from the opposite pole.

It seems to me that the distinction
between these two spindle designs

need not be profound. It will depend on the distribution of lengths of the microtubules and the interactions between them. I therefore propose that all such microtubules should be called "polar" microtubules or PMTs, reflecting their probable place of origin.
"Continuous" and "overlapping" MTs are then used as terms descriptive of their length.

M. Osborn, Göttingen, F.R.G.
I just want to say that we also feel it may be very important to determine the polarity in different types of microtubules. Since isolated tubules have a helical structure with a deformed "handedness" the axis of a tubule should have a fixed polarity. Thus the direction in which a tubule grows in vivo should be preserved in the final polarity of the finished structure. Furthermore, the cell may use such a system to guide the intracellular movement or transport of material in a unidirectional manner (for fuller discussion, see Osborn and Weber, Exp. Cell Res. 103, 331, 1976). The difficulty usually is to try to design experiments that might reveal this polarity directly in microtubules in situ in the cell.

A.S. Bajer, Eugene, Oregon, USA
The polarity of the microtubules is undoubtedly a very important problem. It is difficult, however, to answer this question using the "standard spindle" of higher organisms. Some lower organisms are more promising. I feel also that there are other equally important problems and I can visualize experiments which would give us the answer. Such questions are, e.g., 1) Do microtubules differ in their mechanical resistance (different classes of microtubules? single microtubule along its length? etc.), 2) Do the properties of microtubules change with their age, position in the spindle, etc? It is important to understand such and many similar questions for the understanding of the constantly changeable structure and function of the spindle.

Chairman:
Though there is clear evidence for a polarity of microtubules in different systems,
e.g., flagella, this remains an open question for the mitotic spindle. Though the
answer would offer possibilities to know more about the growth and function of the
spindle, no experiments have thus far been designed to solve this problem.
Some data and experimental designs, however, are available for the problem of the
formation of the spindle. Some data favor the kinetochore as initiator of spindle
formation; on the other hand, it is an established fact that at least part of the
spindle is formed around and from the centrioles and from other centers.

A.S. Bajer, Eugene, Oregon, USA
During the beginning of spindle formation (newt, plant endosperm) microtubules are
formed at random, i.e., are arranged in a very irregular fashion. Microtubules are
then rearranged (disassemble and/or change their position).
At a later stage, but still before the breaking of the nuclear envelope, asters
(if present) are instrumental in microtubule rearrangement. After breaking of the
nuclear envelope kinetochores play a major role in microtubule arrangement.

U.-P. Roos, Zürich, Switzerland
In rat kangaroo cells, kinetochores are present on prophase chromosomes, but they
are structurally unlike metaphase or anaphase kinetochores, for they are not
triple-layered plaques. We do not know if prophase kinetochores are active, because
it is not clear whether there is tubulin in prophase nuclei or if so, whether it
is competent for assembly. Certainly, there are no microtubules on prophase kinetochores.
As Dr. Fuge pointed out, the nuclear envelope begins to disintegrate in the vicinity
of the asters that form during prophase and are quite large at the beginning of
prometaphase. Microtubules then apparently penetrate into the nucleus through the
gaps in the nuclear envelope, but other microtubules seem to be assembled at the
kinetochores, presumably because tubulin becomes available by diffusing into the
"nucleus".

A.S. Bajer, Eugene, Oregon, USA
I would like to stress that some aspects of this problem can only be studied in
large spindles. In the spindle of *Haemanthus* or the newt, microtubules rearrange
or come together and form sheets or bundles. Their lateral movements are clearly
seen in the polarized light (the northern light phenomenon - Inoué and Bajer,
Chromosoma 12, 48, 1961, and especially in the film of Sato and Izutsu). Microtubules
"coming out" of kinetochores diverge and become more parallel at a very short distance toward the poles. Close to the poles kinetochore microtubules again become
more divergent and more disorganized. It is not known wath brings microtubules
closer together; but microtubules often become slightly bent during this process.
Thus, the simplest assumption is that random lateral interaction between microtubules is an important factor during differentiation of the spindle into spindle
fibers.

Chairman:
In the discussion it was stated that during reconstruction of the daughter nuclei
microtubules remain incorporated within the nuclear envelope. Are microtubules,-
or is at least, tubulin present in the interphase nucleus?

M. DeBrabander, Beerse, Belgium
We have conflicting evidence for the presence of tubulin and/or microtubules
within the interphase nucleus of mammalian cells. The immunocytochemical results
(Poster 25) point to the absence of antigenically detectable tubulin from the
nucleus. However, this failure to detect it can be due to many factors. On the
other hand we have demonstrated that rat mast cell nuclei do contain, under certain circumstances, structures that are identical, from a morphologic point of
view, to normal cytoplasmic or spindle microtubules.

M. Hauser, Bochum, F.R.G
Microtubules are often present in interphase nuclei of many lower eukaryotes,
e.g., in *Paramecium bursaria*, where a large part of the telophase spindle persists

during interphase. This is also true in various other cases. Sometimes microtubular material persists as a paracrystalline body, e.g., in *Paramecium* micronuclei.

C. Petzelt, Heidelberg, F.R.G
If we speak about the formation of the spindle, do we merely describe the first events when the spindle is formed or do we try to find some causes for the formation of the spindle inside the cell during mitosis? I guess the latter is true. Can anybody make a comment on the causes of the formation of a mitotic apparatus? Can we induce experimentally a spindle or a similar apparatus in a nonmitotic cell?

D. Mazia, Berkeley, California, USA
There is evidence that the same conditions or factors that determine the condensation of chromosomes also determine the formation of the spindle. A lot of work has been done on premature chromosome condensation; in at least one case I have shown that premature chromosome condensation is accompanied by premature spindle formation.

M. Girbardt, Jena, G.D.R.
During the first steps of basidiomycetous mitosis cytoplasmic microtubules radiate from the globular entities of the nucleus-associated organelle. They never enter the nucleus. After the nuclear envelope has fragmented (fenestrated), other microtubules are formed by the same organelle that forms the spindle. This spindle might be composed of only one type of microtubule.

R. Dietz, Tübingen, F.R.G
There is another point we should consider. I mentioned that I can induce the formation of separate chromosome spindles independently from the asters in tipulids. Normally, however, the asters are close to the nucleus and then the astral rays grow toward the chromosomes. So you have a wide-spread phenomenon whereby the astral rays grow and form a spindle. Astral fibers fail to grow in these flattened cells if they are far away from chromosomes. My feeling is that each chromosome surrounds itself with a field in which polymerization occurs, and if astral rays extend into this field the subunits are simply added at the seeds. However, this growth has nothing to do with a real aster growth; it is induced by the kinetochore and by the field that the kinetochore has formed. If the rays are far away, you only have the chance to see that the spindle is formed around each individual chromosome.

R.B. Nicklas, Durham, North Carolina, USA
The extent of the chromosomal involvement in spindle formation is not yet clear experimentally. This is, if you like, a point of view.

M. DeBrabander, Beerse, Belgium
When mitotic cells in culture are treated with nocodazole, microtubules are completely absent. After the compound has been washed out, microtubules reappear initially at the kinetochores. This adds evidence for the possible role of kinetochores as tubule organizing centers.

R.B. Nicklas, Durham, North Carolina, USA
In my opinion, a problem that remains to be solved in this and similar work, in in vitro experiments is to show by electronmicroscopy that there are no visible microtubule fragments remaining at these kinetochores after treatment. It is difficult to show that there might not be aggregates of tubulin that were formed in the presence of the spindle and that they serve as nucleation sites on these kinetochores that are in fact, not present on normal prophase kinetochores.

H. Fuge, Kaiserslautern, F.R.G.
We should at least discriminate between kinetochore microtubules and nonkinetochore microtubules when we are talking about spindle formation. Are both types induced from "outside," or only the nonkinetochore microtubules?

A.S. Bajer, Eugene, Oregon, USA
I have two questions addressed to Dr. Borisy or Dr. McIntosh: 1. Can we distinguish on a morphologic basis brain microtubules from spindle microtubules when brain microtubules are incorporated into the spindle?

2. If brain tubulin is added to other microtubules, can we distinguish where growth of microtubules begins?

Chairman:
Thus far it is not possible to distinguish, at least morphologically, micro-tubules from different sources.
In publications dealing with the structure of the spindle, different terms for the same part of the spindle can be found. Therefore, during the discussion some efforts were made to find a common terminology.

J.R. McIntosh, Boulder, Colorado, USA
A pole is the region at each end of the spindle toward which the chromosomes will move at anaphase.
A half-spindle is the portion of the spindle running from one set of kinetochores to and including the pole it faces.
The interzone is the portion of the spindle that lies between the sister sets of kinetochores.
Thus, during anaphase, the half-spindle shortens and the interzone lengthens.
These regions are not well defined until metaphase. The term "spindle fiber" is a light-microscopic term; spindle fibers contain clusters of microtubules and additional material.

H. Sato, Philadelphia, Pennsylvania, USA
Rather than to define a half-spindle from a morphologic standpoint, I prefer to define it in physiologic terms. For instance, upon prolonged treatment with 10^{-1} M mercaptoethanol, we can induce "half-spindles" - sea urchin blastomeres that possess fully functional centrosomes and kinetochores. Due to the lack of centriole replication or possible segregation, however, they become half-spindles and in most occasions fail to show anaphase chromosome movement.

A.S. Bajer, Eugene, Oregon, USA
The term "polar region" is more precise than "the pole."
The term "interzone," understood as a region between sister kinetochores, applies to anaphase. It becomes an "abstract" term if applied to a region of the spindle in metaphase of mitosis, especially if there is no regular metaphase plate.

T.E. Schroeder, Friday Harbor, Washington, USA
I object to an unqualified use of the term "half-spindle" as Dr. McIntosh has just defined it - that is, including one of the cell poles. I would prefer to restrict the term to the array of kinetochore microtubules alone. Cell poles are highly varied structures and, often, it is difficult to clearly identify which of several structural elements represents the pole as it might relate to an array of kineto-chore microtubules.

A.S. Bajer, Eugene, Oregon, USA
To avoid misunderstanding and confusion, different terms should be used for electron-microscope and light-microscope structures. E.g., continuous fibers are clearly seen in the polarizing microscope, as well as chromosomal fibers. In the electron microscope the existence of continuous fibers (microtubules) is not clear, and a chromosomal fiber is composed of kinetochore and nonkinetochore microtubules. Thus, at the electron-microscopic level, we could speak about kinetochore and chromosomal fibers.

H. Sato, Philadelphia, Pennsylvania, USA
If we are talking about a general terminology for the spindle structure, then I think we should stick to the term "continuous spindle fiber." If we are talking especially about the microtubules that are oriented from pole to pole, we should call them pole-to-pole microtubules. I think this is more precise.

U.-P. Roos, Zürich, Switzerland
I should like to propose the terms "central spindle" for the interpolar shaft of microtubules and "chromosomal spindle" for the ensemble of kinetochore microtubules.

S. Ghosh, Calcutta, India

Because the presence of an interzone area in all cells is controversial, can we not use the term interkinetochore spindle fibers?

In a paper of Giménez-Martin and coworkers (Risueño et al., J. Microscopie 26, 1976) I read about mitotic cells with persistent nuclear envelope in onion roots treated with cordycepin. Perhaps Dr. Lopéz-Saéz could describe the spindle fibers in these cells.

J.F. Lopéz-Saéz, Madrid, Spain

The statement "spindle can be formed from outside and/or inside the prophase nucleus" is not affected by, or incompatible with the organization of a mitotic apparatus without the breaking of the nuclear envelope, such as occurs in meristem cells treated with ethidium bromide (50-100 µg/ml). These cells are probably blocked by the inhibition of RNA synthesis and, in an interesting way, these prophases, after about a mitosis time, develop a cell plate and the new wall catches and furrows the mother nucleus.

T.E. Schroeder, Friday Harbor, Washington, USA

It seems to me to be an exercise in futility to attempt to find generally accepted terms for parts of the mitotic apparatus and different populations of microtubules. There is an inherent variability in mitotic apparatus organization from organism to organism; the architecture of any given mitotic apparatus changes as a dynamic structure in time; different methods of observation reveal different levels of organization; the literature is burdened with historically important terms - which cannot be uniformly simplified or standardized. I believe that nearly any reasonable set of terms is acceptable as long as we are careful to specify and realize that each term is conditional upon the nature of our specimen and method of observation. Any more rigid definition of terms fails to do justice to nature's variability and the technical conditions of our observations.

D. Mazia, Berkeley, California, USA

Should we not discuss the components of the mitotic apparatus other than microtubules and chromosomes! These other components make up most of the mass of the mitotic apparatus. Sometimes one has the impression that we picture the mitotic apparatus as a harp immersed in a bathtub; surely there is more to it.

A.S. Bajer, Eugene, Oregon, USA

We should pay more attention to the vesicles in the spindle. I would not be surprised if some of them are not preserved by the present standard electron-microscopic techniques. Vesicles can, and probably do have different properties, contents, etc., in various regions of the spindle and aster and may play a very important role in the regulation of the spindle structure and function.

H. Fuge, Kaiserslautern, F.R.G.

Membraneous vesicles, as, e.g., found in sea urchin spindle, are *not* common in all spindles. Hence, a general physiologic function of vesicles for spindle mechanism should be discussed with caution.

Chairman:

There is not only controversy in relation to the existence of vesicles in the spindle but, in particular, to their function. Therefore, more experimental data are necessary to draw final conclusions.

Session IV

Microtubule-Organizing Centers of the Mitotic Spindle

G. G. BORISY and R. R. GOULD, Laboratory of Molecular Biology, University of Wisconsin, Madison, Wisconsin 53706, USA

Discussions of the regulation of microtubule assembly generally have had a dual aspect. Firstly, they have been in terms of the extent of assembly; that is, in terms of whether and to what degree assembly is favoured or not. Secondly, and of equal significance to the amount of microtubules, discussions have focused on their spatial distribution within the cell and the control of their orientation. Clearly, an understanding of both temporal and positional control mechanisms will be required for an explanation of the role that these cytoplasmic structures play in cell behavior.

The essential properties of mitosis are the development of a bipolar spindle and the subsequent movement of the chromosomes. Both polarity and motility are equally essential. Movement of chromosomes without an ordered array of microtubules to define the directions of movement could lead to aberrant segregation and nonviable progeny cells. Hence, the mechanisms specifying positional information for microtubules are of crucial importance in understanding mitosis. In this paper, we will address the problem of the spatial regulation of microtubule assembly, in particular as it applies to the microtubule organizing centers of the mitotic spindle, namely the centrosomes and kinetochores.

A. The Spatial Regulation of Microtubule Assembly

The general consensus of the literature (see Olmsted and Borisy, 1973; Bardele, 1973; Roberts, 1974; Soifer, 1975; and Snyder and McIntosh, 1976 for reviews) is that the regulation of the assembly of cytoplasmic microtubules appears not to be due to differential protein synthesis, but rather to post-translational control mechanisms. Thus, assembly draws largely upon pre-existing pools of tubulin subunits.

A decade ago, Porter (1966) noted that to influence the shapes of cells, microtubules must be distributed according to some prescribed pattern, be anchored at one end and free to grow at the other end, and be oriented, that is, given a direction in their growth from a point. These considerations led Porter to postulate that there exists in the cytoplasm a complex of tubule-initiating sites that follow a spatial and temporal program that determines the distribution and initiation of microtubules and ultimately the form and shape of the cell. Implicit in the notion that microtubule formation is dependent upon tubule-initiating sites is the idea that in vivo microtubules do not spontaneously self-assemble in the absence of these sites. If this were not so, tubulin molecules would be expected to polymerize to form variable numbers of randomly oriented tubules. A random distribution of microtubules is in fact obtained for polymerization in vitro (Fig. 1), suggesting again that some mechanism is required in vivo to specify positional information.

In a previous paper (Borisy et al., 1976) we considered the proposition that cytoplasmic microtubules may be placed into two categories. One category refers to tubules that form spontaneously by self-assembly and

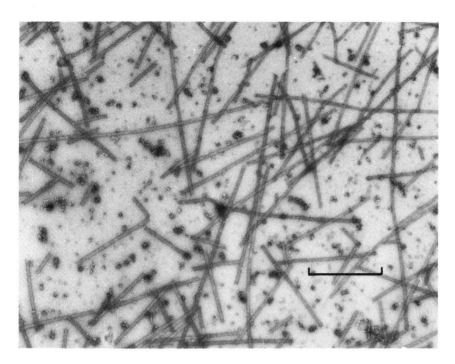

Fig. 1. Polymerization of microtubule protein in vitro. Electron micrograph of microtubules polymerized from microtubule protein purified from porcine brain tissue by two cycles of a reversible assembly-disassembly procedure. Samples were applied to a carbon and Formvar coated grid and negatively stained with 1 % uranyl acetate. Note the random orientation of tubules. *Solid line* indicates 1 μm

hence are free, that is, not attached to or originating from a particular structure. The other category represents tubules whose formation is dependent upon initiating centers, hence we refer to these as site-initiated tubules. The control of these two classes of tubules may be different. If it is important for cells to position their tubules with respect to other cell structures, then it may also be important to specify where microtubules shall not form.

In Figure 2 the self-assembly and site-initiated assembly of microtubules is diagrammatically represented. As indicated in the figure, both modes of assembly involve nucleation. The difference between the two modes is that in site-initiated assembly, the nucleating element is a nondiffusible structure which in vivo is generally located in a particular cytoplasmic region. The nature of this nucleating element is unknown, but we require that the orientation and direction of growth of the microtubule be determined in some way by the orientation of the nucleating element.

The diagram also serves to clarify the distinction between a microtubule-nucleating center and a microtubule-organizing center (Pickett-Heaps, 1969). A microtubule-nucleating center is considered to be an entity that initiates the formation of a single microtubule. The term microtubule-organizing center should be reserved for an entity that determines the overall spatial array of microtubules. Accordingly, the organizing center would then be identified with an interconnected

SELF-ASSEMBLY

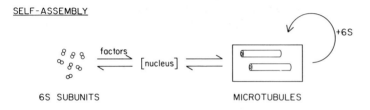

6S SUBUNITS   MICROTUBULES

SITE-INITIATED   ASSEMBLY

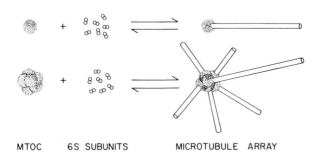

MTOC   6S SUBUNITS   MICROTUBULE ARRAY

Fig. 2. Schematic diagram of microtubule assembly. Self-assembly requires tubulin and under some conditions nontubulin factors, and proceeds through a nucleation stage to form microtubules which subsequently elongate by the biased polar addition of 6S tubulin subunits. In site-initiated assembly, microtubule formation is initiated by a nondiffusible nucleating element, represented as a *single stippled ball*. A microtubule organizing center (MTOC) is then represented by a constellation of nucleating elements and gives rise to a microtubule array

group of nucleating elements. The connections between the nucleating elements would bond them together into a larger cohesive structure. Upon initiation of the microtubules by the nucleating elements, the tubules would remain anchored at their origins by these same connectives. The anchorage of the tubules is probably of importance for mechanical reasons, such as supporting loads under compression, or transmitting forces under tension.

The essential points of this model are that the capacity to nucleate the formation of microtubules resides in nondiffusible nucleating elements; that the elements are structured so as to specify the direction of microtubule growth; and that connectives exist between the elements. Thus, the threefold fundamental properties ascribed by Porter to microtubule-organizing centers, namely, nucleation, orientation, and anchorage of microtubules, follow in a natural way from these properties of constellations of nucleating elements.

A further consequence of the activity of microtubule-organizing centers is that the resulting microtubules become positioned in an array with varying degrees of order. Tucker (1977) has discussed the factors which determine the pattern of microtubule packing and the degree of regularity in the resulting arrays. The pattern of microtubule packing seems to be established in two main ways. In one way, pattern is determined by a pre-existing arrangement of nucleating

elements. In the second way, pattern is established by self-linkage
of microtubules by intertubule bridges after microtubule formation.
In addition, both mechanisms for establishing pattern may operate
together in varying degrees.

In lower eukaryotes, a great diversity of microtubule-organizing
centers have been described (Pickett-Heaps, 1969; Tucker, 1977). However, in higher cells, the microtubule-organizing centers of greatest
importance are the centrosomes and the kinetochores of chromosomes.
Since these are the principal structures involved in the formation of
the mitotic spindle, the balance of this paper will focus on them.

B. Microtubule-Organizing Centers of the Mitotic Spindle

When a mammalian cell prepares to divide, it constructs an elaborate
mechanism to move the chromosomes to opposite poles of the cell. This
mechanism, called the mitotic spindle, is a spindle-shaped array of
fibers, some of which radiate from the poles of the cell, while others
are attached to the chromosomes. Although generations of observers
have been absorbed by the drama of cell division, the fundamental problem of how the spindle fibers are assembled and oriented, let alone
how they function, has remained unsolved. In fact, despite the modern
repertory of analytical techniques, our current understanding of spindle assembly is surprisingly and deeply indebted to some of the earliest
observations of dividing cells.

C. Centrosomes

A key participant in the assembly of the spindle is a structure called
the centrosome, which received its name because it was the central body
of the array of fibers present at each pole of the dividing cell (Van
Beneden, 1883; Boveri, 1888). The centrosome consists of a pair of
centrioles surrounded by a fibrous substance called the "pericentriolar
material". In mammalian cells, the centrosome serves as a microtubule-
organizing center in two distinct ways. During interphase, it resides
near the nucleus and is apparently the origin of a network of cytoplasmic microtubules that are thought to help maintain the cell's shape
(de Thé, 1964; Porter, 1966; Brinkley et al., 1975; Osborn and Weber,
1976). During cell division, the centrosomes, one at each pole of the
cell, give rise to a spindle-shaped array of microtubules that forms
the framework of the mitotic apparatus (Wilson, 1928; Bernhard and de
Harven, 1958; McIntosh et al., 1975). Although the centrosome has been
studied for a century, and although it is a prototype for other microtubule-organizing centers, the mechanism by which it nucleates, orients
and anchors microtubules has remained a mystery.

In their studies of cell division (Van Beneden, 1883; Boveri, 1888;
Wilson, 1928), the early cytologists first posed the crucial question,
which part of the centrosome gives rise to the spindle fibers - the
centrioles or the pericentriolar material? The centrioles were the most
conspicuous and persistent feature of the centrosome (Van Beneden,
1883; Boveri, 1888), but it was the pericentriolar material, seen in
the light microscope as a clear zone, that was observed to wax and
then wane as cell division progressed (Wilson, 1928).

Ironically, the advent of electron microscopy deepened, rather than
resolved, the problem. On the one hand, the centrioles and spindle
fibers were both revealed in thin sections to be bundles of microtubules (de Harven and Bernhard, 1956; Bernhard and de Harven, 1958),

leading to suggestions that the centrioles might "spin out" or nucleate
the microtubules of the spindle (Stubblefield and Brinkley, 1967). On
the other hand, it was soon apparent that while a few spindle microtubules might contact the centrioles directly, the great majority of
them originated in the densely staining, apparently amorphous material
surrounding the centrioles (Robbins et al., 1968). Hence the old di-
lemma, instead of being dispelled, reappeared in a new guise.

Recently, the ability of the centrosomes to nucleate microtubule assembly was directly tested in vitro by incubating lysed, dividing
cells with purified microtubule protein (Weisenberg and Rosenfeld,
1975; McGill and Brinkley, 1975; Snyder and McIntosh, 1975). These
experiments confirmed that the centrosome can nucleate microtubules
in vitro, and in one report it was suggested that the material associated with the centrioles was the nucleating material (Weisenberg
and Rosenfeld, 1975). Again, however, electron microscopy failed to
distinguish unequivocally which component of the centrosome nucleated
microtubules.

In our own studies (Gould and Borisy, 1977) we investigated the structure and function of the centrosomes in vitro, isolated from other
cell components, as obtained from lysed Chinese hamster ovary (CHO)
cells. The centrosomes from both interphase and dividing cells consisted of pairs of centrioles, a fibrous pericentriolar material, and
a group of virus-like particles which were characteristic of the CHO
cells and which served as markers for the pericentriolar material
(Fig. 3).

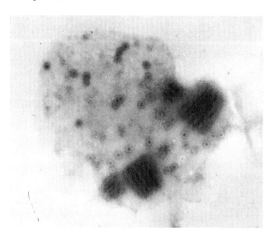

Fig. 3. A centrosome from a Chinese
hamster ovary cell. Lysates were
prepared from cells using the detergent Triton X-100, sedimented
onto grids and negatively stained
with 2 % phosphotungstic acid. The
centrosome consists of a pair of
centrioles embedded in a fibrous
matrix. Virus-like particles are
also associated with the matrix.
Each centriole bears an immature
daughter centriole at right angles
to itself. x 25,000

Interphase centrosomes anchored up to two dozen microtubules when cells
were lysed under conditions which preserved native microtubules. When
colcemid-blocked mitotic cells, initially devoid of microtubules, were
allowed to recover for 10 min, microtubules formed at the pericentriolar
material, but not at the centrioles. When lysates of colcemid-blocked
cells were incubated in vitro with microtubule protein purified from
porcine brain tissue (Fig. 4), 100 to 250 microtubules assembled at
the centrosomes, similar to the number of microtubules that would normally form at the centrosome during cell division. A few microtubules
could also be assembled in vitro onto the ends of isolated centrioles
from which the pericentriolar material had been removed, forming
characteristic axoneme-like bundles. In addition, microtubules assembled
onto isolated fragments of densely staining, fibrous material which
was tentatively identified as pericentriolar material by its association

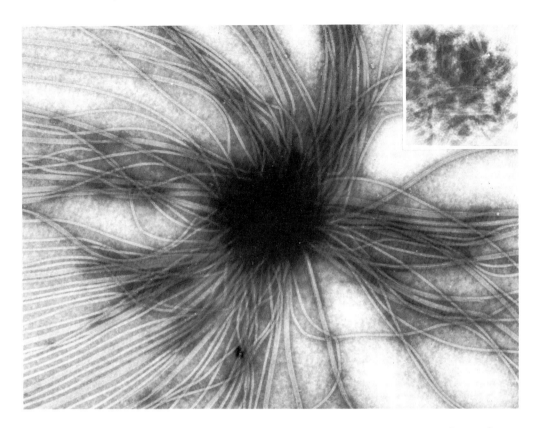

<u>Fig. 4.</u> Centrosome-initiated microtubule assembly in vitro. A centrosome from col-
cemid-blocked mitotic Chinese hamster ovary cells after exogenous incubation with
porcine brain tubulin in vitro showing microtubules polymerized at the centrosome.
Over 125 microtubules emanate from the densely staining center. The stain conceals
a centriole pair, shown at higher magnification and the same orientation in the
inset. x 20,000; inset, x 37,500

with the virus-like particles. Thus, we have evidence that the pericentriolar material is actively associated with the formation of
microtubules in vivo, and that it nucleates microtubule assembly in
vitro. We conclude that it is the pericentriolar material, and not
the centrioles themselves, that nucleates the microtubules of the
mitotic spindle.

D. Kinetochores

The second important participant in the organization of the spindle
is a structure called the kinetochore, which is found at the constriction of the chromosome, and to which the chromosomal spindle fibers
are attached (Metzner, 1894). Despite reports that individual spindle
fibers were formed at the kinetochore (e.g., Hughes-Schrader, 1924),
its possible role as an organizing center for spindle fibers was over-
shadowed by its better established role in the movement of chromosomes
(hence the term kinetochore - "movement place"; Sharp, 1934).

The organizing capacity of the kinetochores was explicitly noted by Pease (1946), who showed that after chromosomal fibers were depolymerized by high hydrostatic pressure, they regenerated from the kinetochore region. Inoué (1964) showed that the integrity of the kinetochore region was essential for regeneration of chromosomal fibers that had been damaged by ultraviolet microbeam irradiation.

The kinetochore was revealed by electron microscopy to be a three-layered plate (Nebel and Coulon, 1962; Jokelainen, 1967; Brinkley and Stubblefield, 1966) to which the chromosomal microtubules are attached. However, like the pericentriolar region, the kinetochore was sufficiently featureless in this sections to yield no clues as to how it might initiate microtubules.

Recently, the microtubule-organizing capabilities of kinetochores was directly tested in vitro by the incubation of either lysed, dividing cells or isolated chromosomes with purified microtubule protein (McGill and Brinkley, 1975; Telzer et al., 1975; Gould and Borisy, unpublished results). Microtubules were found to assemble specifically at the kinetochores, confirming that these sites serve to nucleate microtubules. In our own studies, the microtubule-nucleating capacity of chromosomes was tested in vitro in lysates of Chinese hamster ovary cells using conditions previously employed for centrosomes. Figure 5 shows the microtubule growth onto a chromosome in a lysate incubated with exogenous porcine brain tubulin. Under the conditions employed, greater than 98 % of the chromosomes gave rise to microtubules at their kinetochore regions thus unequivocally demonstrating that chromosome are competent to initiate microtubule formation.

E. Discussion

In vitro polymerization experiments have now demonstrated that centrosomes and kinetochores serve as microtubule-organizing centers. The centrosome has been shown to be an organelle in itself, which can be isolated intact at any stage of the cell cycle, and which consists of a pair of centrioles and pericentriolar material. Thus, the early cytologists who first used the term "centrosome" (Van Beneden, 1883; Boveri, 1888; Wilson, 1928) were quite right in hinting at its discrete nature. In recent years, the distinction between the centrioles, and the larger centrosomes of which they are a part, has been blurred (Snyder and McIntosh, 1975), and occasionally the two terms have been used interchangeably (McGill and Brinkley, 1975). However, the centrioles and pericentriolar material are two functionally distinct parts of the centrosome, and it is important to maintain the distinction between them. From our studies we conclude that it is the pericentriolar material and not the centrioles which is primarily responsible for nucleating and anchoring microtubules at the poles of the mitotic spindle.

The studies on chromosomes are less advanced and it has not yet been possible to test whether the organizing activity of the chromosome is separable from the bulk chromatin as the pericentriolar material is separable from the centrioles. However, an intriguing possibility is that the microtubule organizing centers of both centrosomes and chromosomes are similar in their molecular composition if not structural arrangement. Evaluation of this suggestion will require the isolation and biochemical characterization of microtubule organizing material. The ultimate goal of these studies will be to understand the basic strategy by which a cell constructs its mitotic spindle in terms of the positioning and control of microtubule nucleating material at the centrosomes and kinetochores.

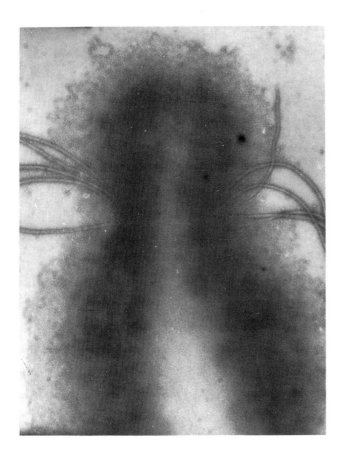

Fig. 5. Chromosome-initiated microtubule assembly in vitro. A chromosome from col-cemid-blocked mitotic Chinese hamster ovary cells after incubation with exogenous porcine brain tubulin in vitro showing microtubules polymerized at the kinetochore region. The stain conceals the kinetochore structure. x 24,800

Acknowledgment. This work was supported by National Institutes of Health grant GM21963 to G.G. Borisy and by National Institutes of Health Postdoctoral Fellowship CAO2357 to R.R. Gould.

References

Bardele, C.: Struktur, Biochemie und Funktion der Mikrotubuli. Cytobiologie 7, 442-488 (1973)

Beneden, E. Van: Recherches sur la maturation de l'oeuf. La Fécondation et la Division Céllulaire. Paris: Masson 1883

Bernhard, W., Harven, E. de: L'ultrastructure du centriole et d'autres éléments de l'appareil achromatique. In: 4th Intern. Conf. Electron Microscopy. Bargmann, W., Petirs, D., and Wolpers, C. (eds.). Berlin, Heidelberg, New York: Springer 1958, Vol. II, pp. 217-227

Borisy, G.G., Johnson, K.A., Marcum, J.M.: Self-assembly and site-initiated assembly of microtubules. In: Cell Motility. Goldman, R., Pollard, T., and Rosenbaum, J. (eds.). Cold Spring Harbor Lab. 1976, pp. 1093-1108

Boveri, T.: Zellen-Studien. Jena: Fischer 1888, 2, 68

Brinkley, B.R., Fuller, G.M., Highfield, D.P.: Cytoplasmic microtubules in normal and transformed cells in culture: analysis by tubulin antibody immunofluorescence. Proc. Nat. Acad. Sci. US 72, 4981-4985 (1975)

Brinkley, B.R., Stubblefield, E.: The fine structure of the kinetochore of a mammalian cell in vitro. Chromosoma 19, 28-43 (1966)

Gould, R.R., Borisy, G.G.: The pericentriolar material in Chinese hamster ovary cells nucleates microtubule formation. J. Cell Biol. 73, 601-615 (1977)

Harven, E. de, Bernhard, W.: Etude au microscope êléctronique de l'ultrastructure du centriole chez les vertebrates. Z. Zellforsch. 45, 378-398 (1956)

Hughes-Schrader, S.: Reproduction in *Acroschismus wheeleri* (Pierce). J. Morphol. 39, 157-205 (1924)

Inouê, S.: Organization and function of the mitotic spindle. In: Primitive Motile Systems in Cell Biology. Allen, R.D., Kamiya, N. (eds.). New York: Academic Press 1964, 549-594

Jokelainen, P.T.: The ultrastructure and spatial organization of the metaphase kinetochore in mitotic rat cells. J. Ultrastruct. Res. 19, 19-44 (1967)

McGill, M., Brinkley, B.R.: Human chromosomes and centrioles as nucleating sites for the in vitro assembly of microtubules from bovine brain tubulin. J. Cell Biol. 67, 189-199 (1975)

McIntosh, J.R., Cande, Z., Snyder, J., Vanderslice, K.: Studies on the mechanism of mitosis. Ann. N.Y. Acad. Sci. 253, 407-427 (1975)

Metzner, R.: Beiträge zur Granulalehre. Arch. Anat. U. Physiol. 309, 1894. In: The Kinetochore of Spindle Fiber Locus in *Amphiuma tridactylum*. Schrader, F. (ed.). Biol. Bull. 70, 484-498 (1934)

Nebel, B.R., Coulon, E.M.: The fine structure of chromosomes in pigeon spermatocytes. Chromosoma 13, 272-291 (1962)

Olmsted, J.B., Borisy, G.G.: Microtubules. Ann. Rev. Biochem. 42, 507-540 (1973)

Osborn, M., Weber, K.: Cytoplasmic microtubules in tissue culture cells appear to grow from an organizing structure towards the plasma membrane. Proc. Nat. Acad. Sci. US 73, 867-871 (1976)

Pease, D.C.: Hydrostatic pressure effects upon the spindle figure and chromosome movement. II. Experiments on the meiotic divisions of *Tradescantia* pollen mother cells. Biol. Bull. 91, 145-169 (1946)

Pickett-Heaps, J.: The evolution of the mitotic apparatus: An attempt at comparative ultrastructural plant cytology in dividing plant cells. Cytobios. 3, 257-280 (1969)

Porter, K.R.: Cytoplasmic microtubules and their functions. In: Principles of Biomolecular Organization. Wolstenholme, G.E.W., O'Connor, M. (eds.). London: J. and A. Churchill Ltd. 1966, pp. 308-356

Robbins, E.L., Jentzsch, G., Micali, A.: The centriole cycle in synchronized HeLa cells. J. Cell Biol. 36, 329-339 (1968)

Roberts, K.: Cytoplasmic microtubules and their functions. In: Progress in Biophysics and Molecular Biology. Butler, A.J.V., Noble, D. (eds.). Oxford: Pergamon Press 1974 Vol. 28, pp. 271-420

Sharp, L.W.: Introduction to Cytology. New York: McGraw Hill 1934

Snyder, J.A., McIntosh, J.R.: Initiation and growth of microtubules from mitotic centers in lysed mammalian cells. J. Cell Biol. 67, 744-760 (1975)

Snyder, J., McIntosh, J.R.: Biochemistry and physiology of microtubules. Ann. Rev. Biochem. 45, 699-720 (1976)

Soifer, D. (ed.): The biology of cytoplasmic microtubules. Ann. N.Y. Acad. Sci. 253, 848 (1975)

Stubblefield, E., Brinkley, B.R.: Architecture and function of the mammalian centriole. Symp. Intern. Soc. Cell Biol. 6, 175-218 (1967)

Telzer, B.R., Moses, M.J., Rosenbaum, J.L.: Assembly of microtubules onto kinetochores of isolated mitotic chromosomes of HeLa cells. Proc. Nat. Acad. Sci. US 72, 4023-4027 (1975)

Thê, G. de: Cytoplasmic microtubules in different animal cells. J. Cell Biol. 23, 265-275 (1964)

Tucker, J.B.: Shape and pattern specification during microtubule bundle assembly. Nature (London) 226, 22-26 (1977)

Weisenberg, R.C., Rosenfeld, A.C.: In vitro polymerization of microtubules into asters and spindles in homogenates of surf clam eggs. J. Cell Biol. <u>64</u>, 146-158 (1975)

Wilson, E.B.: The Cell in Development and Heredity. 3rd Ed. New York: Macmillan 1928

Discussion Session IV: Tubulin and Microtubules
Chairman: H. PONSTINGL, Heidelberg, F.R.G.

C. Petzelt, Heidelberg, F.R.G.
Dr. Borisy, how did you isolate and separate the pericentriolar material?

G.G. Borisy, Madison, Wisconsin, USA
The centrosomes are isolated from cells lysed in a low ionic strength polymerization buffer with 10 mM PIPES, 0.25 mM Mg^{2+}, 0.1 % Triton X-100, 1 mM EGTA. Under these conditions the centrosome pops out of the cell and is fairly stable. If this is incubated in the absence of EGTA at elevated temperature for 15 min, then the pericentriolar material works loose and we have the centrioles.

U.P. Roos, Zürich, Switzerland
Dr. Borisy, did you expose your isolated chromosomes to tubulin in suspension or to chromosomes fixed to a grid? Have you fixed and embedded chromosomes with newly assembled microtubules and have you sectioned such?

G.G. Borisy, Madison, Wisconsin, USA
The answer to the first question is that the tubulin was reacted with the chromosomes in suspension, and the complexes were subsequently attached to a grid. The answer to the second question is: Yes, we have embedded and thin-sectioned chromosomes prepared from the lysed cells. We have seen the kinetochore morphology as it was before addition of tubulin. But it is not so easy to trace the tubules in the sections. From the limited section data that we have, I cannot give you quantitative data on the number of tubules initiated nor of the length of the tubules.

L. Hens, Brussels, Belgium
Is there any evidence for chromosomal microtubule organizing centers other than the centromere? From Dr. Borisy's pictures of the assembly of microtubules on CHO chromosomes, it was not clear if all microtubules started from the centromere. Is it possible that a few microtubules start from the chromosome arms?

G.G. Borisy, Madison, Wisconsin, USA
As I mentioned in the talk, the staining of the chromosomes is so dense that it did not permit one to see the origins of the tubules as they come in. Some of the tubules originate from the kinetochores, but then come up over the arms of the chromosomes. Then, we only see them when they emerge from the chromosomes. They apparently arise from an arm of the chromosome. However, on the plates it is possible to trace these tubules to the point of origin, and although sometimes there are tubules that we cannot see going to the kinetochores, the overwhelming majority, more than 90 % of them, are traceable to the kinetochore region.

C. Petzelt, Heidelberg, F.R.G.
Dr. Borisy, have you tried to isolate centrioles from interphase cells and checked on their capacity to assemble microtubules?

G.G. Borisy, Madison, Wisconsin, USA
We have tried to isolated centrosomes from interphase cells and to test them for initiation capacity, and we do have some initiation. But our results are not clean enough to tell you how this compares to the initiation capacity in the mitotic cells.

M. Hauser, Bochum, F.R.G.
Have you done the same experiments with isolated basal bodies? Was it the same result as with the isolated centrioles liberated from the dense material?

G.G. Borisy, Madison, Wisconsin, USA
No, but Snell and others in Rosenbaum's laboratory did such experiments with basal bodies isolated from *Chlamydomonas* and obtained biased directional assembly (Snell, W.J. et al., Science 185, 357, 1974), as we see here on the centrioles. Also McGill and Brinkley (J. Cell Biol. 67, 189, 1975) described biased directional assembly on centrioles in lysates of HeLa cells.

A.M. Lambert, Strasbourg, France
The nuclear envelope may be a nucleating center for microtubule polymerization in plant cells. During meiosis prophase (pachytene) in the moss *Mnium hornum*, I described microtubules that were clearly associated with the nuclear envelope. On the surface of such intact nuclei, microtubules are embedded in dense material and distributed regularly. A detailed description of such association and its hypothetical role is given in: Lambert, A.M., Abst. Intern. Meet. of Electron Microscopy, Grenoble, Volume III, 1970, and C.R. Acad. Sci. Ser. D, 1970.

W. Herth, Heidelberg, F.R.G.
Immunofluorescence investigations of *Leucojum* endosperm cells with a monospecific antibody against tubulin from porcine brain show accumulations of diffuse positive reaction at the nuclear periphery and at the poles in prophase to prometaphase. Our control experiments with total IgG of nonimmunized rabbits and with omission of the tubulin antibody rule out a nonspecific binding of the tubulin antibody (Franke, W.W., et al., Cytobiologie 15, 24, 1977).

D. Mazia, Berkeley, California, USA
I wanted to call attention to the fact that a problem in mitosis is the problem of the spindle fiber. There are bundles as units of connection and motion. There would be no reason to expect that analysis of the assembly of the microtubule itself would provide an explanation of the formation of bundles.

T.E. Schroeder, Friday Harbor, Washington, USA
I have had the impression from micrographs of some kinetochores and their associated bundles of microtubules that the degree of parallel alignment (vs. divergence) correlates with the flat, planar nature of the kinetochore (vs. convex form). If this relation holds, it suggests a remarkable control of microtubule bundling by the kinetochore itself.

U.-P. Roos, Zürich, Switzerland
In rat kangaroo cells, one can often observe a bundle of quite parallel microtubules connected to a very convex kinetochore. Thus, the comment made by Dr. Schroeder does not apply to this situation, for the organization of the microtubules does not reflect the convexity of the kinetochore.
However, it is true that in prometaphase, especially at the beginning of this stage, kinetochore microtubules are not tightly bundled, but they do become more tightly bundled as prometaphase progresses. However, this is not related to a cessation of chromosome movements, because the chromosomes continue to move throughout late prometaphase and metaphase.
I also at one time looked at some sections of *Orthoptera* spermatocytes, whose bivalents have kinetochores resembling those of *Haemanthus*, i.e., they look like a ball in a cup. However, as far as I can recall, the kinetochore fibers were bundles of quite parallel microtubules, not divergent as in *Haemanthus*, but this may need to be confirmed by a more thorough investigation.

D. Mazia, Berkeley, California, USA
Do we know whether the microtubule-organizing bodies can regulate the length of microtubules? We do see such regulation as a real happening which may be very important. An example is the unequal divisions of some eggs (or later blastomeres) which we can identify with critical steps of early differentiation. The unequal division is brought about by the formation of mitotic apparatus with large asters at one pole and small asters at the other. This shifts the equator of the spindle and therefore the plane of division. The difference in size of the asters at the two poles obviously results from a difference in the lengths of the astral microtubules. One would think that the phenomenon is telling us to look for differences in the centers that regulate the different lengths of the microtubules.

G.B. Borisy, Madison, Wisconsin, USA
What that would require at the molecular level, it seems to me, is not an exact determination of the length of the tubule but rather the value of the modal length of a population of tubules. One could still have a stochastic process which determines the distribution of tubule length. If one accepts that, then I think one can speak about the physical chemistry. It has been shown in a number of laboratories that for tubules that have formed by self assembly, the length distribution

depends upon the ratio of nontubulin factors to tubulin. That is, if one makes mixtures of tubulin with low concentrations of factors, then one gets few tubules that are very long. If one takes tubulin with high concentrations of factors, one gets many tubules which are relatively short. And, as I showed in the histograms of tubule length obtained by self-assembly, there was a rather paucidisperse distribution with a well-defined mode. So, perhaps it is conceivable to think of some diferences locally in the cell in the ratio of initiating factors to tubulin, which in turn determines the value of the mode.

J.R. McIntosh, Boulder, Colorado, USA
What Dr. Borisy suggested permits the prediction that the large aster is going to have longer tubules, but fewer of them. Is that a fact?

H. Sato, Philadelphia, Pennsylvania, USA
Unequally developed asters in a mitotic spindle, which will lead the cell later into an unequal division, can be observed at the polar body formation stage in mature starfish oocytes or in the 4th division in sea urchin embryos.
From birefringence measurements on both intact and isolated spindles, it has been found that both the smaller and the larger asters could hold the same amout of microtubules. At least we could not detect any significant difference in these asters except their size.

J.G. Carlson, Knoxville, Tennessee, USA
The spindle of a cell treated with an appropriate concentration of colchicine or dose of 225 nm ultraviolet radiation just before breakdown of the nuclear membrane will develop a spindle much smaller than normal - perhaps half its normal diameter and length. Its general shape and capacity to separate the daughter chromosomes normally at anaphase, however, will be unimpaired. Is it likely that its shorter length is due to decreased polymerization of tubulin during its formation and its smaller diameter to the presence of a smaller number of microtubules?

P. Harris, Eugene, Oregon, USA
Kawamura reported studies on asymmetric divisions in grasshopper neuroblast cells (Exptl. Cell Res. 21, 1, 1960; 106, 127, 1977). The asymmetric mitosis occurred only when the neuroblast was associated with several companion cells, but divided symmetrically when these cells were removed. It suggests that there can be a topical effect increasing microtubule growth locally.

H. Sato, Philadelphia, Pennsylvania, USA
Application of low concentrations of colchicine or colcemid will reduce the concentration of polymerizable tubulin in living cytoplasm. As a result, a small and weakly birefringent spindle is formed in the cell. However, 366 nm near-ultraviolet light irradiation on the spindle region will convert tubulin-bound colchicine to lumi-colchicine, the inactivated form of colchicine, and thus release the polymerizable tubulin. Eventually, the spindles grow and regain the size and birefringence of spindles, reflecting the available amount of polymerizable tubulin (Aronson, J, J. Cell Biol. 51, 579, 1971 and Sluder, G., J. Cell Biol. 70, 75, 1976).

R. Dietz, Tübingen, F.R.G.
In the crane fly spermatocyte, two of the six autosomes are oriented toward one pole, and four autosomes toward the other. In metaphase the spindle is highly asymmetric: The half spindle, in which only two chromosomal spindle fibers have been organized, is considerably longer than the half spindle in which four chromosomal spindle fibers have been assembled. The anaphase velocity is also different: The two chromosomes in one anaphase group move significantly faster than the four chromosomes in the other. Since both poles are occupied by asters, the asymmetry of the spindle is obviously caused by competition between kinetochores. The diffusion of tubulin protomers apparently is highly restricted in living cells. This may explain that approximately the same subpools are available for the assembly of four chromosomal spindle fibers within one half spindle, and for the assembly of two (correspondingly longer) chromosomal spindle fibers in the other. Sequestering of the protomeric tubulin pool may also explain the division mentioned by Dr. Mazia. If the two asters lie within cell regions in which different amounts of protomeric tubulin are sequestered, one aster may grow much larger than the other.

A.S. Bajer, Eugene, Oregon, USA
Microtubules form spontaneously around the spindle, and only then on the randomly
arranged microtubules organizing centers; begin to exert their influence; microtubules are then reorganized.

G.G. Borisy, Madison, Wisconsin, USA
Dr. Bajer raises a very good point. In my talk I focused exclusively on the site-
initiated tubules just to sharpen the issues. However, from Dr. Bajer's work and
also that of Dr. McIntosh and Dr. Fuge, it is clear that many free tubules are
present in the spindle, tubules not associated either with the centrosome or the
chromosomes. Yet many of these tubules in the mature spindle are arranged parallel
to the axis of the spindle. We have the very interesting question of how these
tubules came to be in this position, and I would agree with Dr. Bajer that they
may form randomly. I cannot see any reason why some self-assembly should not occur
in cells and that under the influence of organized tubules they may come to take
up positions parallel to the site-initiated tubules. These free tubules, if they
interact laterally with the site-initiated microtubules, may be part of the spindle
fibers, either centrosomal spindle fibers or chromosomal fibers.

G. Blumenthal-Kasbekar, Washington, DC. USA
Has any attempt been made to determine the minimum size of the microtubule fragments necessary for the in vitro elongation?

G.G. Borisy, Madison, Wisconsin, USA
In our experiments in which we made microtubule seeds, the modal length was 1 µ.
There were tubules that were shorter, extending down to several tenths of a µ.

C. Petzelt, Heidelberg, F.R.G.
There is a certain amout of tubulin polymerized in the mitotic spindle, and every-
body knows that there is much more tubulin in the cell. What prevents the tubulin
from polymerizing - say at mitosis?

M. Osborn, Göttingen, F.R.G.
At the moment it is dangerous to interpret quantitatively immunocytochemical data
on pools of tubulin because the relative affinities of the antibodies for microtubules versus soluble tubulin is not known. However, immunofluorescence microscopy with tubulin antibodies (Brinkley's lab, our lab) suggests that at the beginning of mitosis a large fraction of cytoplasmic microtubules may be depolymerized,
ant that as the next cell cycle begins, cytoplasmic microtubules are repolymerized.
We assume that at least some of the depolymerized tubulin molecules from the cytoplasmic microtubules is cycled into spindle microtubules, since metaphase cells
show a very bright spindle against a dark background.

E. Schnepf, Heidelberg, F.R.G.
In young leaflets of the moss, *Sphagnum*, the overall amount of cytoplasmic microtubules before mitosis is about 1000 µm per cell. In mitosis, all cytoplasmic microtubules have disappeared; the amount of microtubules in the spindle then is, again,
about 1000 µm (Schnepf, F.: Protoplasma 78, 145, 1973). In both cases, however,
there is a certain pool of free tubulin, as can be shown with heavy water, which
induces the formation of additional microtubules (Schnepf et al., Cytobiologie 13,
341, 1976).

L. Schimmelpfeng, Köln, F.R.G.
We looked for an in vivo situation similar to mitotic cells concerning the equilibrium between spindle and cytoplasmic microtubules. We found that freshly trypsinized cells after cold treatment showed a shape recovery (thought to be a function of microtubules) that was clearly not inhibited by colchicine. On the other
hand, we found that attached cells kept for a given time at 14°C, a temperature
known to inhibit polymerization of tubulin in vitro, had a mitotic activity of
at least 25 %. Thus we considered cytoplasmic microtubules to be in some way different from those of the spindle, because of their different reaction upon cold
treatment and colchicine.

J.R. McIntosh, Boulder, Colorado, USA
There is some indication that the stability of microtubules varies simply with
the extent of their bunching. For example, the early anaphase interzone tubules

are untouched and are labile. A little later, what appear to be the same tubules are bunched, and they become the most stable tubules in the spindle.

P.F. Baker, London, England, UK

What information is available on the sensitivity of microtubules to their ionic environment? There seems to be some difference between isolated microtubules and microtubules inside cells, and I wonder whether any systematic attempt has been made to alter the internal ionic environment of cells and examine any concomitant effects on microtubular organization.

The normal levels of cation in mammalian cells are approximately Na, 10 mM; K, 140 mM; Mg, 10 mM; and Ca, 1 mM. Results with ion-selective electrodes and dyes suggest that the bulk of the Na and K is ionized, about 1/3 to 1/2 of the Mg is ionized (Baker-Crawford; J. Physiol. $\underline{227}$, 855), but less than 0.1 % of the Ca is ionized; and ionized Ca inside cells seems to be 0.1 µM or less (Baker, Prog. Biophys. Mol. Biol. $\underline{24}$, 179).

C. Petzelt, Heidelberg, F.R.G.

Is there any reliable method for measuring the ionized Ca^{2+} in situ in the cell?

P.F. Baker, London, England, UK

The most accurate measurements of ionized Ca have been made in large cells, and methods are not yet available for measuring ionized Ca accurately in small cells. The bulk of intracellular Ca is bound either within organelles, such as mitochondria, and derivatives of the endoplasmic reticulum, such as sarcoplasmic reticulum, or to protein. Nothing is known about Ca binding in the nucleus. Injection of Ca into cells results only in very localized change in ionized Ca because of the powerful binding system. The precise localization of Ca is difficult, and most techniques need to be treated with caution. Anything altering the effectiveness of binding systems can have very dramatic effects on intracellularly ionized Ca. The total amount of Ca available for intracellular binding is controlled by Ca pumps located in the surface membrane, and alteration in the activity of these pumps can lead to changes both in total and ionized Ca inside cells.

Intracellular ionized Ca can be increased by exposure to the ionophore A23187; but as with cytochalasin, this ionophore has many effects. It can alter the permeability of both surface and intracellular membranes to Ca and Mg, and flooding the cytosol with Ca tends to activate any energy-dependent Ca binding systems leading to a severe depletion of ATP.

Nevertheless, used with care, it can provide valuable information. Ionophores selective for Na or K can also be used to regulate the intracellular levels of these ions.

M. DeBrabander, Beerse, Belgium

We have some evidence for a role of Ca in microtubule polymerization in intact cells. This is primarily the work of Dr. Borgers in our laboratory. Intranuclear microtubules appear in peribronchiolar mast cells in rats in which an anaphylactic shock was produced. This coincides with the disappearance of Ca from the nucleus of these cells, as was demonstrated with the pyroantimonate cytochemical technique. This technique, however, is not quantitative and is subject to several criticisms.

J.R. McIntosh, Boulder, Colorado, USA

Kiehart and Inoué have injected small drops of millimolar Ca^{2+} solutions into metaphase marine eggs. They see a local reduction in spindle birefringence. The volume of the zone of reduced birefringence is about the size of the injected drops, so the local free Ca^{2+} concentrations are probably high. It is interesting that the zone of reduced birefringence does not spread, as one might expect if the Ca^{2+} were free to diffuse. Probably the Ca^{2+} is rapidly bound and/or sequestered into vesicles in the egg cytoplasm.

I.B. Heath, Downsview, Ontario, Canada

When considering the possible role of ions such as Ca^{2+} in controlling mitosis, one must remember that in a number of cells one gets concomitant polymerization and depolymerization in closely adjacent microtubules.

C. Petzelt, Heidelberg, F.R.G.
What indication do we have that there are real differences between the brain tubulin and tubulin from other sources?

H. Ponstingl, Heidelberg, F.R.G.
Pfeffer et al. (J. Cell Biol. $\underline{69}$, 599-607, 1977) measured the colchicine binding to tubulin from chick brain and sea urchin mitotic spindles. They found the K_{Ass} to be lower by one order of magnitude in the mitotic spindle tubulin. But this may as well reflect differences between species.

T. Bibring, Nashville, Tennessee, USA
We have detected differences between tubulins from different microtubules of a single organism by gel electrophoresis, chromatography, and peptide analysis. All the tubulins we looked at were different - tubulins from doublet microtubules of sperm flagella, mitotic apparatus microtubules, and the A-tubule of cilia in sea urchins. The in vitro studies of polymerization suggest that mitosis could proceed with brain tubulin, but our findings reopen the possibility that specific tubulins are selected from heterogeneous pools by polymerization in vivo.
A separate question is the heterogeneity of tubulin from any one microtubule system. Here we find as many as four subunits, two α subunits and two β subunits, by isoelectric focusing in gels containing urea.

H. Ponstingl, Heidelberg, F.R.G.
How did you obtain the tubulins, and how did you know they were tubulins?

T. Bibring, Nashville, Tennessee, USA
The flagellar and ciliary tubulins were obtained from isolated flagellar doublet microtubules and A-tubules of cilia. The tubulin was extracted with organic mercurial, which is quite selective for tubulin. The mitotic apparatus tubulin was obtained by isolation of mitotic apparatus and extraction with mercurial. The tubulins were further purified by DEAE-cellulose chromatography.

M. Little, Heidelberg, F.R.G.
Organic mercurials could modify and extract any proteins which contain sulfhydryl groups. Are you sure that the proteins which you are looking at are all tubulins?

T. Bibring, Nashville, Tennessee, USA
They were similar, and tubulin-like, in gel electrophoresis, chromatography on DEAE and on hydroxylapatite, amino acid composition, and end-group analysis.
The mercurial was removed by dialysis against mercaptoethanol, followed by carboxymethylation. We have no formal proof that we did not introduce heterogeneity during this procedure, but we got reproducible amounts of the subunits. If the heterogeneity was artifactual, one might expect the subunits to be present in variable amounts. Even if the heterogeneity were artifactual, though, it would not explain the reproducible differences we obtained between the different tubulins, which would have to arise from intrinsic differences.

H. Ponstingl, Heidelberg, F.R.G.
How many peptide differences are there between the tubulins?

T. Bibring, Nashville, Tennessee, USA
It seems to be several, but we are not prepared to quantitate that yet.

B. Jokusch, Basel, Switzerland
Dr. Bibring, did you prepare antibodies against the three different kinds of tubulin you found in one cell type?

T. Bibring, Nashville, Tennessee, USA
We made an antibody against the flagellar doublet tubulin, and it reacted with the tubulin from mitotic apparatus. We have not yet tried it with the ciliary tubulin.

E. Mandelkow, Heidelberg, F.R.G.
Could someone summarize which kinds of tubulins have been shown to copolymerize with others?

G.G. Borisy, Madison, Wisconsin, USA
There are different kinds of copolymerization which have been achieved in vitro.
Pig tubulin has been incorporated into spindles of several different sorts: rat
kangaroo spindles, clam oocyte spindles, *Chaetopterus* spindles.

E. Mandelkow, Heidelberg, F.R.G.
Did this tubulin add to existing mitotic tubules or did it form new tubules?

G.G. Borisy, Madison, Wisconsin, USA
It is not clear in the cases where tubulin is incorporated into the spindles.
In addition tubulin has been added onto basal bodies, flagella, sperm tails,
spindle pole bodies, and centrioles from several different kinds of cells. Pig
brain tubulin has also been reported to copolymerize - subunits are incorporated
together in the same polymer - with yeast and with HeLa.

G. Wiche, Vienna, Austria
Tubulin preparations from 3T3-fibroblast cells can be incorporated into microtubules formed from extracts of rat glial C_6-cells.

E. Mandelkow, Heidelberg, F.R.G.
Have you checked whether the microtubules grown from the organizing centers were
typical, i.e., do they have 13 protofilaments and the 3-start helical arrangement?

G.G. Borisy, Madison, Wisconsin, USA
I have not looked in detail. One can see protofilaments, and the tubules look
superficially indistinguishable from brain tubules, but I have not done any dif-
fraction experiments.

K.H. Doenges, Heidelberg, F.R.G.
Dr. Borisy, what do you think about the function and role of rings during in
vitro polymerization of neural tubulin? Are these rings important for tubulin
assembly?

G.G. Borisy, Madison, Wisconsin, USA
We do not view them as obligatory intermediates.

G. Wiche, Vienna, Austria
"Rings" with the physical appearance of Kirschner's 36 S material have been observed in preparations of rat glial C_6-cell microtubules at 4°C in rare number.
The extent of ring formation could be increased by the addition of 1 mM Ca^{2+} or
hog brain τ-factor preparations. In preparations from other cell lines we have
not studied the formation of ring-like structures in particular. However, it
seems unlikely that the formation of these structures occurs only in C_6-cell microtubule preparations, since our studies with preparations from other cell types
(including mouse fibroblast cells) did not reveal differences in any of the other
features of the polymer formation process we have studied so far.

M. DeBrabander, Beerse, Belgium
Has anybody ever seen rings in cells?

H.P. Erickson, Gif-sur-Yvette, France
Rings have not been identified in vivo, but this is not really surprising since
they are composed of only one or two tubulin protofilaments, 5 nm thick. Such a
small filamentous structure would be very difficult to visualize in typical fixed
and embedded specimens, especially if the sections were thicker than a ring diameter, 40 nm. The main problem is that the contrast of a single filament would
probably be much lower than the background from cytoplasmic protein.

G.G. Borisy, Madison, Wisconsin, USA
I would like to comment on rings, just to indicate the complexity of the phenomenon that is being considered here. Rings are not one entity; There are different
kinds of rings and I think it would be useful to list the kinds of rings that
have been reported. We have reported a ring which we measured to have a sedimentation coefficient of 30 S. Kirschner reported a ring which had a sedimentation
coefficient of 36 S, Doenges reported a ring which is 20 S and Timasheff, 42 S.
We have tried to repeat this work and we agree with these findings. What is more,

we find another particle which we thought was a 20 S ring but we call it now an 18 S particle which is probably not a ring. So we have some complexity in this system, and we have to be careful in what we are dealing with. The 30 S particle we now feel to be a double ring or a short helix or lock washer. We have evidence from shadowing experiments that the height of this ring corresponds to two turns of dimers, and that this contains high-molecular-weight molecules which add drag to this particle. The calculated sedimentation coefficient for a single ring is 20 S, and for a double ring as I have described here, it would be about 40 S. I suggest that the high-molecular-weight components, which we have demonstrated to be part of this structure, add drag more than they add mass, and they slow down the calculated 40 S particle and bring it to 30 S. I think Kirschner's particle may be similar to this, but he has less of this material and the drag is less. The 20 S particle is probably a single ring consisting of τ and tubulin. When we have purified the τ fraction and have added it back to tubulin, we have made a 20 S ring. Timasheff's particles are Mg-induced and have no factors, neither τ nor high-molecular-weight.

K.H. Doenges, Heidelberg, F.R.G.
I want to comment on the high-molecular-weight (HMW) bands found in mammalian brain. We have been able to assemble in vitro tubulin from nonneural Ehrlich ascites tumor (EAT) cells and never detected HMW proteins. The only nontubulin proteins we found had mol wts of about 70,000, 49,000, 45,000, and 30,000. We conclude from these results that HMW proteins are not essential for the in vitro polymerization of EAT cells. We also never detected ring-like structures (36/30 S or 20 S components) and thus suggest that these tubulin aggregates are neither intermediates nor required nucleation centers in EAT tubulin assembly.

G.G. Borisy, Madison, Wisconsin, USA
The data that I presented only have to do with brain tubulin, and who is to say what kind of accessory proteins will be present in tissue culture cells? I would be surprised if the same ones would be present in all cells. I do expect that some of the proteins in tissue culture cells that are associated with tubulin will turn out to be of high molecular weight, because one can see bridges or filamentous structures associated with the walls of tubules; assuming that such structures are represented by big molecules, then we look for HMW proteins as a subset of the proteins associated with tissue culture tubules.

J.R. McIntosh, Boulder, Colorado, USA
J. Snyder in my laboratory prepared a rabbit antiserum to porcine brain high-molecular-weight microtubule-associated protein (MAP), eluted from an SDS gel. She finds with indirect immunofluorescence that these antibodies stain interphase and mitotic cells to give an image that looks like cells stained with antitubulin.

G. Wiche, Vienna, Austria
Microtubule preparations obtained from extracts of rat glial C_6-cells contain components which exhibit a molecular weight similar to that of their counterparts from hog brain; this was concluded from analysis of the preparations on SDS-polyacrylamide slab gels. However, the relative amount of these protein components in C_6-cell microtubule preparations is considerably lower than that in brain tissue preparations. In preparations from C_6-cells these components seem to undergo a fast decay and therefore are sometimes missing completely in aged preparations. Preparations from mouse fibroblast cells have similar characteristics.

M. DeBrabander, Beerse, Belgium
Dr. McIntosh, does the antiserum against HMW proteins which Dr. Snyder has been using stain the midbody? Is the osmiophilic material in the midbody related to HMW proteins?

J.R. McIntosh, Boulder, Colorado, USA
None of the antisera we have used on spindles - antitubulin, actin, tropomyosin, or α-actinin - will stain the zone of overlapping microtubules found at the center of the telophase midbody.

K.H. Doenges, Heidelberg, F.R.G.
Dr. Borisy, why do you get HMW proteins in your tubulin preparations and Kirschner's group does not detect them? Does the presence of glycerol in the assembly broth play an essential role?

G.G. Borisy, Madison, Wisconsin, USA
He uses a different procedure and we have repeated it exactly as he describes it and we have essentially gotten his results, which is that τ is a large component of the nontubulin proteins present. However, we found that there were also HMW proteins in his preparation, and he now also says that there are HMW proteins. But they are a minor component of the nontubulin proteins. I do not know why his procedure gives that result. I do not know if it is an effect of glycerol. But there are other differences in the procedure. He uses a different buffer, different pH, a different method of homogenization. I have not explored all of these parameters to determine what is the cause of the difference.

E. Mandelkow, Heidelberg, F.R.G.
Do the cross-bridges that some people see in the spindles have about the same dimensions as your HMW proteins?

J.R. McIntosh, Boulder, Colorado, USA
It was initially tempting to think that the HMW MAP from neurotubulin might correspond to the arms and bridges seen on microtubules in situ. But we have been unable to see any indication that purified neurotubules bridge to one another, though they are well decorated by HMW.

G.G. Borisy, Madison, Wisconsin, USA
We have also tried to determine whether or not these molecules are bridges for brain microtubules. We have found that the distribution of nearest neighbor distances of tubules, that have been packed together in the presence of these factors, is random.

E. Mandelkow, Heidelberg, F.R.G.
The variability of distance between adjacent microtubules could be due to the wrapping of the HMW proteins around the microtubule cylinder, rather than sticking out straight. I believe Linda Amos used this argument in her recent analysis of the HMW proteins associated with brain microtubules (J. Cell Biol. 72, 642, 1977).

J.R. McIntosh, Boulder, Colorado, USA
If anything, the HMW protein seems to be involved in keeping the microtubules apart: in vitro tubules polymerized by the Borisy procedure essentially never touch wall-to-wall, even if you pack them in a centrifuge. After polymerization with polyethylene glycol or sometimes with glycerol, we observe many tubules whose walls touch. Since the yield of polymer per mg microtubule protein is higher when we do this polymerization, we may well be making some tubules with lower amounts of HMW proteins on their surface, and that may account for their ability to touch one another.

G. Wiche, Vienna, Austria
When a large excess of hog τ-factor preparations was added to depolymerized rat glial C_6-cell microtubule preparations and polymer formation was induced by incubation at 37°C, polymers predominantly in the form of double tubules (two tubules, sticking wall-to-wall to each other) have been observed.

G.G. Borisy, Madison, Wisconsin, USA
I agree with Dr. McIntosh's statement that, when tubules are polymerized with the HMW factors, pelleted, and the nearest neighbor distances are observed, there is an excluded zone around each tubule, and tubules do not approach each other closer than by a distance equal to the length of the HMW molecule. We have cleaved off this protein with trypsin and then pelleted the tubules. Under these conditions the tubules approach each other far more closely. So we would conclude that the HMW molecule prevents tubules from touching each other.
If the activity of the HMW protein is due to its polycationic character, this character is restricted to a very small part of the molecule, because again with trypsin cleavage it is possible to dissect the molecule into a large fragment of

255,000 daltons, which does not bind to the microtubules, and a small fragment of about 35,000 daltons, which does bind to tubulin, promotes ring formation, and stimulates microtubule assembly.

H.P. Erickson, Gif-sur-Yvette, France

Dr. Borisy has demonstrated that, in his standard reassembly buffer (0.1 M MES, 5 mM Mg, 0.1-1 mM GTP, pH 6.5-7), microtubule assembly requires the presence of the microtubule-associated protein. I would like to recall the demonstration in our lab that a variety of basic proteins or polycations, such as RNase A, histones, and DEAE dextran, can replace these MAPs and stimulate assembly of microtubules from purified tubulin. The in vitro assembly system of Lee and Timasheff (Biochemistry 14,5183, 1975), in which microtubules are assembled from highly purified tubulin in a buffer containing 5 to 15 mM Mg plus 3 M glycerol, is even more dramatic evidence that the MAPs are not essential for microtubule assembly. I would like to make two points based on these observations: First is the hypothesis that MAPs, polycations, and magnesium may all promote microtubule assembly by a similar mechanism, i.e., neutralization of negative charges on the tubulin subunits, which allows or promotes the aggregation of the tubulin. One can imagine further that a MTOC (microtubule organizing center) might consist of a cluster of cationic proteins that could initiate microtubule assembly by the same mechanism. The second point is more negative and cautionary. Since we know that a variety of cationic macromolecules can replace the MAPs in the assembly reaction, it will be necessary to distinguish a true regulatory factor from the mixture of proteolytic fragments and nonspecific proteins that may be included in the MAP fraction. Likewise we might imagine that any clump of cationic proteins could initiate microtubule assembly and appear as a MTOC. In summary, I feel that the hypothesis that microtubule assembly in vivo is regulated by accessory protein factors is very attractive, but a simple demonstration of stimulatory activity in the in vitro system may be misleading.

I would also like to point out one difference in assembly with the native mixture of MAPs and that induced by polycations or by magnesium and glycerol. In the latter two systems the microtubules are usually organized into bundles of up to 50 parallel and closely packed microtubules. The microtubules assembled with MAPs never appear to interact with each other, and thus always appear isolated or individual.

Session V

Nontubulin Molecules in the Spindle

J. W. SANGER, Department of Anatomy, University of Pennsylvania,
School of Medicine, Pennsylvania, USA

Die weitgehende Ähnlichkeit zwischen der Kon-
traktion der Zellmodelle und Muskelmodelle ver-
leiht der Vorstellung erhöhtes Gewicht, daß die
Fundamentalprozesse der Zell- und Muskelmotili-
tät sehr nahe verwandt sind.

> H. Hoffman-Berling and H.H. Weber
> (1953)

The great similarity between the contraction
of the cell model and the muscle model gives
added weight to the idea that the fundamental
processes of cell and muscle motility are very
closely related.

A. Introduction

Interest in the topic of nonmuscle motility has been greatly stimulated
recently by the realization that the contractile proteins of muscle
cells are probably ubiquitous in eukaryotic cells. The contractile
protein present in greatest concentration in nonmuscle cells is actin.
Actin has been identified biochemically and ultrastructurally in many
animal cells and in a few plant cells (Pollard and Weihing, 1974).
There is now even evidence that an actin-like protein is present in
Escherichia coli (Minkoff and Damadian, 1976). The question is no longer
whether or not actin is present in a cell but what it is doing and how.

It has generally been assumed that the actin filaments seen in nonmuscle cells by themselves do not exert a motive force but, in a manner analogous to muscle contraction, interact with cytoplasmic myosin.
Even before the discovery of actin filaments in nonmuscle cells, both
actin and myosin had been extracted from a number of these cells
(Loewy, 1952; Bettex-Galland and Luscher, 1959). Cytoplasmic myosins
have been isolated that have ATPase activities similar to those of
muscle myosins and that interact with actin filaments in a characteristic fashion (Adelstein et al., 1972). Filaments similar in diameter
to muscle myosin filaments have been observed in several nonmuscle
cells known to contain myosin (Nachmias, 1964; Behnke et al., 1971;
Allera and Wohlfarth-Bottermann, 1972). In addition to actin and myosin, other proteins, tropomyosin, troponin, and actinin, present in
thin filaments of muscle cells have also been found in nonmuscle
cells (Cohen and Cohen, 1972; Fine et al., 1973; Tanaku and Hatano,
1972; Tilney et al., 1973). More recently, proteins that bind to actin in nonmuscle cells, but that are unknown in muscle cells, have
been reported (Stossel and Hartwig, 1975; Kane, 1975; Pollard, 1974).
The properties and distribution of these proteins are still under active study.

The finding of these cytoplasmic "contractile" proteins has naturally
led to the assumption that there is a common chemical basis for
muscle contraction and cell motility (Huxley, 1963). The amount of

myosin in nonmuscle cells is very small compared with actin, however. In striated muscle there are about 3.5 mol of actin per mol of myosin, while in smooth muscle there are 31 mol of actin per mol of myosin (Murphy et al., 1974). In nonmuscle cells the actin-myosin ratio is 100:1 (Pollard and Weihing, 1974). Tilney and various collaborators, in fact, presented evidence that actin alone can cause movement in echinoderm sperm heads (Tilney et al., 1973). Additional work in Tilney's laboratory (Tilney and Detmers, 1975) has indicated that actin is involved in the structural reinforcement of the red blood cell membrane. Thus, the large amount of actin in nonmuscle cells may be involved in both motility and in a cytoskeletal role. To begin to understand nonmuscle cell motility, we need to know, therefore, not simply that actin or myosin or other contractile proteins are present in a cell but how these contractile proteins are distributed within the cell, especially with respect to the motile regions of the cell. The task of localizing these proteins is complicated by the fact that the site of contraction changes with time, as for example, when pseudopods extend first in one direction and then in another.

Actin has been localized within cells using several techniques. If the heavy meromyosin (HMM) fragment of myosin is added to fibrous actin, the actin filaments become decorated with arrowheads (Huxley, 1963). Ishikawa et al. (1969) were the first to localize actin in nonmuscle cells with this technique. They found actin filaments in the cytoplasm of cartilage, fibroblast, nerve, and epithelial cells. Muscle HMM has never been found to bind with microtubules, neurofilaments, intermediate filaments (100-Å filaments), tonofilaments, or even bacterial flagella (Ishikawa et al., 1969; Pollard and Weihing, 1974). The binding to actin is moreover, prevented by the presence of Mg-ATP. Thus the distinctive ultrastructural test for the localization of actin is the exposure of HMM to the cell models and the localization of arrowheads. Perry et al. (1971) using newt eggs, Forer and Behnke (1972) using insect spermatocytes, and Schroeder (1973) using cultured HeLa cells showed that the thin filaments in the cleavage furrow formed arrowheads with HMM and were thus actin.

Ultrastructural localization of actin using HMM permits very small areas of a cell to be examined at one time. To reconstruct the total distribution of actin in a cell using thin sections would be a difficult task. To undertake the detection of any changes in actin distribution during a cell cycle in a wide variety of cells would truly be a Herculian task using the ultrastructural technique of serial sectioning and scoring for arrowhead complexes. Luckily, it is possible to put a fluorescent tag on heavy meromyosin (Aronson, 1965; Sanger, 1975a) and use it to localize actin in nonmuscle cells on a light-microscopic level (Sanger, 1975a-c; Sanger and Sanger, 1976).

The fluorescently tagged HMM does not lose its ability to bind reversibly to actin. We have demonstrated its specificity by binding it to a variety of myofibrils: rabbit, chick, lobster, and scallop. The staining was intense in the I-bands and decreased in the overlap region of the A-bands where actin filaments overlap with myosin. No staining was observed in the H-zones or in the Z-bands. The decreased staining in the overlap region is due to the binding of actin in that region to the native myofibrillar myosin. If myofibrils are exposed to an 0.8 M KCl solution the A-band myosin is removed. When these "ghosts" (myofibrils lacking A-bands) were stained with fluorescent HMM, the full length of the thin filaments stained. No staining of the myofibrils was obtained if the fluorescent HMM was used in the presence of MgATP. The specificity of the fluorescently tagged HMM was further confirmed by the localized staining of acrosomal caps of sperm, intestinal brush borders, ruffled membranes, nerve and chondrogenic cells, and smooth, and cardiac muscle fibers, all known to contain actin (Sanger, 1975c; Sanger and Sanger, 1976).

A variety of interphase cells can be demonstrated to contain bundles of cytoplasmic fibers when exposed to fluorescent HMM and S-1 fragment of myosin (Fig. 1). Motile interphase cells possess not only

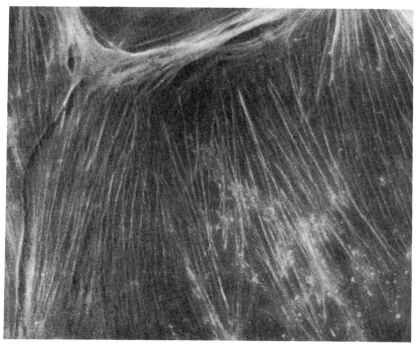

Fig. 1. Rat kangaroo cells opened with nonidet and stained with fluorescein-labeled S-1 fragment of myosin

fluorescent bundles of actin but also have fluorescent staining in the pseudopodial regions. This interphase staining pattern undergoes a dramatic change when the cells enter mitosis. When fibroblasts enter mitosis they round up and the fibrous actin bundles cannot be detected. When the cells rounded up, the small spikes holding the cell to the substratum were fluorescent, but there was a diffuse fluorescence over the rest of the cell. A change from this diffuse pattern of actin staining was observed in telophase when the furrow region was found to be densely stained. At the end of telophase the midbody is only slightly fluorescent with strong fluorescence in the cytoplasm adjacent to the midbody. It should be noted that there is also some staining outside the furrow region. This indicated that not all of the actin was located in the furrow region. After completion of cytokinesis, staining was now concentrated in the distal ends of the cells corresponding to the developing pseudopod regions. The increase in nonfurrow pole staining was accompanied by a concomitant decrease in the former furrow regions of the two daughter cells. When the separated daughter cells began to flatten the fluorescent staining was concentrated along the perimeter of the cells. Fully flattened cells exhibited the continuous fluorescent fibers typical of interphase cells. The total time period for this division is 1 h. In this short period, four different patterns of actin localization are observed. These changes in actin staining demonstrate large amounts of actin present in the cell as well as assembly and disassembly of actin during mitosis.

Using indirect fluorescent antibody technique, Lazarides and Weber (1974), Lazarides (1976), Weber (1976) localized not only actin but also tropomyosin, α-actinin, and myosin in several nonmuscle cells. In interphase cells Lazarides and Weber (1974) noted that the actin was present in bundles of filaments. Lazarides (1975, 1976) demonstrated that tropomyosin was distributed along these same bundles in a periodic fashion. Weber and Groeschel-Stewart (1974) reported that myosin was also distributed in a fibrous pattern similar to that of actin.

The major problem with all of these protein localization techniques, however, is that cells must be fixed, i.e., a cell model made, before the localizing agent can be introduced. We have to ask how closely the fixed cell models resemble the living cell. By using different techniques and comparing the results they give, we can reduce the chances that we will be fooled by artifact, but we must always be concerned that our cell models are not living. Let us now proceed to analyze protein localization in mitotic spindles.

B. Actin and Myosin in the Spindle Apparatus - An Overview

Actin and myosin have not only been localized in the cytoplasm of a wide variety of cells, but also seem to be present in the mitotic and meiotic spindle apparatus. Actin filaments decoratable with HMM have been identified in electron-microscopic analyses of thin sections of the spindle of glycerinated cells. Fluorescent-labeled myosin fragments (HMM and S-1) have also been used to identify the localization of actin in the chromosomal spindle fibers. Using a completely different experimental approach, actin has also been localized on the chromosomal spindle fibers using the indirect immunofluorescent method. If an actin-myosin system is involved in chromosomal movement, one would expect to find myosin localized where actin is. Indeed, by the direct immunofluorescent method myosin has been shown to be localized between the chromosomes and the pole where the chromosomal spindle fibers reside.

C. Actin in the Mitotic Spindle - Electron Microscopic Localization

Within two years of the publication of the work of Ishikawa et al. (1969), two groups using the HMM technique reported the presence of actin and "arrowhead" in spindles. Behnke et al. (1971) reported the presence of actin in meiotic spindles (crane fly testes), and Gawadi (1971) reported actin in mitotic spindles of locust spermatogonia.

Both groups of workers expanded upon their original observations in later papers. Forer and Behnke (1972) demonstrated clearly that thin filaments capable of binding HMM in an arrowhead fashion were present in the spindle regions. Gawadi (1974) expanded her original report of 1971 and showed that there were two types of thin filaments within the mitotic spindle. One group not associated with the microtubules readily bound to the HMM and formed arrowheads. However, there was a second group of thin filaments that were cross-bridged to microtubules and did not bind to HMM as detected by the absence of arrowheads in these filaments. However, Gawadi (1974) was able to trace a few of these cross-linked noncoated thin filaments away from their associations with microtubules. She found that as these filaments coursed away from the associated microtubules arrowheads could be detected. She thus deduced that the filaments were actin but that their linkage to microtubules prevented their binding to HMM. Gawadi proposed not that the

With both glycerol and detergent preparation of cells for staining, the possibility that actin was moving from its in vivo localization was of great concern. In view of the considerable amount of actin in cells, this is the greatest potential source for error when trying to determine the cellular site of relatively small amounts of actin such as are found in the spindle. To avoid this problem cells were plunged into cold acetone to "precipitate" or fix the actin in place. They were then air-dried and rehydrated with standard salt. The whole procedure took 10 min, after which the cells were stained for 20-60 min and washed for 10-15 min. Again the same unique chromosomal fiber staining pattern was obtained.

Another variation used was to fix the cells in formaldehyde and then open them with acetone. A final variation was to fix in formaldehyde but open the cells with detergents. Whatever the fixation method used the spindle staining pattern was always the same. I should point out that the interphase actin bundles and cleavage rings still stained after these various procedures.

In addition to fluorescein isothiocyanate, tetramethylrhodamine iso-thiocyanate (TMRITC) was also coupled to HMM. The staining results were the same. We are currently using TMRITC-HMM and FITC-AB against tubulin to localize actin and tubulin in the same cells. We have also used the S-1 fragment of myosin labeled with a fluorescent dye to stain cells. Again the same specific results were obtained. We have added unlabeled rabbit HMM to mitotic cells and then stained them with a fluorescein-labeled antibody against rabbit myosin (Dr. Frank Pepe's generous gift). When the cell is pretreated with HMM, staining is obtained in chromosomal spindle fibers. If no HMM is added to the cell models, no antibody staining can be seen in the chromosomal fibers.

What is the possibility that HMM induces actin filaments to form in the spindle? Heavy meromyosin can polymerize a solution of monomer actin even when the monomer concentration is very low (Yagi et al., 1965; Kikuchi et al., 1969). Actin in unactivated sperm heads is un-polymerized (Tilney et al., 1973); yet if unactivated sperm are gly-cerinated and stained with HMM, decorated filaments are sometimes seen. It is possible, then, that unpolymerized cytoplasmic actin may move into the spindle when we stain and be polymerized by the fluores-cent HMM. If this is so, it is curious that the actin staining is con-fined to the chromosomal spindle fiber region. We must yet show, however, that astral and interzonal microtubules are still present after our staining procedure.

It is also possible that HMM stabilizes actin filaments that are al-ready in the spindle and protects them from osmium destruction. The whole question of proper fixation has historical analogies in the cases of microtubules, smooth muscle myosin filaments, and the transverse tubular system of muscle, all of which were initially not recognized because of inadequate fixation (Figs. 4 and 5).

What is the possibility that any fluorescently tagged molecule will attach to dense structures in the cells? The following experimental results indicate that the staining reactions discussed above are spe-cific. Fluorescently tagged antibody against rabbit tropomyosin that stains the I-bands of rabbit and scallop cross-striated muscles also stains PtK$_2$ cells (Fig. 6). The antibody staining in the Ptk$_2$ cells is concentrated around poles and in bundles perpendicular to the cleavage furrow. The furrow itself does not stain with the tropomyosin antibody nor is the staining in the spindle observed in chromosomal spindle fibers. Furthermore, unlike fluorescent myosin fragments, the

Fig. 4. Rat kangaroo cells viewed after normal glutaraldehyde and osmium fixation in the electron microscope. Note the vesicles around the polar region. x 28,000

antibody does stain a filamentous network in the nonspindle area of the cytoplasm (Fig. 7). This pattern of staining was so unexpected that we examined the cells in the electron microscope. Bundles of keratin-like filaments are present in cell areas stained by the tropomyosin antibody. They are especially prominent adjacent to the outer nucleus membrane. During mitosis they are present within the spindle, around the centrioles, and also in a scroll-like network in the cytoplasm (matching the interphase Ab-pattern). The antibody-stained filaments present during mitosis cannot be actin since they do not bind the myosin fragments. In some instances, arrowheads were observed on thin filaments associated with microtubules in the spindle of PtK$_2$ cells but the keratin-like filaments were smooth in appearance. That

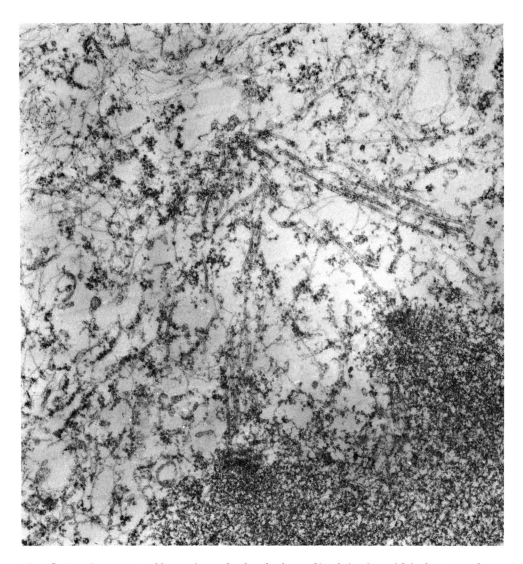

Fig. 5. Rat kangaroo cell at the end of telophase fixed in formaldehyde, opened with detergent, and then stained with fluorescein-labeled HMM. After washing out the unbound stain, the cells were fixed in glutaraldehyde and osmium and then processed for electron microscopy. Note the abundance of coated microfilaments distributed about the polar region and associated microtubules. In comparing this Figure with Figure 4, note the marked extraction of the matrix material that has taken place. x 35,000

the two fluorescently labeled stains give distinctly different patterns argues against nonspecific adherence of fluorescent label to fibrous systems in the cell.

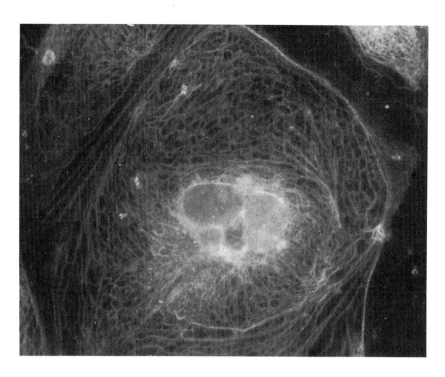

Fig. 6. Rat kangaroo cell (PtK$_2$) stained with directly labeled fluorescein anti-rabbit tropomyosin. Note the scroll-like pattern of stain throughout the cytoplasm. Compare this pattern with the typical interphase actin pattern in Figure 1

II. Actin and Myosin Antibodies

The indirect fluorescent antibody technique has been used successfully by Lazarides and Weber (1974), Weber and Groeschel-Stewart (1974), Lazarides (1976), and Weber (1976) to localize actin, tropomyosin, α-actinin, and myosin in nonmuscle cells. Recently, Cande et al. (1977) extensively studied the localization of actin and tubulin in the mitotic spindle of the rat kangaroo cell (PtK$_1$). They demonstrated that actin is found only in the chromosomal spindle fibers. It was found continuously from the pole to the attachment on the chromosome at the kinetochore. Using conditions identical to those used for the actin antibody staining they added antibody against tubulin to sister cultures. The tubulin antibody stained not only the chromosomal spindle fibers, but also the astral and continuous spindle fibers. This clearly demonstrated that the two antibody staining reactions were specific and again counters the criticism that fluorescent staining of cells is due to the sticking of the fluorescent compounds to dense fibrous networks in the cell.

Of even greater interest is the agreement of results of indirect actin antibody staining with those of fluorescent myosin fragment staining. That both stains should be confined to the chromosomal spindle fibers to the exclusion of the interzonal fibers and astral rays greatly strengthens the argument that actin is localized in vivo in this region. Furthermore, tubulin indirect antibody staining shows that microtubules have not been destroyed in any particular region of the spindle. Thus, if actin moved into the spindle as a result of the

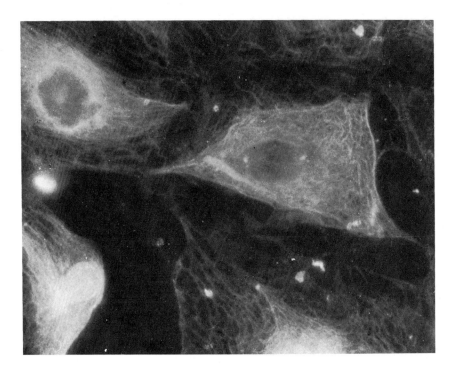

Fig. 7. A mitotic PtK$_2$ rat kangaroo cell stained with the same antibody as shown in Figure 6. Though the poles are stained, there is no chromosomal spindle fiber staining. There is an extensive fiber system in the cytoplasm. Contrast this with the absence of cytoplasmic actin bundles in the mitotic cell in Figure 4

staining technique, it seems unlikely that it would only become aligned along the chromosomal fibers.

With the exception of sperm and red blood cell membranes, wherever actin has been found so also has myosin. For actin to play a contractile role in mitosis one would expect myosin to be present within the spindle also. Since actin molecules do not shorten or contract by themselves, the absence of myosin from the spindle would seem to rule out a contractile role for actin. On the other hand, the finding of myosin within the spindle has not proved that actin and myosin are responsible for chromosome movement. The recent paper by Fujiwara and Pollard (1976) clearly demonstrates the localization of myosin in the chromosomal spindle fiber area of HeLa cells. Their directly labeled antibody against human myosin did not stain mitotic apparatus of salamander cells or of rat kangaroo cells. This latter experiment is further evidence that fluorescent agents are not unspecifically staining dense structures in cells.

E. Chromosome Movement

All living cells demonstrate some form of motility. The two main systems responsible for cell movement are (1) tubulin-dynein as seen in cilia, and (2) actin-myosin as in muscle. Which one of these systems is used for chromosome movement?

The work of Inoué and Sato (1967) clearly indicates the importance of microtubules in mitosis. If the microtubules are destroyed there is no movement of the chromosomes to the pole (anaphase movement). Sanger and Sanger (1976) and Cande et al. (1977) demonstrated that if mitotic cells are exposed to colcemid and then stained for the presence of actin, none is found. Thus, the orientation of actin in chromosomal spindle fibers is dependent on the presence of microtubules.

Ris (1943, 1949) demonstrated that chromosomal separation has two components: 1) the movement of the chromosomes to the poles accompanied by the shortening of chromosomal spindle fibers; and 2) the separation of the poles due to spindle elongation. The independence of the two components was demonstrated by Ris using chloralhydrate. This compound prevented spindle elongation but did not affect the anaphase movement of the chromosomes. Data obtained by micromanipulation and microbeaming studies indicate that the chromosomal spindle fibers pull the chromosomes to the poles (see review by Nicklas, 1971). The specific localization of actin in chromosomal spindle fibers, both by the indirect immunofluorescence technique (Cande et al., 1977) and by the use of fluorescently tagged myosin fragments (Sanger, 1975a; Sanger and Sanger, 1976) and now the localization of myosin in the chromosomal spindle fiber area as well (Fujiwara and Pollard, 1976) suggest an actin-myosin interaction as the force-producing mechanism for chromosomal movement in anaphase.

The movements of the poles apart from one another is thought to be due to the activity of the continuous spindle fibers. Actin and myosin have not been localized in these fibers. Thus the separation of the poles could be caused by a sliding (McIntosh and Landis, 1971) and/or polymerization (Inoué and Sato, 1967) of microtubules.

If actin and myosin are involved in chromosomal movement, one would expect some interaction and cooperation between the two systems. In tissue culture cells colchicine will inhibit forward translational movement of a cell but not the ruffling of the periphery of the cell. When the drug is removed, translational movement resumes. It is thought that this results from the interaction of microtubules with actin filaments. Gawadi (1974),in fact, has shown direct connections between actin filaments and microtubules in another system. I would like to see more cases of this, so we can have greater assurance of this type of interaction. The shortening of chromosomal spindle microtubules could control the speed with which an actin-myosin system pulls a chromosome (Forer, 1974). Perhaps the microtubule system is like a floating raft that provides a skeletal surface for the actin and myosin to attach, contract, and to reorganize, and to contract again. So little is known about any interaction between microtubules and actin-myosin that one could propose any number of possibilities. The simplest model would be to have the poles held in place by continuous microtubules and the chromosomes attached to the pole by chromosomal fibers composed of microtubules - actin and myosin. The pole could be a Z-band to which the thin filament attaches directly or indirectly. The chromosomes could be held at the metaphase plate by the pushing of the chromosomal spindle microtubules. Anaphase movement then starts when the chromosomal microtubules begin to depolymerize and the actin-myosin system pulls the chromosome toward the pole. Some support for the polarized direction of actin in chromosomal spindle fibers has been obtained by Gawadi (1974), who demonstrated the pointing of the arrowheads away from the pole. This is the same relationship the Z-band has with the thin filaments in the sarcomere.

There is always the possibility that the presence of actin and myosin within the spindle has nothing to do with chromosome movement. There are other types of movements associated with the spindle that have nothing to do with chromosome movement (Rebhun, 1963; Bajer, 1967). Perhaps this is the role of actin and myosin. What we need now are direct tests on cell models of mitosis.

I. Ca^{2+} Regulation in Mitotic Spindles

If actin and myosin function in a contractile system in the spindle, there must also be a system for regulating the level of Ca ion concentration within the spindle. Harris (1962) observed an extensive vesicular system among the microtubules and around the centriolar regions of sea urchin spindles. Robbins and Jentzsch (1970) reported a similar distribution in HeLa cells. This vesicular system has also been observed in a meiotic spindle by Friedlander and Wharman (1970) and in the plant mitotic spindle by Bajer and Molè-Bajer (1969) and in rat kangaroo cells (Fig. 4). Could it be the equivalent of a sarcoplasmic reticulum? The membranous system and the ATPase could be used to release Ca^{2+} to break down the chromosomal microtubules and activate the actin-myosin system. The slow speed of chromosome movement would be due to the rapid activity of Ca^{2+} release and capture. This would imply that the chromosomal microtubules are much more sensitive to free calcium concentration than astral or continuous microtubules. We do not even yet know if these vesicles contain Ca ions.

F. Conclusions

I have tried to review the evidence that actin and myosin are localized in the area of the chromosomal spindle fibers of the mitotic apparatus. I have tried to indicate the problems encountered when cell models must be used instead of living cells. The large amount of actin present in cells would make it very easy for some of the actin to be translocated during staining procedures, and although myosin is present in much smaller amounts, it too might be moved post vivo. I feel, however, that the agreement of the several methods used for actin and myosin localization are compelling enough to affirm that actin and myosin are to be found in the area of the chromosomal spindle fibers.

The micromanipulative and microbeaming experimental evidence clearly indicate that chromosomes are pulled to the poles. Microtubules are well known to serve a cytoskeletal role and in some cases to exert a pushing force. Actin and myosin in a vast number of systems are involved in shortening actions. Therefore, by analogy, one can envision an actin-myosin system pulling the chromosomes to the poles and the dynein-tubulin system pushing the poles apart.

The force needed to move chromosomes to the poles is so small (Taylor, 1965; Nicklas, 1971), however, that there may well be some unique process such as is postulated in Inouè's dynamic equilibrium model, which can account for this movement. We must not let chromosome movement obscure from us the other movements that are seen in or near the spindle (Rebhun, 1963; Bajer and Molè-Bajer, 1972). Actin and myosin may be associated with this movement and not with chromosome movement.

Finally, the idea that actin and myosin are present in the spindle is a relatively new one and thus must be treated with caution. We would like to feel that, as some of the observations of Hoffman-Berling have been shown to be correct after initial scepticism 25 years ago, that 25 years from now some of our present observations will be incorporated into a working model of chromosome movement.

References

Adelstein, R.S., Conte, M.A., Johnson, G., Pastan, I., Pollard, D.: Isolation and characterization of myosin from cloned fibroblasts. Proc. Natl. Acad. Sci. U.S. 69, 3693 (1972)

Allèra, A., Wohlfarth-Bottermann, K.E.: Extensive fibrillar protoplasmic differentiations and their significance for protoplasmic streaming. Cytobiologie 6, 261 (1972)

Aronson, J.F.: The use of fluorescein-labelled heavy meromyosin for the cytological demonstration of actin. J. Cell Biol. 26, 293-298 (1965)

Bajer, A.: Notes on ultrastructure and some properties of transport within the living mitotic spindle. J. Cell Biol. 33, 713 (1967)

Bajer, A., Molè-Bajer, J.: Formation of spindle fibers, kinetochore orientation and behavior of the nuclear envelope during mitosis in endosperm. Fine structural and in vitro studies. Chromosoma 27, 448 (1969)

Bajer, A., Molè-Bajer, J.: Spindle dynamics and chromosome movements. In: Int. Rev. Cytol., Suppl. 3. New York: Academic Press 1972, pp. 2-71

Behnke, O., Forer, A., Emmersen, J.: Actin in sperm tails and meiotic spindles. Nature (London) 234, 408-410 (1971)

Bettex-Galland, M., Luscher, E.F.: Extraction of an actomyosin-like protein from human thrombocytes. Nature (London) 184, 276 (1959)

Cande, W.Z., Lazarides, E., McIntosh, J.R.: A comparison of the distribution of actin and tubulin in the mammalian mitotic spindle as seen by indirect immunofluorescence. J. Cell Biol. 72, 552-567 (1977)

Cohen, I., Cohen, C.: A tropomyosin-like protein from human platelets. J. Mol. Biol. 68, 383 (1972)

Fine, R., Blitz, A., Hitchcock, S., Kaminer, B.: Tropomyosin in brain and growing neurones. Nature (New Biol.) 245, 182 (1973)

Forer, A.: Possible roles of microtubules and actin-like filaments during cell division. In: Cell Cycle Controls. Padilla, G.M., Cameron, I.L., Zimmerman, A.M. (eds.). New York: Academic Press 1974, pp. 319-336

Forer, A., Behnke, O.: An actin-like component in spermatocytes of a crane fly (Nephrotoma sutermalis Loew). 1. The spindle. Chromosoma 39, 145 (1972)

Forer, A., Jackson, W.T.: Actin filaments in the endosperm mitotic spindle in a higher plant, Haemanthus katherinae Baker. Cytobiologie 17, 199-214 (1976)

Forer, A., Zimmerman, A.: Characteristics of sea-urchin mitotic apparatus isolated using a dimethyl sulphoxide glycerol medium. J. Cell Sci. 16, 481 (1974)

Friedlander, M., Wahrman, J.: The spindle as a basal body distributor. A study in the meiosis of the male silkworm moth, Bombyx mori. J. Cell Sci. 1, 65-89 (1970)

Fujiwara, K., Pollard, T.D.: Fluorescent antibody localization of myosin in the cytoplasm, cleavage furrow, and mitotic spindle of human cells. J. Cell Biol. 71, 848-875 (1976)

Garrels, J.I., Gibson, W.: Identification and characterization of multiple forms of actin. Cell 1, 793-805 (1976)

Gawadi, N.: Actin in the mitotic spindle. Nature (London) 234, 410 (1971)

Gawadi, N.: Characterization and distribution of microfilaments in dividing locust testis cells. Cytobios 10, 17 (1974)

Harris, P.: Some structural and functional aspects of the mitotic apparatus in sea urchin embryos. J. Cell Biol. 14, 475-488 (1962)

Hinkley, R., Telser, A.: Heavy meromyosin-binding filaments in the mitotic apparatus of mammalian cells. Exp. Cell Res. 86, 161 (1974)

Huxley, H.E.: Electron microscope studies on the structure of natural and synthetic protein filaments from striated muscle. J. Mol. Biol. 7, 281 (1963)

Huxley, H.E.: Muscular contraction and cell motility. Nature (London) 243, 449-455 (1973)

Inoué, S., Sato, H.: Cell motility by labile association of molecules: the nature of mitotic spindle fibers and their role in chromosome movement. J. Gen. Physiol. 50, 259 (1967)

Ishikawa, H., Bischoff, R., Holtzer, H.: Formation of arrowhead complexes with heavy meromyosin in a variety of cell types. J. Cell Biol. 43, 312 (1969)

Kane, R.E.: Preparation and purification of polymerized actin from sea urchin extracts. J. Cell Biol. 66, 305 (1975)

Kikuchi, M., Noda, H., Maruyanna, K.: Interaction of actin with H. meromyosin at low ionic strength. J. Biochem. 65, 945 (1969)

Lazarides, E.: Tropomyosin antibody-specific localization of tropomyosin in non-muscle cells. J. Cell Biol. 65, 549-561 (1975)

Lazarides, E.: Actin, alpha-actinin and tropomyosin interaction in structural organization of actin-filaments in non-muscle cells. J. Cell Biol. 68, 202-219 (1976)

Lazarides, E., Weber, K.: Actin antibody: the specific visualization of actin filaments in non-muscle cells. Proc. Natl. Acad. Sci. U.S. 71, 2268-2272 (1974)

Loewy, A.G.: An actomyosin-like substance from the plasmodium of a myxomycete. J. Cell Comp. Physiol. 40, 127 (1952)

McIntosh, J.R., Cande, W.Z., Snyder, J.A.: Structure and physiology of the mammalian mitotic spindle. In: Molecules and Cell Movement. Inoué, S., Stephens, R.E. (eds.). New York: Raven Press 1975, pp. 31-76

McIntosh, J.R., Landis, S.C.: The distribution of spindle microtubules during mitosis in cultured human cells. J. Cell Biol. 49, 468 (1971)

Minkoff, L., Damadian, R.: Actin-like proteins from *Escherichia coli*: concept of cytolomus as the missing link between cell metabolism and the biological ion-exchange resin. J. Bacteriol. 125, 353-365 (1976)

Müller, W.: Elektronenmikroskopische Untersuchungen zum Formwechsel der Kinetochoren während der Spermatocytenteilungen von *Pales Ferruginea* (Nematocera). Chromosoma 38, 139 (1972)

Murphy, R.A., Herlihy, T., Megerman, J.: Force-generating capacity and contractile protein content of arterial smooth muscle. J. Gen. Physiol. 64, 691-705 (1974)

Nachmias, V.T.: Fibrillar structures in the cytoplasm of *Chaos chaos*. J. Cell Biol. 23, 183 (1964)

Nicklas, R.B.: Mitosis. In: Advances in Cell Biology. Prescott, D.M., Goldstein, L., McConkey, H. (eds.). New York: Appleton 1971, p. 225

Perry, M.M., John, H.A., Thomas, N.S.T.: Actin-like filaments in the cleavage furrow of newt eggs. Exptl. Cell Res. 65, 249-252 (1971)

Pollard, T.D.: The role of actin in the temperature dependent gellation and contraction of extracts of *Acanthamoeba*. J. Cell Biol. 68, 579 (1974)

Pollard, T.D., Weihing, R.: Actin and myosin and cell movement. CRC Crit. Rev. Biochem. 2, 1 (1974)

Rebhun, L.I.: Saltatory particle movements and their relation to the mitotic apparatus. In: The Cell in Mitosis. Levine, L. (ed.). New York: Academic Press 1963, pp. 67-106

Ris, H.: A quantitative study of anaphase movement in the aphid *Tamalia*. Biol. Bull. 85, 164-179 (1943)

Ris, H.: The anaphase movement of chromosomes in the spermatocytes of the grasshopper. Biol. Bull. 96, 90-106 (1949)

Robbins, E., Jentzsch, G.: Ultrastructural changes in the mitotic apparatus at the metaphase-to-anaphase transition. J. Cell Biol. 40, 678-691 (1970)

Sanger, J.W.: Changing patterns of actin localization during cell division. Proc. Natl. Acad. Sci. U.S. 72, pp. 1913-1916 (1975a)

Sanger, J.W.: Presence of actin during chromosomal movement. Proc. Natl. Acad. Sci. U.S. 72, pp. 2451-2455 (1975b)

Sanger, J.W.: Intracellular localization of actin with fluorescently labelled heavy meromyosin. Cell Tissue Res. 161, pp. 431-444 (1975c)

Sanger, J.W., Sanger, J.M.: States of actin and cytochalasin-B. J. Cell Biol. 67, 381a (1975)

Sanger, J.W., Sanger, J.M.: Actin localization during cell division. In: Cold Spring Harbor Conferences on Cell Proliferation 3, pp. 1295-1316 (1976)

Schroeder, T.: Actin in dividing cells. Contractile ring filaments bind heavy meromyosin. Proc. Natl. Acad. Sci. U.S. 70, 1688 (1973)

Stossel, T.P., Hartwig, J.H.: Interaction between actin, myosin, and an actin binding protein from alveolar macrophages. Alveolar macrophages myosin Mg^+ ATPase requires a cofactor for activation by actin. J. Biol. Chem. 250, 5706 (1975)

Szamier, P.M., Pollard, T.D., Fugiwara, K.: Tropomyosin prevents the destruction of actin filaments by osmium. J. Cell Biol. 67 (2 P +2): 424 (1975)

Tanaka, H., Hatano, S.: Extraction of native tropomyosin-like substances from myxomycete plasmodium and the cross reaction between plasmodium F-actin and muscle native tropomyosin. Biochim. Biophys. Acta 257, 445-451 (1972)

Taylor, E.W.: Brownian and saltatory movements of cytoplasmic granules and the movement of anaphase chromosomes. In: Proc. of the 4th Internatl. Cong. on Rheology. Copley, A.L. (ed.). New York: Interscience 1965, p. 175

Tilney, L.G., Detmers, P.: Actin in erythrocytes ghosts and its association with spectrin. Evidence for a nonfilamentous form of these two molecules in situ. J. Cell Biol. 66, 508 (1975)

Tilney, L.G., Hatano, S., Ishikawa, H., Mooseker, M.S.: The polymerization of actin: its role in the generation of the acrosomal process of certain echinoderm sperm. J. Cell Biol. 59, 109 (1973)

Weber, K.: Visualization of tubulin-containing structures by immunofluorescence microscopy: cytoplasmic microtubules, mitotic figures and vinblastone induced paracrystals. In: Cold Spring Harbor Conferences on Cell Proliferation 3, 403-417 (1976)

Weber, K., Groeschel-Stewart, U.: Antibody to myosin: the specific visualization of myosin - containing filaments in non-muscle cells. Proc. Natl. Acad. Sci. U.S. 71, 4561-4564 (1974)

Whalen, R.G., Bulter-Browne, G.S., Gros, F.: Protein synthesis and actin heterogeneity in calf muscle cells in culture. Proc. Natl. Acad. Sci. U.S. 73, 2018-2022 (1976)

Yagi, K., Mase, R., Sakakibara, I., Asai, H.: Function of heavy meromyosin in the acceleration of actin polymerization. J. Biol. Chem. 240, 2448 (1965)

Discussion Session V: Nontubulin Molecules

Chairman: E. MANDELKOW, Heidelberg, F.R.G.

J.R. McIntosh, Boulder, Colorado, USA
I would like to reinforce the caution Dr. Sanger expressed about possible artifacts
in the current studies showing actin and related molecules in the spindele. The
specificity of the tracers, either heavy meromyosin or antibodies, is well
documented, but the need for some form of lysis to get the tracer into the cell
is hard to evaluate. Electron microscopy of the cells used by Cande et al. (J. Cell
Biol. 1977) shows them to be strongly extracted when viewed in the fluorescence
microscope. Thus we must face the problem of loss or rearrangement of important
molecules. I think that in this situation one must use essentially esthetic criteria
as well as controls to evaluate the likelihood that a structure is correctly seen.
When we compare the images of the spindle seen after staining with antiactin and
antitubulin, we see that they are clearly different. Thus, the simple idea of entrap-
ment of actin in the spindle as an artifactual cause of the image is difficult to
accept. One can readily imagine a model to account for the difference, but the situation becomes more complex. For instance the chromosome might sweep the actin out
of the interzone during anaphase, so entrapped actin is not found in that part of
the spindle. One must look for other kinds of evidence.

B.M. Jockusch, Basel, Switzerland
Dr. Hauser and I have repeated Dr. McIntosh's result with antiactin in mitotic
spindles on meiotic grasshopper spindles. Our results confirm staining of the spindle
pole regions and spindle fibers. The pictures are not as pretty as with mitotic
PtK cells because the cells are smaller and not as flat. Also, since cytokinesis
sets in, in early anaphase meiotic cells, the central region (interzone) in anaphase
is stained with antiactin, but this is not so in mitotic cells. However, we would
like to stress that the findings of Dr. McIntosh and Dr. Sanger in mitotic cells
can be repeated in meiotic cells: Antiactin binds to spindle pole areas and chromosome fibers.

A.S. Bajer, Eugene, Oregon, USA
I am not surprised that actin is present in the spindle. In fact, it would sur-
prise me, if actin were not there in cells where it is present in the cytoplasm.
Elimination of actin from the spindle would require the existence of specific "active" transport. Furthermore, the studies of transport phenomena in the spindle
(behavior of small granules, etc.) combined with the studies on fine structure,
predict that actin would not be distributed uniformly in the spindle. I would expect that it would be accumulated between or along kinetochore fibers. The presence
of actin does not tell us, however, what the actin does. It would be important to
find out whether actin has specific polarity or whether its polarity is random.

Chairman:
The question of polarity of microtubules was raised yesterday, and I want once more
to draw attention to the poster of Dr. Wegner on how a linear polymer can grow in
a polar fashion. He showed that due to the hydrolysis of ATP during the assembly
of actin filaments, the association of a G-actin-ATP monomer to the end of the filament is not the reverse of the dissociation of a G-actin-ADP monomer, and thus
F-actin grows mainly on one end and shortens on the other (head-to-tail polymerization, J. Mol. Biol. 1977)

J.R. McIntosh, Boulder, Colorado, USA
Lazarides and I have used antiserum directed against α-actinin to look for the
presence of this molecule in mammalian spindles. We found that metaphase spindles
stain well with this antiserum. The cells may be lysed before fixation, but lysis is
not necessary, and cells fixed in formaldehyde from the living state and then either
detergent or cold-acetone treated, show a spindle that stains brightly, relative
to the surrounding cytoplasm. Anaphase and telophase cells stained with α-actinin
antiserum are fluorescent not only between chromosomes and poles where the antiactin binds, but also in the interzone. Such staining is quite graphic at telophase,

and it is clear that the α-actinin image is different from the actin image. This observation lends further esthetic weight to the immunofluorescent evidence for muscle proteins in the spindle. While one might argue that a general staining is not an indication of specificity, the most likely origin of nonspecific staining for α-actinin would be actin binding. Thus the difference between α-actinin and actin images suggests something interesting.

R.B. Nicklas, Durham, North Carolina, USA
Concerning the application of esthetic and functional criteria to judging the results of attempts to localize molecules within the spindle, how do you interpret the apparently generalized distribution of α-actinin?

J.R. McIntosh, Boulder, Colorado, USA
The functions of α-actinin in muscle are not well understood. It is known to be localized in the Z line, but its actin-binding properties are temperature sensitive, its behavior in solution is complex, and functional statements would be premature. Nonetheless an analogy between muscle and spindle would predict α-actinin staining of poles and kinetochores. This is *not* observed. The straightforward model is thus not supported, and the immunofluorescent image leads me to believe that spindle α-actinin is serving a general linking role to couple actin to spindle tubules.

G. Wiche, Vienna, Austria
I would like to emphasize the necessity of very careful controls for studies using the indirect immunofluorescence method, since we have found that IgG preparations obtained from two different preimmune sera of rabbits contained components that gave rise to the visualization of cellular structures in 3T3 fibroblasts and other cell lines, which very closely resembled those observed with a highly specific tubulin antibody preparation, using the identical method. We believe that these components in preimmune sera represent naturally occurring antibodies against structural proteins in the preimmune sera of the tested animals.

M. DeBrabander, Beerse, Belgium
In addition to direct and indirect immunofluorescence the Sternberger immunocytochemical technique can be used with some advantage for localization of fibrous structures (see Poster 25). This technique uses only immunologic links between the antigen and the marker, which enhances the specificity and sensitivity. Moreover the marker is visible both at the light microscopic and at the ultrastructural levels. This allows sequential investigations on the same cells.

M. Osborn, Göttingen, F.R.G.
At least in the interphase cell three different antibodies distinguish three different fibrous systems. In a PtK2 cell treated with antiactin antibody, which stains the microfilament bundles, the predominant arrangement is of straight fibers moving across the cell, with some fiber bundles moving around the edge of the cell. Another cell has been treated with monospecific antitubulin antibody that stains the microtubules. The microtubules are of uniform thickness and have a rather radial display. The third cell has been treated with a third antibody, which in these cells stains a third fibrous system running in thick wavy fibers all over the cell. The three fibrous systems are clearly different on esthetic grounds. However, more importantly they can also be clearly distinguished if the same cells are treated with drugs known to inhibit specifically a particular fibrous system. The microfilament bundles of cells treated with cytochalasin B are reduced to "star-like heaps." As shown here the microtubule and intermediate fiber systems are not affected by cytochalasin B treatment. Cells treated with colcemid do not show any fibers when treated with tubulin antibody. The microfilament system remains intact, and the intermediate fiber system seems only slightly affected. Thus those three systems are distinguishable in the same cell line, not only on esthetic grounds, but also after drug treatment.
The visualization of this third thick wavy fibrous system with the immunofluorescence microscope led us also to look at this cell line by electron microscopy, in collaboration with Dr. Franke. It is very clear from our sections of this cell line, which are similar to those shown by Dr. Sanger, that PtK2 contains large wavy bundles of "pre-keratin-like" fibers.

M. DeBrabander, Beerse, Belgium
Did anyone ever determine experimentally the minimal size of a fiber that can be seen with immunofluorescence microscopy?

M. Osborn, Göttingen, F.R.G.
One has to choose one's system carefully. Clearly good antibodies, flat cells, and good contrast are all important. Perhaps it is worth saying that any linear structure below 1000-2000 Å in diameter will result in a line of equivalent thickness in the microscope. Thus, bundles of small numbers of microtubules and a single microtubule (assuming it to be visible) could appear as lines of equivalent thickness in a fluorescent micrograph. Stained rat sperm tails treated so as to spread out the outer doublet microtubules show staining with our monospecific antitubulin antibody. In some specimens we can see 3-4 fluorescent fibers. Thus we think we may be getting very close to being able to stain individual microtubules.

M. DeBrabander, Beerse, Belgium
Using the PAP-immunocytochemical method we have actually studied sequentially the microtubule system in whole cells with the light microscope and the ultrastructural distribution of stained microtubules after embedding the very same cells. It is our belief that individual microtubules can possibly be discerned with the light microscope under ideal circumstances in the thin peripheral lamellae. In thicker parts of the cells, e.g., in the center, this is not possible due to the presence of bundles, microtubule overlap, and diffuse staining. The light microscopic picture can thus only be regarded as representing the general pattern. Certainly it cannot be used to determine the presence and distribution of individual microtubules whenever the cell shape is drastically altered, such as rounding after detachment. The immunofluorescence technique yields pictures identical to those obtained using the PAP technique. Thus, the previous remarks probably apply also to the immunofluorescence technique.

W.W. Franke, Heidelberg, F.R.G.
Demonstration of individual microtubules and comparison of light microscopic images using indirect immunofluorescence with electron microscopy is certainly facilitated in the lamellipodia of tissue culture cells. In addition to the very useful technique of Porter and Buckley, flat sections of cells fixed and embedded on the coverslips after indirect immunolabeling (work done in cooperation with Drs. Weber and Osborn from Göttingen) indicates that a correlation is in principle possible but is somewhat hindered by the partial interruptions (disintegrations) of microtubular profiles that obviously have resulted from the insufficient fixation.

M. Girbardt, Jena, G.D.R.
In addition to the examples mentioned, vegetative hyphae of *Polystictus versicolor* would be a test cell for proving that single or at least 2-3 microtubules can be demonstrated by immunofluorescence. The cell contains a constant amount of microtubules, varying from the top to the basis. Near the septum of the 400 - 700 μm long cell (diameter 5 μm) there are only 1-5 microtubules per cross section.

H. Ponstingl, Heidelberg, F.R.G.
I wonder whether instead of just identifying and localizing spindle components, one could find functional tests by injecting specific inhibitors or modifiers. For instance perhaps phalloidin would be a useful reagent for actin.

J.R. McIntosh, Boulder, Colorado, USA
We have used phalloidin, kindly donated by Dr. Wieland, to look for actin filaments with the electron microscope in lysed, phalloidin-treated mammalian cells. The spindles of such cells contain many thin filaments, but since phalloidin induces the polymerization of actin in solution, it is perfectly possible that the filaments are induced in the spindle by the treatment.

H. Ponstingl, Heidelberg, F.R.G.
Dr. Sanger, I am somewhat surprised by what you say about colchicine-treated cells. If there is no effect of colchicine on actin filaments, why should they vanish from chromosomal fibers?

J.W. Sanger, Philadelphia, USA
I did not say that colcemid depolymerized fibrous actin. Colcemid has no effect on
our ability to detect interphase actin bundles in our cells. Rather I pointed out
in my lecture that both Sanger and Sanger (1976) and Cande et al. (1971) demonstrated
that cells arrested in metaphase with colcemid do not have actin bundles in the
spindles. This tells us that the localization of actin in the chromosomal spindle
fiber is dependent on the presence of chromosomal spindle microtubules, and raises
two possibilities: 1) Actin can be artifactually located on only the chromosomal
spindle microtubules. 2) The actin must be linked to the microtubule. If they are
linked in vivo, then it is not surprising that destroying the skeleton disrupts
the localization of actin in spindles.

J.R. McIntosh, Boulder, Colorado, USA
Dr. Ponstingl has raised an interesting and important point. One does not expect
an effect of colchicine directly upon the spindle actin, so the destruction of
the metaphase actin image by colchicine implies an indirect effect and thus that
the organization of the actin depends upon the spindle tubules. During interphase
the antisera show that the tubule and actin systems are largely independent. I infer that there is a change, as a function of the cell cycle, that provides a situation where actin and tubulin of the spindle will interact.

J. Robertson, München, F.R.G.
How sticky is actin? There is a sticky organelle, the ribosome, which might bind
actin and protect it from depolymerization. Since ribosomes are dispersed throughout the spindle, ribosomal-bound actin microfilament fragments could appear in
fluorescence studies as one filament continuous through the spindle. As the chromosomes move toward the poles, the ribosome and its actin would be either swept
aside or pushed toward the poles. Does anyone have any evidence regarding binding
of actin to ribosomes?

B.M. Jockusch, Basel, Switzerland
I have tried to find out, in one system, whether ribosomes do bind actin and found
that *Physarum* ribosomes do not bind F-actin. However, it is not clear whether that
actin which we see aligned to chromosome fibers with indirect immunofluorescence
is actually F-actin.

D. Mazia, Berkeley, California, USA
Should we not be impressed by the fact that cytochalasin has no effects on mitosis
in the very same cells in which it stops cytokinesis completely? Of course, we
now tend to dismiss cytochalasin as an unreliable tool - yet, if an effect of cytochalasin on mitosis had been found, we would have regarded it as a brilliant demonstration of a role of actin in chromosome movement.

J.W. Sanger, Philadelphia, Pennsylvania, USA
I agree with Dr. Mazia that many people would believe that actin and myosin are
responsible for chromosomal movement if cytochalasin stopped the movement. However,
it is by no means clear that cytochalasin does act on actin. Cytochalasin stops
some cells (sea urchin) from cleaving but not others (HeLa, L-fibroblast, etc.).
Cytochalasin injected into *Nitella* stops protoplasmic streaming. However, when the
drug is injected into ameba, there is no inhibition of movement. Both of these experiments were performed by the same experimenter (Dr. Richard Cotby).
Sanger and Sanger (1975) also thought that cytochalasin B might affect non-fibrous
actin. We therefore exposed sperm of brittle stars and sea urchins to the drug and
activated them. Acrosomal filaments formed, indicating that the drug did not affect
the polymerization of "protofilamentous actin" to fibrous actin.

J.R. McIntosh, Boulder, Colorado, USA
Stossel and Hartwig and Weihing have published evidence in the Journal of Cell
Biology that cytochalasin acts on the gelation of actin mediated by high-molecular-
weight, actin-binding proteins. Thus, a lack of effect of cytochalasin B on mitosis
simply implies these gels are not important for chromosome movement.

M. Hauser, Bochum, F.R.G.
In my own work on mitosis of the nuclear apparatus of ciliates, which I undertook
to study a highly polyploid nucleus with multiple cell nuclear components - es-

pecially spindle material - I found not only masses of actin-like filaments but also numerous thick 150-180 Å filaments, without using glycerination or any other conventional fixation treatment. Their actomyosin-like nature is evident by their reaction with ATP (they become parallel and show cross-bridging) and their positive reaction in immunofluorescence. Finally, extracts of homogenized isolated nuclei show a protein pattern corresponding to that of actin and myosin in disc electrophoresis.

Chairman:
Is there any evidence for a control system regulating the formation and activity of microfilaments and microtubules?

J.W. Sanger, Philadelphia, Pennsylvania, USA
If the vesicles found in some spindles are analogous to sarcoplasmic reticulum (SR), it is possible to have highly localized release and uptake of Ca2+ in the spindle. In skeletal muscle, Ca ions are stored in the terminal cisternae, which when stimulated, release the Ca2+ and cause contraction. Almost immediately, the longitudinal elements of the SR take up the Ca ions, causing relaxation (provided that no more Ca ions are released). In muscle it is possible to stimulate a single sarcomere while surrounding sarcomeres are unaffected. In the spindle no one has yet demonstrated calcium in spindle vesicles. If this can be shown (one might expect two different types of vesicles) in the spindle, Ca ions may be thought to control microtubule polymerization and depolymerization and/or perhaps actin-myosin interaction.

P.J. Harris, Eugene, Oregon, USA
In a thin section of a prometaphase sea urchin egg, the aster and spindle appear as a clear area from which yolk is excluded, and this clear area, especially in the asters, is primarily a mass of vesicles. Elongated vesicles are arranged radially from the centers to form the astral rays, along with the microtubules. The continuous growth of the aster and the centrosphere, or aster center, as seen in a thick section of isolated mitotic apparatus, suggest the pattern one sees in a trigger wave system as shown in the film of the Belousov reaction. This is a preparation in which successive bands of an oxidized and reduced state move outward from a central pacemaker or point of perturbation. If calcium release and sequestering occurs in a similar way in the cell, the necessary control in both space and time could be accomplished.

J.R. McIntosh, Boulder, Colorado, USA
The vesicles at the poles are attractive candidates for a system analogous to the SR, as suggested by Dr. Harris. It is well to remember, however, that many asters associate with vesicles and granules visible in living cells and then move them. Rebhun has described colchine-sensitive saltations of vesicles in the living cell. Many of these are saltations of vesicles toward the central region of the aster. Spindles themselves will transport nonchromosomal material, and the motions of vesicles described by several workers is toward the spindle poles. Thus there is an accumulation of vesicles near the mitotic centers of many cells, and the *presence* of the vesicles at the spindle poles may be a fortuitous result of spindle and astral transport properties.

Steinhardt, Berkeley, California, USA
We have measured free calcium in sea urchin eggs with aequorin and at the same time considerably changed the intracellular pH with molecules such as ammonia. At our levels of sensitivity of the detection of free calcium, we see no change in intracellular ionic calcium with large changes in intracellular pH. We have been able to detect the changes in ionic calcium at fertilization, and intracellular pH electrodes have been used to measure directly the intracellular pH transient at fertilization and in other conditions.

P.J. Harris, Eugene, Oregon, USA
Measurements of luminescence made on the entire cell do not really tell us what may be happening in a limited region. Again I refer to the fertilization reaction in Medaka demonstrated by Ridgway. The narrow band of aequorin luminescence shows that calcium release and perhaps high concentrations of calcium ion may exist for a short period followed by sequestering.

Steinhardt, Berkeley, California, USA
It is true we cannot inject as much aequorin into sea urchin eggs as into Medaka eggs. However, we can easily follow the peak transient release of calcium at fertilization in the sea urchin egg. We have measured the pH changes in the same system and we have changed the pH and looked for changes in free calcium.

P.F. Baker, London, U.K.
The only experimental data I know where an attempt was made to fix intracellular Ca at different levels are the data of Baker and Warner (J. Cell Biol. 53, 579, 1972). They found that injection of the Ca^{2+}-chelating agent EGTA into *Xenopus* eggs blocked cleavage, but, nuclear division still took place. Levels of Ca^{2+} in the range 0.2-1.0 μM did not block either nuclear division or cleavage. Injection of the Ca-sensitive protein into eggs revealed a small increase in light output (and presumably ionized Ca^{2+}) associated with cleavage. These observations suggest that nuclear division can take place at very low levels of ionized Ca^{2+} but the cleavage requires a somewhat higher level of calcium.

Chairman:
Regarding the affinity of a possible calcium pump, I have been checking with the muscle calcium pump people and they say their pump has an affinity of micromolar or a tenth of micromolar. We have heard from Dr. Petzelt that his calcium pump has an affinity of millimolar. It remains unclear how this pump would work in an environment like the spindle.

C. Petzelt, Heidelberg, F.R.G.
There are several possible explanations. Either we have destroyed the high Ca^{2+}-affinity of the mitotic ATPase by the purification procedure or we have lost some cofactors necessary for the enzyme's function in the cell. Additionally, since we are still working with a protein mixture in which the enzyme is not yet completely purified, the presence of a Ca^{2+}-binding protein might lower the actual concentration in the enzyme assay down to the micromolar range. This has to be worked on.

T.E. Schroeder, Friday Harbor, Washington, USA
It may be relevant to note that there is some potentially important negative evidence for an important role of calcium in cell division in sea urchin eggs. The "Ca-ionophore" A23187, applied under a variety of conditions, fails to alter the short-term course of mitosis or cell cleavage. This compound appears to increase free calcium in unfertilized eggs of the same species, so presumably it is doing something similar in fertilized eggs.

Steinhardt, Berkeley, California, USA
I would like to back up Dr. Schroeder's remark. A23187 is active on both unfertilized and fertilized sea urchin eggs. We have measured the burst in O_2 uptake at fertilization and shown that it is dependent on free calcium ion. In an egg which has been fertilized you can follow the release and sequestering of calcium by the time course of O_2 uptake. Using fertilized eggs which have released and re-pumped up their calcium, we get another release of intracellular Ca^{2+} and another respiratory burst with the ionophore A23187. So the ionophore is active in both unfertilized and fertilized sea urchin eggs, and the calcium store can be recharged in a few minutes.
On the other hand, it is entirely possible that in large lipid-filled eggs, such as the sea urchin or *Xenopus*, that A23187 never reaches the center of the egg, which, of course, is an important reservation.

J.M. Mitchison, Edinburgh, U.K.
Duffus has found that fission yeast can be synchronized by treatment with calcium ionophore.

J.F. Lopéz-Saéz, Madrid, Spain
An important observation on the role of Ca^{2+} in plant cell division was reported by Paul and Goff (1973). They also proposed that the inhibiting effect of caffeine on cell plate formation might be related to a calcium-caffeine antagonism. Our group has confirmed this antagonism by showing a decrease in the caffeine efficiency in the presence of 10 mM calcium. We found a magnesium-caffeine antagonism and even

moreso, a clear calcium-magnesium potentiation.
The action of caffeine on cell division depends markedly on the drug concentration, as is true with the vast majority of inhibitors. Thus, 5 mM caffeine is able to inhibit cytokinesis without disturbing mitosis morphologically, while 50 mM caffeine shows a clear c-mitotic action with metaphase arrest of the chromosomes and an increasing chromosome condensation with time.

P.J. Harris, Eugene, Oregon, USA
Some years ago in Mazia's lab sea urchin eggs were blocked in mitosis by mercaptoethanol, and when allowed to recover after the controls had divided the second time, they divided directly from one to four cells. Likewise, if caffeine is used to block division in sea urchin eggs, a one-to-four division follows on recovery. In cells blocked just before metaphase, the spindle shortens and the centrioles separate to form four separate poles connected by spindle fibers and with chromosomes randomly distributed. Perhaps mercaptoethanol and caffeine are both acting in the same way by somehow affecting the calcium pump.

C. Petzelt, Heidelberg, F.R.G.
In a recent paper, Nath and Rebhun showed that caffeine could block mitosis and they showed at the same time that caffeine blocks or inhibits the Ca-ATPase.

T.E. Schroeder, Friday Harbor, Washington, USA
As I recall these experiments, however, the effects of caffeine on eggs are long-term ones. Caffeine has surface-active properties, as well as effects upon cyclic AMP metabolism, and it may be rash to assume that caffeine effects on mitosis are related to calcium-release activity of sarcoplasmic reticulum from muscle treated with caffeine.

Session VI
Mitosis in Differentiation, Morphogenesis, and Cancer

F. DUSPIVA, Zoologisches Institut der Universität Heidelberg, Fachrichtung Physiologie,
Im Neuenheimer Feld 230, 6900 Heidelberg, FRG

A. Introduction

Mitotic cell division is undoubtedly the mechanism by which the im-
mensely complex adult multicellular organism originates from the apparently simply organized zygote. The question arises whether the
only purpose of mitosis is to distribute, as perfectly as possible,
the genome from the mother cell to the daughter cells, or has mitosis
additional functions essential for differentiation, e.g., when daughter
cells exhibit features and behavior lacking in the mother cell. Have
we reached with this concept the main point of the process called differentiation?

The terms differentiation and morphogenesis are nearly 100 years old.
They derive from a time when the only disciplines available to biologists
were anatomy and histology. In many contemporary textbooks differentiation is still defined as the variety that originates from determination, topogenesis (a process of movement in relation to morphogenesis), histogenesis, and growth, whereas determination refers to all
of the invisible events in a cell, in a blastema, or in its neighborhood that are expressed as morphogenesis. Today we know that these developmental processes are invisible because they take place at a molecular level and that they can be explored with the available biochemical methods. With this widening of our biologic horizon, the classical terms differentiation and morphogenesis have become too narrow.

In the search for a new definition for differentiation, applicable
also to processes of molecular organization, many biologists found
that the changes in structure and behavior of cells during a developmental step are correlated with a sudden excessive rise in synthesis
of various macromolecules (enzymes, contractile proteins, hemoglobin),
thus expressing the phenotype of these cells. If we try to define the
term differentiation in terms of the ability of cells to synthesize
"working" molecules (unfortunately called "luxury" molecules), we have
to distinguish these from all the other ubiquitous molecules necessary to support fundamental cell metabolism. The latter group should
be called "essential" or "housekeeping" molecules (Weiss, 1973).

Apart from the fact that only the fully mature, functionally efficient
cell contains such "luxury" molecules, this definition involves many
difficulties. In many cases it is not possible to isolate and characterize specific luxury molecules, especially where it is not the
material itself but its form and structure that constitute differentiation.

An example of this is the development of the wing scales of Lepidopters
(*Ephestia kühniella*), a process that may undoubtedly by described as a
process of differentiation from a morphologic viewpoint. From a scale
stem cell both a socket mother cell and a scale mother cell derive by
differential mitoses. The scale, a cuticular layer with a highly complicated ultrastructure, is derived from the scale mother cell (Stossberg, 1938). There is no luxury molecule to be found; chitin is used
to cover all other epidermis cells (Fig. 1). In this case it is not
the substrate that comes into question but the ultrastructure. Can

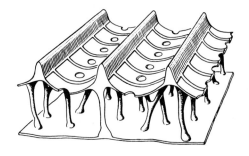

Fig. 1. Chitin structures of wing scales from butterflies, constructed after electron-microscopic pictures. (From Kühn, 1965)

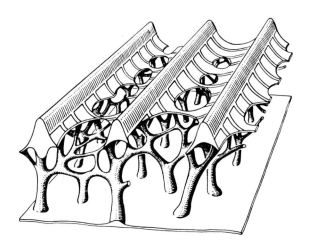

this process be included under the term differentiation? Perhaps the scale structure depends on the synthesis of special proteins that assemble to form an intimate space structure for the later chitinization. Such an interpretation is absolutely speculative; we know too little about the mechanisms necessary to construct subtle structures such as shells and diverse forms of skeletal elements in *Radiolaria* and *Foraminifera*. In many cases, it is the outfitting with certain proteins (hemoglobin, basic substance of chondrus) that determines the morphologic and physiologic features of cells.

Given the stage of our present knowledge, it is difficult to formulate an unambiguous definition for the term determination that is also useful at all levels of the structural hierarchy of life; therefore it was hoped that at least the main problem could be resolved by investigating a variety of phenomena seen during the development of animals and plants. It seemed clear a priori that the information for the entire developmental process and life cycle of an individual is enclosed in the genome of the zygote. Thus an attempt was made to trace back the very different phenotypes of differentiated cells to a genotype common to all. On the basis of all presently available research, collected by Davidson (1968), it seemed plausible that the mechanism of selective gene activation is the principle underlying differentiation and morphogenesis. If we consider that in a single process of differentiation many hundreds, probably thousands of genes must be regulated selectively, it seemed plausible to search for certain overlying mechanisms that are able to simultaneously activate or inhibit groups of "structure" "action" genes. Such a mechanism of gene regulation brings about a drastic restriction of activation steps. A model of gene regulation based on our present knowledge was proposed by Britten and Davidson (1969) and Davidson and Britten (1971).

B. Orientation of Spindle Axis and Its Role in Differentiation

To become acquainted with the mechanism by which cells differentiate, there is perhaps no better object in nature than the Spiralian egg. The typical members of this group of animals (Annelida, Mollusca) have drastically shortened the developmental process to achieve, as early as possible, an independent viable larva. During evolution this group of animals attained the minimal number of developmental steps necessary for the formation of the trochophora larva from the egg (Cather, 1971).

It is astonishing that already during the first division steps of the egg cell, the direction of the spindles is not meridional or latitudinal as usual but oblique (Fig. 2). Therefore the cleavage is somewhat spiral in nature. The divisions are typically unequal: The first cleavage divides the egg into two unequal blastomeres AB and CD. After the second division A, B, and C attain the same size, while the D-blastomer becomes substantially larger. The third cleavage divides each of the four blastomeres into four small micromeres and four still larger macromeres. After fertilization the ooplasm is not at all homogeneous but exhibits a complicated architecture oriented toward the pole (segregation of "pole plasma"). Therefore, during cleavage, the blastomeres receive qualitatively and quantitatively different plasmas. The exact programming of the course of division results in the distribution of cleavage nuclei into a very specific ooplasmic environment. It will be explained later how by this situation, important determinative steps have taken place that are of fundamental importance for the organization of the future larva.

It has been observed in many other systems that asymmetric cell division occurred together with the differentiation events. In scale formation of the butterfly wings, one cell that originates from the epidermic unity is transformed into a scale mother cell. This cell divides unequally: A small cell lies on the inside and a large cell on the outside. The small cell subsequently dies while the large cell is transformed to a scale cell and a socket cell by one oblique differential mitosis (Stossberg, 1938) (Figs. 3,4).

Such a process reveals many problems:

1. Why must the normal epithelial cell go through several cell generations before it can be transformed to a scale cell?

2. By what mechanism does the polar distribution of determinants occur in the mother cell? What is the molecular nature of these determinants and what is the mechanism of "differential" mitosis governing the change in the phenotype of daughter cells relative to the mother cell?

3. How is the orientation of the spindle in the cell controlled, so that after an unequal differential mitosis the daughter nuclei come into predestinated areas of maternal cytoplasm equipped with a specific pattern of determinants?

I want to discuss these questions in reversed order. The plant kingdom also has many examples of differential cell divisions, whereby

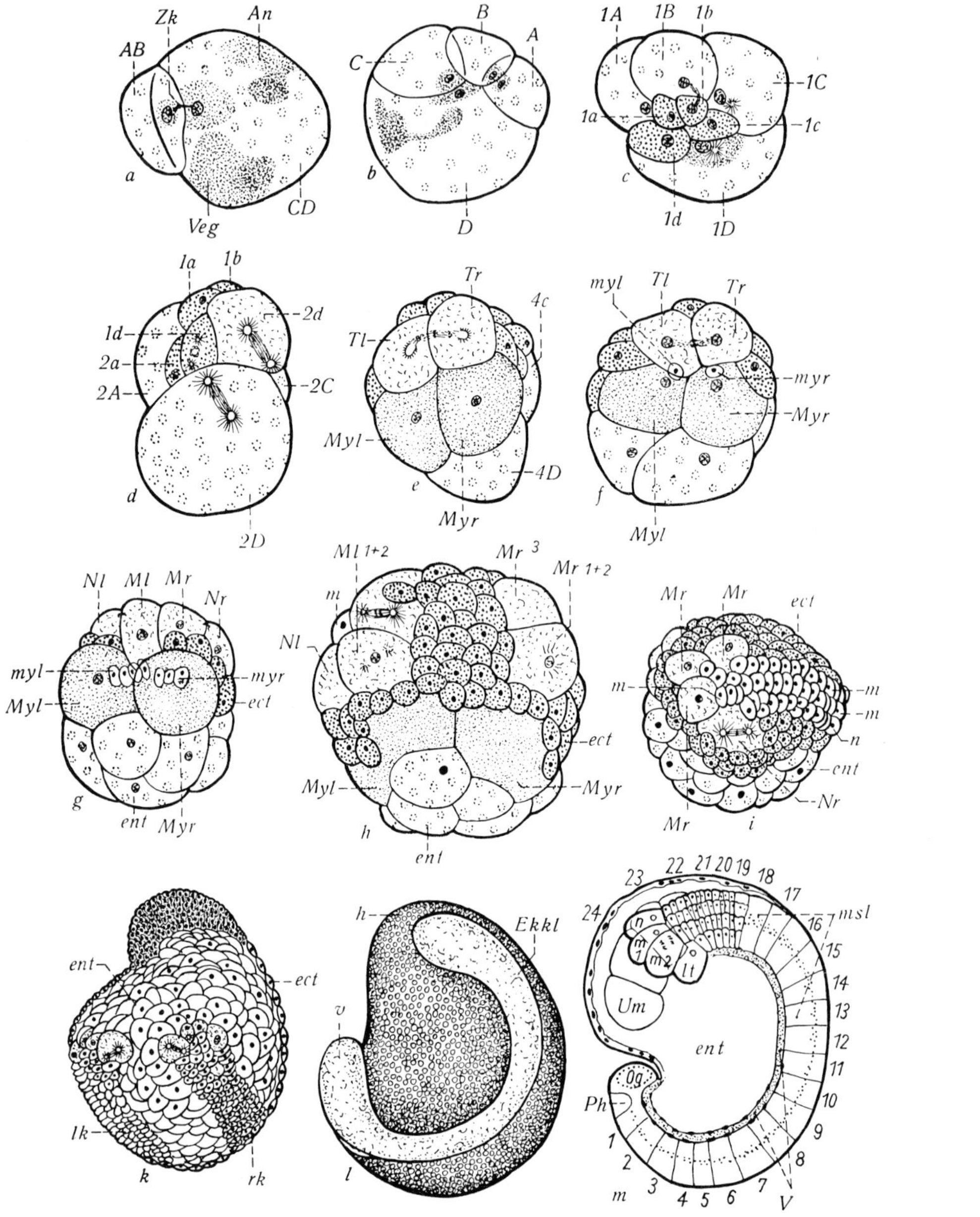

Fig. 2a-m. Development of *Tubifex* (Annelida). (a) 2-cell stage; (b) 4-cell stage;
(c) 8-cell stage; (d) 12-cell stage; (e) stage with both mesentoblast cells *(Myr,
Myl)*; (f) stage with the primary teloblasts *(Tr)* and *(Tl)* and both mesentoblast
cells *(Myr, Myl)*; (g) stage with the primary myoblasts *(Mr, Ml)*, the neuroblasts
(Nr, Nl), and 4 pairs of mesoblast cells *(myr, myl)*; (h) stage with 3 teloblasts
(Mr 1+2, Mr 3, Nl, Ml 1+2); (i) stage with 4 teloblasts; (k) stage with germ bands
(lk, rk); (l) germ bands joined together in their anterior part; (m) *An, Veg,* animal and vegetal pole-plasma; *ect,* ectoblast; *ent,* endoblast; *m,* cells budded from
myoblasts; *Myr, Myl,* mesentoblast cells; *lk, rk,* left and right half of germ bands;
ph, pharynx; *Og,* supraesophageal ganglion; *V,* dorsal blood vessel; *msl,* lateral line
of circular muscle; *n,* neuroblast; *m₁/m₂,* myale teloblasts; *lt,* lateral teloblast;
Um, mesentoblast cell. (From Pflugfelder, 1962)

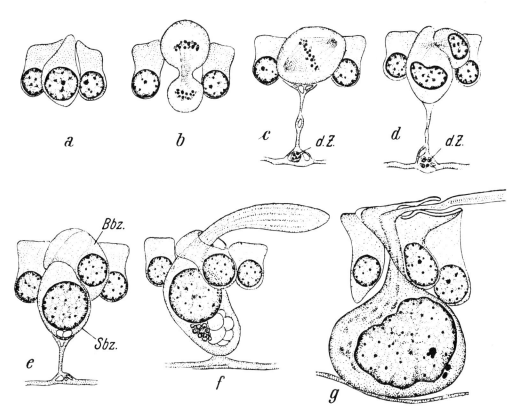

Fig. 3a-g. Development of wing scales in the butterfly *Ephestia kühniella*: (a) scale
stem cell I; (b) its 1st differential mitosis; (c) its 2nd differential mitosis;
(d-f) scale mother cell *(Sbz)* and socket mother cell *(Bbz)*; (g) *d.Z.*, degenerating
cell. (From Stossberg, 1938)

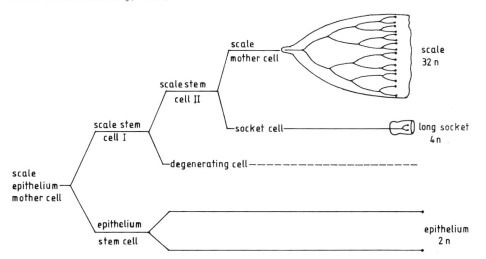

Fig. 4. Scheme of differential cell divisions and steps of endomitosis during development of wing scales. (After Kühn, 1965)

polar distribution of cytoplasm and its contents in the mother cell
precedes the determination of daughter cells. Classical examples of
asymmetric differential cell division include the pattern of differentiation in the leaves of the peat moss *Sphagnum* (Zepf, 1952) and
the differentiation of root hairs and the formation of stomata in
the leaves (Bünning and Biegest, 1953). Polarization of the cell must
always precede unequal cell division. In many examples (eggs of *Ascidia*
or *Ctenophora*) immediately after fertilization various vitally stain-
able components are distributed in the egg in a gradient-like manner;
they look as if their contents (yolk platelets) had been stratified
by electrophoresis or centrifugal force (Spek, 1938). The mechanism
for the phenomenon is unknown. Robinson and Jaffe (1975, 1976) recently showed that the *Fucus* egg forms a polar axis some hours after
fertilization, visible as a bulging out of the cell wall at one point
where the rhizoid begins to grow while at the other end the thallus is
formed. These eggs drive an electric current through the plasma. The
electric field is created by ions, especially Ca^{2+}, which enter more
quickly at the prospective growing pole and leave the egg faster at
the opposite pole. Is it possible that the idea of an electrophoretic
mechanism as the basis of polarity is not so wrong after all? The
mechanism of spindle orientation in a polarized plasma is still unknown.
Early experiments by Hörstadius (1928) on the sea urchin egg showed that
single blastomeres isolated from blastula divide as though they were
in their cell community (Fig. 5). If one of the mitoses is suppressed
by short-term inhibition of cytokinesis, it is not subsequently under-
taken. At the next division step the orientation plane of the spindle
is determined by the time interval since the moment of fertilization.

We can conclude from this only that an autonomous process in the cytoplasm determines the positions of the spindles, just as if a biologic
clock started at fertilization. The workings of this mechanism are
unknown. We suppose that such a mechanism is located in the cell cortex, because a lesion of the cortical layer caused by a prick with a
thin needle is sufficient to produce an abnormality of the mitotic
apparatus and of chromosomes during cleavage(Brachet and Hubert, 1972).
Unfortunately the nature of the cortex and of the plasma layer beneath
it is not sufficiently clarified. Recent investigations with Annelidian
and sea urchin eggs have revealed that fibrils run from the cortex
into the plasma. Evidently this fibrous network can hold minute cell
particles very tightly, which therefore cannot be transposed by centrifugation. This effect explains an old phenomenon. The developmental
biologists knew a long time ago that centrifugation of an egg cell did
not change the organization of the embryo. They made a good guess in
assuming that the cortical layer is highly viscous and therefore retains in place important determinants during centrifugation. The fibrous network clarifies this situation. Whether a still unknown cortical structure has any influence upon the orientation of spindles is
still an unresolved matter (Rappaport, 1971).

We can be sure that during differential mitosis the deviation in the
phenotype of daughter cells from the mother cell or among the daughter
cells themselves is dependent on the diverse distribution of determinants in the cytoplasm and not on the difference in quality of the cell
nuclei. I want to recall the classical experiment of Carlson (1952),
who succeeded in turning the spindle with its set of chromosomes about
180° in an insect neuroblast dividing unequally, so that each daughter
cell obtained those chromosomes that were normally intended for its
sister. This operation had no effect whatsoever on the subsequent development.

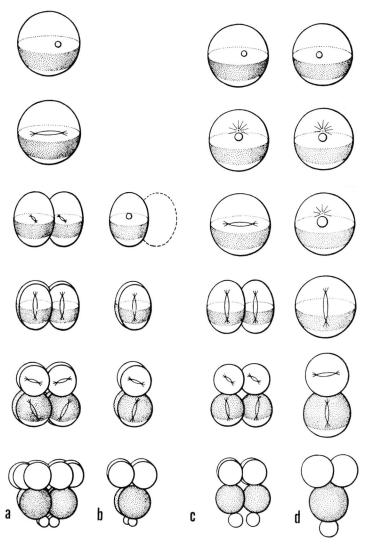

Fig. 5a-d. Schemes of normal (a) and altered cleavages in sea urchin eggs; (b) cleavage of a 1/2 blastomere; (c) the first mitosis suppressed; (d) the first and the
2nd mitoses suppressed. (From Hörstadius, 1928)

C. Mode of Action of Determinants in Morphogenesis

The characterization of determinants and of their mode of action is a
fascinating problem that has been discussed for a long time and that
is still relevant. The well-known classical experiments on cell lineages during the development of Spiralian egg (Annelida, Mollusca,
Insecta) led to the idea that early in development, segregation of
morphogenetic potency takes places that gives rise to regional specific localization of cytoplasmic information in the egg space. Taken up
by the stem cells, this information directs a number of diverse developmental pathways, which, after a relatively long cell lineage, lead
to different results out of which organ-anlagen are formed.

An extraordinary system in which to study "visible" morphogenesis is the egg of some mollusk species in which, during the first three division steps, so-called polar lobes are seen for a short time on the D-quadrant, which contains vegetative cytoplasm but no nucleus (Cather, 1971). The material of the polar lobes is taken up exclusively by the D-quadrant. During separation of the second micromere quartet, some parts of the vegetative and of the whole animal pole plasma join together to form a 2d-cell, named the *first* somatoblast, which then undergoes determination to a stem cell via this step. This is an unique example in nature demonstrating the origin of a stem cell in an embryo. From this stem cell a cell lineage originates, that produces as its terminal members epidermis, myoblasts, and neuroblasts. The remaining parts of the vegetative pole-plasma of 2D are included in a 3D-cell, which after the next mitosis is completely transformed into a 4d-cell, named the second somatoblast. This is the stem cell for the mesoderm anlage. The formation of stem cells for different cell lineages is a fundamental process of morphogenesis (Figs. 6,7).

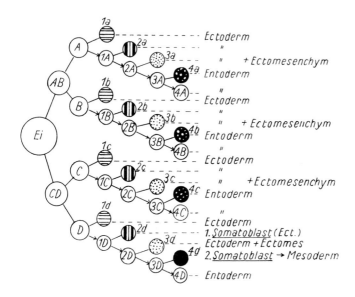

Fig. 6. Cleavage model of *Spiralia*. (From Pflugfelder, 1962)

What is the role of polar lobe material? After removal of the first polar lobe, all the specifities of the D-quadrant have been lost (Fig. 8). Cleavage continues and gastrulation takes place; even a trochophora forms but shows many deficiencies: It lacks an apical organ and the whole posttrocheal region out of which the body of the adult animal forms. There is no mesoderm development of its derivatives. Consequently bilateral symmetry is also absent and a larva results showing radial symmetry. After removing the second polar lobe, a trochophora grows with an apical organ but it still lacks a caudal region. The factors responsible for its formation have been lost with the second polar lobe. However, embryos without polar lobes still demonstrate a considerable amount of histodifferentiation. After removal of the polar lobes, an embryo developes with normal shape and ciliary rings. The latter are the expression of cell differentiation, a fundamental process of development that remains after removal of the polar lobes; however, the embryo lacks morphogenetic differentiation, which proceeds only when formation of the stem cells is guaranteed. If this is not the case, then an adult animal lacks a foot, shell, velum, and

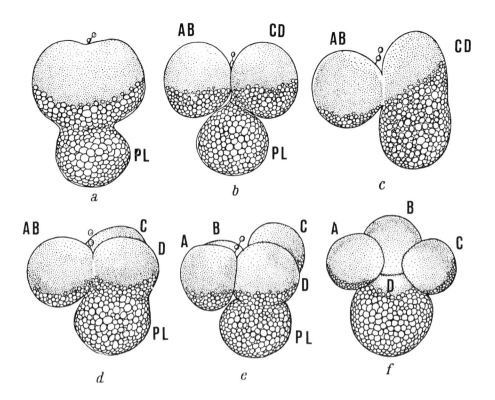

Fig. 7a-f. Formation of polar lobes in the gastropod *Ilyanassa obsoleta*. (a-f) Normal cleavage; (a-c) formation of polar lobe (*PL*) during the first cytokinesis and its transfer into the 1/2 blastomere *CD*; (d-f) polar lobe of 2nd cytokinesis and its fusion with the 1/4 blastomere D. [According to Morgan (1927), from Kühn (1965)]

eyes, typical attributes of gastropodian organization. After this discussion we could conclude that Spiralian eggs develop in a mosaic manner, resulting from systematic distribution of the determinants. Thus the quadrants ABC + 2d should be able to produce shell, since a 2d-cell forms the shell gland in normal embryos. ABC + 2d + 3d should form a foot; but heart and gut derive at least in part from the mesentoblast 4d. Indeed, these organ-anlagen are present under these experimental conditions; however, without the D-quadrant morphogenetic development is never normal - adult organs are present but only partially developed. The organization of velum, foot, and shell must be attributed to an inductive mechanism exerted primarily at the third quartet stage (Cather, 1971). This example illustrates the two most important categories of embryonic determination:

1. Cytoplasmic segregation of determinants (morphogenes) localized in the interior of cells, which are initially distributed unequally among daughter cells (cell differentiation).

2. Embryonic induction, depending upon interaction between cells and tissues by means of diffusible inducing substances represented mainly by peptides of low molecular weight (differentiation of certain cell lineages).

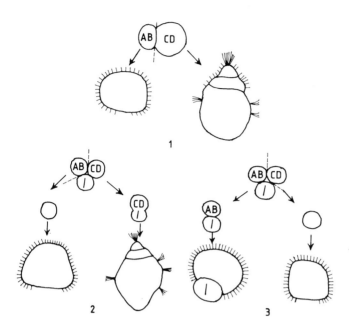

Fig. 8. Larvae of *Dentalium antillarum*. (1) Ciliar globe from *AB* and larva from *CD* blastomere, isolated at the 2-cell stage; (2) larva from blastomere *CD*, isolated at the trefoil stage fused with the polar lobe, and a ciliar globe from *AB*; (3) ciliar globe from blastomere *AB*+polar lobe, isolated at the trefoil stage, polar lobe remained unfused but attached to the globe. (From Verdonk, 1968)

The well-known "hypothesis of variable gene activity" in early development represents an attempt to explain the mode of action of determinants. Briefly, the hypothesis says that the localization of morphogenetic determinants is a fundamental aspect of development and is of universal occurrence. These are molecules, synthesized during oogenesis, whose role is to specify the patterns of gene activation in the early developmental steps. The diverse patterns of gene activity, required for initiation of differentiating cell lineages, are established at least in part through the interaction of these determinants with the totipotent blastomere genomes, which are distributed among them during oogenesis (Davidson, 1968).

Ways of testing this hypothesis are: 1) analysis of nuclear activity after implantation into cells of another differentiated state; 2) removal of genetically or experimentally caused defects in eggs by injection of normal egg cytoplasm; 3) isolation of ooplasmic cell particles containing information for cell development; 4) search for factors causing the remarkable stability of the differentiated state. Are these factors localized in the nucleus or in the cytoplasm?

Cell nuclei from adult frog brain, synthesizing barely detecable amounts of DNA, were injected by Gurdon (1975) into three different cells: 1) mature oocytes synthesizing especially rRNA but not DNA; 2) ovulating oocytes in the stage of meiosis exhibiting condensed chromosomes; and 3) unfertilized eggs, whose nuclei synthesize DNA only after activation but no RNA (Fig. 9). In every case the brain nuclei were induced within a few hours to modify their activity to that characteristic for the cell into which the nuclei were injected:

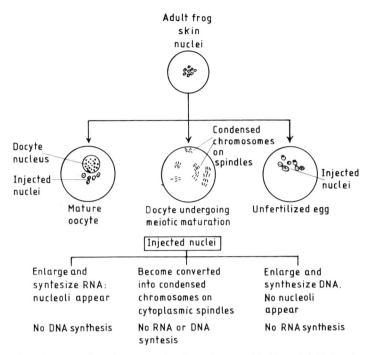

Fig. 9. Cytoplasmic control of nuclear activity. Adult brain or skin nuclei, which do not synthesize DNA or pass through mitosis, change their activity in different ways following exposure to different kinds of oocyte or egg cytoplasm. (From J.B. Gurdon, 1975)

In the oocyte the nucleus swells, nucleoli appear, and RNA is synthesized, but DNA is not reduplicated.

In meiotic oocytes the chromosomes condense and do not synthesize either RNA or DNA. In unfertilized eggs the nuclei swell, DNA is synthesized but not RNA. It follows from this and similar experiments that the egg cytoplasm must contain an inducer for DNA synthesis and a repressor for RNA synthesis that are independent of each other and operate on different gene classes. Further evidence for the action of plasmic factors on the expression of genes was provided by cell fusion experiments. However, it is impossible to give here a detailed discussion.

The following example permits insight into the molecular nature of cytoplasmic factors regulating early development. Axolotl (*Ambystoma mexicanum*) females homozygous for gene o$^-$ (ova deficient) lay eggs that do not develop further than the gastrula stage, even if the sperm carries a normal allel o$^+$. Therefore the effect of o$^-$ is purely maternal. The genetic lesion occurs in the ovary and causes a cytoplasmic deficiency that impairs fundamental events in early morphogenesis. Injection of nucleoplasm from a normal o$^+$ oocyte nucleus (germinal vesicle) into the mutant o$^-$ egg corrects the developmental stop at the gastrula stage. The o$^-$ substance is synthesized in the lampbrush stage of the germinal vesicle. After its breakdown the substance enters the cytoplasm. This factor, probably a high-molecular-weight acidic protein, is found only until the late blastula stage; later it is either destroyed or bound to some cell structures. It is well known that amphibian eggs develop until the late blastula stage without a functional genome. Then gene transcription is necessary to initiate gastrulation. This activity depends on the presence of o$^+$ substance (Briggs and Justus, 1968; Briggs, 1972). Mutant blastulae have little or no RNA; their RNA synthesis is practically eliminated. Brothers (1976) used nucleus transplantation to show that the interaction between o$^+$ protein and blastula nuclei gives rise to a heritable condition of nuclear activation. According to this method, activated nuclei are transplanted into enucleated eggs, which then are incubated until the early blastula stage; afterward nuclei are transplanted for the second time into enucleated mutant eggs, from which a series of transplants are made. These clones yielded normal larvae. These experiments show that nucleus activation, once acquired, represents a stable and heritable condition. More than 30 successive mitoses are not able to abolish this condition. Only nuclei between late cleavage and early blastula stage have the competence to interact with the o$^+$ substance; later on, this capacity is lost. This example can be used as a model to show how a stable differentiated state is established by interaction between nuclei and cytoplasm. It is shown that a regulation protein is effective in this system. Now the question arises, from where does the necessary information come to synthesize such a protein at a given time in development?

According to a widely held belief, morphogenetic information is stored in the ooplasm in the form of stabilized nucleoprotein particles. Such a particle would have most of the properties that embryologists attribute to determinants, such as sensibility to ultraviolet radiation and resistance to centrifugal forces. Many other particles participating in morphogenesis, e.g., yolk granula or mitochondria, are delocated by centrifugation of suitable force; however, this is not true of the very small particles.

The abundant occurrence of such particles in egg cells was demonstrated by a Russian group (Spirin, 1966; Samarina et al., 1968) and by an American group (Gross, 1967). In the telotrophic-meroistic ovary of insects, RNP particles are synthesized in a nurse chamber by polyploidic nurse cells and transported via nurse strands into the growing oocytes. Thus ribosome preparations of nurse cells show the same particle spectrum as eggs immediately before deposition (Fig. 10). These particles are composed of polysomes, monosomes, large and small subunits of ribosomes, and a postribosomal sedimenting particle population. From the latter it is possible to prepare RNA species with different molecular weights (maximum at 7-9 S); they contain a high percentage of poly (A) sequences and have a stimulating influence on the incorporation of radioactively labeled amino acids in a cell-free protein synthesizing system (Fig. 11). Therefore they are recognized as mRNA molecules. The RNA enclosed in these particles is stable. It is well known that the capacity of the oocyte nucleus for RNA synthesis is largely or totally inhibited. However, in the terminal phase of oogenesis, a sudden rise of short-term synthesis of new nonribosomal RNA species takes place. They sediment between 30 and 5 S and possess up to about 57 % poly (A) segments. This RNA is not stable and is translated at once. In eggs, immediately before deposition, only the newly synthesized proteins are demonstrable; mRNA is broken down immediately after translation and is now found as fragments of very low molecular weight. However, the most interesting result of this study is the finding that in newly deposited eggs only the postribosomal sedimenting particles carry some of the proteins synthesized in the late oocyte (Fig. 12). It seems that labeling the RNA particles immediately before deposition of the eggs with certain proteins is of significance for the control of translation later on in development (Duspiva et al., 1973; Winter et al., 1977; Winter, 1974).

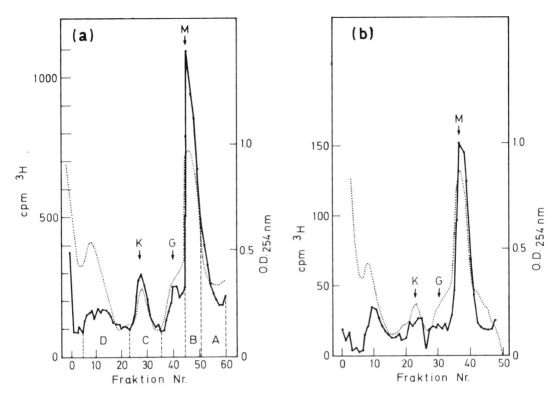

Fig. 10. Profiles of nucleoprotein particles after centrifugation of the ribosome fractions from the nurse chamber and from newly laid eggs of the bug *Dysdercus inter-medius* into a 10-40 % linear sucrose gradient. (a) Nucleoprotein particles from the nurse chamber. The bugs were injected with [³H]-adenosine 2 h before preparation of RNA; (b) nucleoprotein particles from eggs. The insects were injected with [³H]-aden-osine 8 days before preparation of RNA

D. The Morphogenetic Capacity of Nuclei During Development

Many examples show that interaction between nuclei and factors in the cytoplasm plays an important role in differentiation. Is the morpho-genetic capacity of nuclei changed by this interaction?

There are indeed some cases that show that the genome of daughter cells deviates from that of the mother cell. It may be a change in the position of genes or an increase in the number of certain genes. For the first case it has been proved that translocation or inversion of genes can strongly diminish their expression. Such a change in gene sequence is irreversible and cannot be considered as a general mechanism in the development of organisms. On the contrary, amplifi-cation of ribosomal genes occurs often, e.g., the two million ribo-somal gene sets in the mature oocyte nucleus of *Xenopus* or the gene amplification in certain cross-bands of the polytene chromosomes in salivary glands of some Sciaridae species (Crouse and Keyl, 1968; Pavan and de Cunha, 1969). However, this process is also limited and occurs only in certain cases. There are also single cases of chromo-some diminution, e.g., in *Ascaris* and some Diptera and Hymenoptera species. They are limited to the differentiation of soma as opposed to the germ

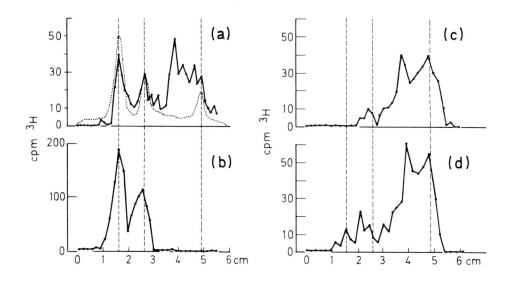

Fig.11a-d. Profiles of RNA species from nucleoprotein particles, which were prepared from nurse chamber of insects as in Figure 9. The profile of the particles was separated into 4 classes. RNA was prepared from each of them and fractionated on 2.5 % polyacrylamide gels. (a) Polyribosomes; (b) monosomes and the larger ribosomal subunit; (c) small ribosomal subunit; (d) postribosomal sedimenting particles

cell lineage (Kühn, 1965). Apart from these extreme examples it is very difficult to prove directly that genes, present only as a single copy in a genome, are not changed during embryonic development. At present, there are not sufficient data to answer this question.

However, one can ask another question. Is a single somatic cell still capable of mitotic division, growth, and production of several new cell types? Recently botanists succeeded in cultivating perfect plants with leaves, flowers, and seeds from a single cell, originating from the parenchyma of carrot root (Steward, 1970). The conditions of culturing were complicated. They included a series of treatments with growth hormones, coconut milk, changes in nitrogen supply, and osmotic pressure. However, no one has yet succeeded in obtaining a multicellular organism from a single animal cell in a similar way. It seems to me that this difference between plants and animals most clearly demonstrates a major problem of differentiation and morphogenesis. To obtain a better understanding of the difference between animal and plant cells, it must be clarified whether the stem cell of the in vitro fully grown carrot plant was really a differentiated cell. Such a cell never produces daughter cells capable of developing gametes. However, if one uses as a criterion of differentiation the ability to synthesize "luxury" molecules in the sense of cell-type-specific proteins, difficulties arise. Plant cells do not store type-specific proteins in such vast excess and quality that their phenotype is determined by them. It could be assumed that specific carbohydrates of the cell wall would eventually characterize the differentiated state of a plant cell. However, the parenchyma cell from carrot root is neither distinguished by certain proteins nor by specific carbohydrates. The ability to produce a perfect plant is probably only de-

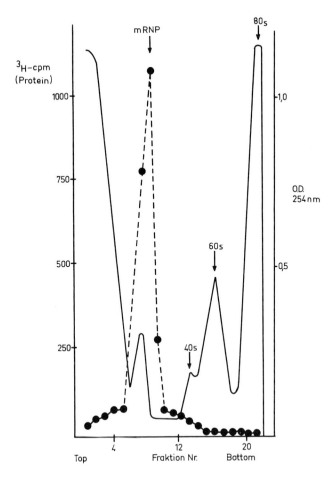

Fig. 12. Profile of nucleoprotein particles from newly laid eggs. During vitello-
genesis bugs were injected with a mixture of [^{3}H]-valine and [^{3}H]-glutamic acid.
Only the postribosomal fraction is labeled

pendent upon the fact that in plant cells neither inhibition nor permanent activation of certain gene sets is the basis of their differentiation.

In contrast to plant cells it is very hard to induce an animal cell
to change its cell type. If a piece of skin from the foot of *Xenopus*
is cultured, cells grow outward from the tissue that fill up with
birefringent keratin-like material one or two weeks after the onset
of growth (Gurdon, 1975). Chondrocytes also exhibit a differentiated
state, detectable as chondroitin sulfate and collagen, during many
generations, the number of which varies according to the culture conditions. However, in a monodisperse culture, they gradually lose
their ability to synthesize cell-type-specific substances and undergo intense proliferation. The change in appearance, similar to fibroblasts, is frequently designated as "dedifferentiated." However, the
cell does not lose its typical character at all; it is only trans-

formed from an overt "differentiated" stage into a covert "determinated" stage. Weiss (1973) proposed the term modulation for this type of
transformation.

An excellent method for studying the stability of a determinated
state is the serial transplantation and culture in vitro of imaginal
disks of *Drosophila* larvae, introduced by Hadorn (1963). Imaginal disks
are embryonic anlagen for a series of organs in the adult fly, which
were determined very early in the blastoderm stage. In the larva, the
disk consists of roundish, strongly basophilic cells with large nuclei
without visible marks of differentiation. These cells were mitotic
and very active in the larva. Differentiation and the initiation of
adult structures take place initially during the pupation period. It
is easy to isolate the imaginal disks from the larvae and to trans-
plant them into the body cavity of a host animal. The disk can be
cultivated for an unlimited time in the abdomen of adult flies; hemo-
lymph is an ideal culture medium (Fig. 13) which makes intensive proliferation of cells possible in a determined state. Because adult flies
lack metamorphotic hormones, terminal differentiation does not take
place. Because flies do not have a long life, the grafts must be transferred into a young fly every 2 to 4 weeks. The result of these experiments is that, as a rule, grafts retain their specific condition
of determination over innumerable generations of cells and over many
years without a detectable change in their determined state. Replan-
ted into the body cavity of a larva, the graft takes part in metamor-
phosis and develops its typical adult structures. This is proof that
the originally acquired determination is maintained during the long
cultivation period. All these experiments demonstrate that the determinated state is exceptionally stable. However, this stability is limited. After many mitotic cell divisions it may be that certain clones
show some deviation in their determined state. This change (transde-

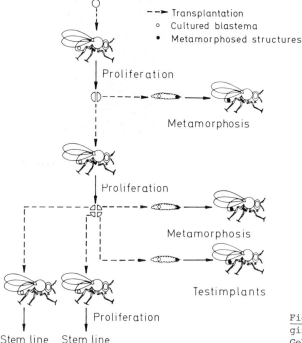

Fig. 13. Method of culturing imaginal disks in vivo. (From W.
Gehring, 1968)

termination) brings about a new, but not at all chaotic determination pattern. Interestingly during transdetermination every cell of the anlage only has the option to change into one of the cell types of the new organ (e.g., cells of antenna-anlage into cells of a leg) whereby a certain probability exists for a specific transition (Gehring, 1968) (Fig. 14). The transdeterminated state also is remarkably stabile. The

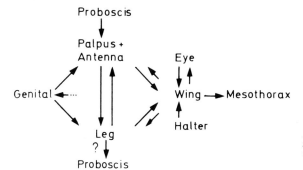

Fig. 14. Scheme of transdetermination sequences. (From W. Gehring, 1968)

real cause of transdetermination is unknown; it seems to be dependent on the mechanism of cell heredity itself. Perhaps some mistakes in the replication of determination quality take place after a prolonged series of mitotic divisions. Hadorn et al. (1966) think that some kind of dilution of the carriers of determination could lead to activation of a new set of genes on the basis of a feedback mechanism. The small number of alternative pathways to direct development leads us to assume that only a few sets of controlling genes (integrator genes in the model of Britten and Davidson) play the main role in differentiation of organs in embryogenesis (Gehring, 1968).

The determinated state of a cell is apparently stable only as long as the nucleus is embedded in a specific system of factors in the surrounding cytoplasm. The well known experiments of Gurdon (1975, 1968, 1970) showed that, when a nucleus of a highly differentiated cell like a keratin-producing skin cell isolated and freed of cytoplasm, is injected into an enucleated egg, it enters at once into mitosis and initiates development, and after three days young tadpoles are formed. This occurs even if serial nucleus-transplantations were interposed.

Now the main question to be answered is: What is the relevance of mitosis to a differentiation step?

Cell divisions and changes of gene expression are in accord with visible processes of differentiating blastemes and tissues during embryonic development. Is DNA synthesis followed by mitosis and cytokinesis, always an obligatory requirement for activating thus far repressed genetic information? One opinion is that cell-type-specific genes can be regulated only once in every cell cycle (Gurdon and Woodland, 1968). The basis of this idea is the conception of a cyclic reprogramming of gene expression (Fig. 15). It is assumed that gene activity depends on the association of regulatory proteins (acidic nonhistone proteins) with particular regions of chromosomes. As they become condensed for mitosis, these proteins are dissociated from chromosomes and are released into the cytoplasm, where they are mixed with other regulatory proteins synthesized during the last interphase. When mitosis is terminated the decondensed chromosomes acquire a portion of the cytoplasmic store of regulatory proteins, which will determine the genes to be activated or inactivated during the next interphase. If these proteins

Daughter cells

Interphase   Mitosis

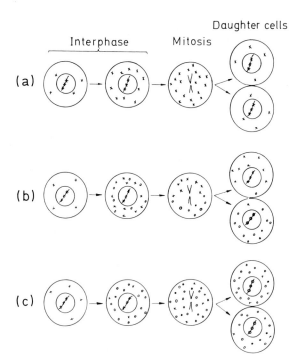

Fig. 15a-c. Hypothetical model of cyclic gene reprogramming and its association with chromosome condensation and with nuclear enlargement that takes place after mitosis. (a) Represents a stable determinated differentiated state, which is propagated through several divisions. Proteins that determine the activity of genes are synthesized throughout interphase and reprogram the chromosomes of the daughter cells; (b) represents an unequal mitosis. In this case a new gene becomes active during interphase, and its products are unequally distributed at mitosis, so that the two daughter cells have their chromosomes reprogrammed in different ways; (c) shows a similar situation to that shown in (b), except that a large number of products of the newly active gene are synthesized but are equally distributed at mitosis; as a result, both daughter cells are programmed for an activity different from that of the parent cell. (From J.B. Gurdon, 1975)

are different from those released at the beginning of that mitosis, then the gene activity of the daughter cell will be different from that of the mother. This may always be the case in early development, when cleavage nuclei occupy different regions of the ooplasm. Gurdon pointed out that the immense enlargement of nuclei and the chromosome dispersal after transplantation into an egg cell, enables the nuclei to acquire new regulatory proteins. A similar process may take place each time unequal cell divisions take place.

E. The Quantal Cell Cycle

Holtzer (1972) has termed a cell cycle that produces daughter cells with options of a new synthesis, not open to the mother cell, a "quantal" cell cycle and points out that the search for an underlying mechanism may be the most essential problem of differentiation. Whether a new program is really activated during only one quantal cell cycle, can be tested experimentally only in biologic systems that

synthesize luxury molecules that are easily detectable in single cells
from among a heterogeneous population of diverse cell types by em-
ploying histochemical techniques. For example, the terminal step of
the myogenetic cell lineage is a suitable object of choice.

Myogenesis has been thoroughly investigated by many authors in vivo
and in vitro mostly in chick embryo. Whether myogenic cell lineage begins during the blastula or gastrula stage is unknown. However, still
later in development there are primitive mesenchyme-containing cells
that are precursors of the myogenic, chondrogenic, and fibrogenic cell
lineages. Myoblasts are cells that have the ability to synthesize and
organize myosin, actin, and tropomyosin into myofibrils. It is possible
to distinguish myoblasts from their precursors using fluorescein-labeled antibodies against these proteins.

To determine whether DNA synthesis or events associated with cell division are obligatory for transition from precursor to mature myoblasts, they were cultivated in a medium containing 10^{-6} M 5-fluorodeoxyuri-
dine (FUdR) for a little more than one cell cycle. Distinct inhibition
of the transfer frequency from precursor to myoblast was observed.
After removing the drug, inhibition proved to be reversible. A similar
effect was obtained with cytosine arabinoside. Mitosis of presumptive
myoblasts can be stopped in metaphase by colcemid; this treatment also
inhibits the transition to the myoblast-stage, the synthesis of myosin,
and the organization of myofibrils. Inhibition of cytokinesis does
not present a transition. These results led Holtzer (1975), (Holtzer
et al., 1972) to suppose that the possibilities for synthesis in the
G1 phase of precursor myoblasts must differ from those in the G1 phase
of the next generation; therefore the transition to the mature myoblast
requires a quantal cell cycle.

The main problem concerning a quantal cell cycle is not answered by
Holtzer's experiments. By definition, synthetic pathways should be
opened for the daughter generation and closed for the mother cell;
this means a transcriptional control. Holtzer saw only the appearance
of luxury proteins, but assumes that during the previous mitosis activation of muscle-specific genes had taken place. It would, of course,
be ideal to show that transcription of muscle-specific mRNA had taken
place. Such investigation would require synchronous cultures of myogenic cells which are not attainable at present. The first step in
this direction was made by Buckingham et al. (1974). Myoblasts from
fetal calf muscle grown in culture first remain stationary. After a
change of culture medium a logarithmic growth phase starts, terminating in a transition phase of growth. This stage passes into fusion,
forming myotubes, without preceding mitosis. The grade of synchrony
was only moderate. Although the cells of myogenic cell lineage contain numerous mRNA species, only a 26 S messenger could be correlated
to myogenesis: It is the messenger for the larger subunit of myosin.
The result of this study was surprising: The 26 S messenger is not
synthesized - as expected - immediately before fusion, but in all
stages of the myogen cell lineage. The genome was programmed to express muscle-specific characters many cell generations before the
terminal myogenic cell, the myoblast, is induced to produce myosin.
The transition of the mononucleated myoblast into fusion is not characterized by new synthesis but by stabilization of mRNA types, including the 26 S messenger. It seems therefore that control of terminal phenotypic expression depends on a cytoplasmic process. The half-
life of messengers increases from 10 h in proliferation precursor
myoblasts to more than 50 h after the transition phase. The mechanism
of stabilization is unknown, but there are some indications that stabilization is attained by enclosure of messengers in cytoplasmic pro-

tein particles, well known under the term RNP particles, to be found in abundance in egg cells.

Erythropoiesis is an excellent system for studying the origin of a luxury molecule. The terminal stage of the red cell lineage contains so much hemoglobin that its protein content is represented by more than 90 % globin. Chick embryos, 40- to 70-h-old, contain hematocyto-blasts that synthesize DNA. They divide and produce first-generation daughter cells (erythroblasts). These already contain globin mRNA in sufficient amounts to synthesize enough hemoglobin (Hb) for detection in single cells by spectrophotometric methods. For initiation of Hb synthesis, DNA synthesis was necessary. To detect the earliest stage at which globin genes are transcribed, Groudine et al. (1974) rehy-bridized highly radioactive labeled complementary DNA, prepared by reverse transcriptase from globin mRNA, with mRNA from various stages of the red cell lineage and from cells outside of this lineage. This study showed that all primary erythroblasts contain globin mRNA, of course in very low concentrations, but that hematoblasts, muscle cells, and fibroblasts lack the messenger. The authors concluded that only members of the red cell lineage are able to transcribe globin genes. However, a negative result often depends on the sensitivity of the method. After further improvement using saturation-hybridization of cDNA with mRNA samples, Humphries et al. (1976) showed, that globin mRNA sequences are found not only in erythroid cells but also in non-erythroid tissues, such as adult brain and liver, sometimes, however, in very low amounts. Because this result was unexpected, a variety of control experiments were conducted. While α- and β-chains of globin are limited exclusively to the differentiated erythroid cell types, globin genes are transcribed by all cells of the animal, the accumu-lation of RNA varying according to cell type. Especially remarkable is the observation that the percentage of globin mRNA in total RNA from one tissue to another or from one cell lineage to another is essentially similar whether found in cell nuclei or in the cytoplasm. In the cytoplasm of erythroid cells, however, the percentage of globin mRNA sequences is always extraordinarily higher than in cells of another cell lineage. These results clearly point to the action of a posttranscriptional control mechanism for the supply of luxury mole-cules in terminal members of a cell lineage. Such a posttranscriptional regulation can be affected either by stabilization of some mRNA species in the cytoplasm or by the breakdown of transcription products that have no significant role in the differentiated cell's metabolism. In both cases an accumulation of certain mRNA classes would take place in time. Aviv et al. (1976) tested this concept by labeling a stem of erythroleukemic Friend cells, induced to differentiate with dimethyl-sulfoxide, with [^{3}H]-uridine and then submitting them to a chase. After a given labeling period 2-4 % of the labeled poly (A)-containing RNA was in globin RNA; it decayed during the chase with a half-life of 16-17 h. The rest of the poly(A) RNA (the main component) was composed of two kinetic populations; 85-90 % decayed with a half-life of about 3 h, while 10 % decayed with a half-life of 37 h. The portion of globin RNA behaved in an unexpected manner during the chase. The percentage of globin RNA increased rapidly at first, reaching a maximum of about 15 % at 20 h, but subsequently declined gradually. A similar obser-vation was published by Lodish and Small (1976): Maturation of reticu-locytes is accompanied by a selective reduction in the synthesis of a polypeptide relative to globin. This reduction is correlated with a reduction in the amount of translatable mRNA for this polypeptide.

F. Differentiation in Monocellular Organisms

In contrast to multicellular organisms, their differentiation takes place within the life cycle of a single cell. Protozoa are both a cell and an organism. In metazoa, differentiation leads to the emergence of some stem cells for cell lineages, terminating in "differentiated" cells synthesizing luxury molecules that are only of use for the whole organism, not for the synthesizing cell itself. It seems that differentiation in metazoa and cell differentiation in protozoa may be different processes not directly comparable.

The life cycle of *Acanthamoeba castellanii*, a small earth ameba, consists of a growth and proliferation phase (trophozoite) as well as an encystment and excystment phase (Weissman, 1976). From a morphologic view point, encystment can be understood as cell differentiation, because we can see a condensation of cytoplasm and formation of a thick cell wall. Biochemically, encystment is characterized by the synthesis of cellulose. This molecule is typical of a certain cell stage, but not of a definite cell type. The term luxury molecule can in this case be used only with reservation. Encystment is brought about by inhibitors of growth (ethidium bromide) or simply by transfer of amebae into a medium free of nutritive material. The question is, whether for the synthesis of a carbohydrate some previously blocked genetic information is made available to the cell in the stage of encystment. Encystment is not a consequence of mitosis. After transfer of amebae into an encystment medium the activity of cellulose synthase both rises and steadily increases and the insertion of cellulose into the wall is initiated. Since actinomycin D and cycloheximide inhibit both functions, Potter and Weisman (1972) concluded that a stage-specific protein synthesis dependent on new mRNA, initiates the formation and construction of cysts. This points to a selective gene activation in the beginning of encystment. Some years ago, Mrs. Jantzen, a member of our staff, investigated the information value of this RNA. Her first results obtained with the DNA-RNA hybridization method seemed to confirm this opinion, because transcription of new RNA species could be found. However, problems arose, because it appeared that many more additional DNA sequences were transcribed during encystment than in the trophozoites (Jantzen, 1973). Furthermore, it could be shown that the amount of transcription was dependent on the method with which encystment was induced; it fluctuated between 90 % of total DNA sequences and none at all, when encystment was induced by ethidium bromide (Fig. 16). Finally, the main part of the excess new RNA species does not belong to the class of poly(A) RNA. These observations demonstrate that there is no causal relationship between excessive increase in transcription and encystment (Jantzen, 1974, 1973). Naturally, very small amounts of specific messengers for cellulose synthase could be present among the newly synthesized RNA. These can be detected only by rehybridization with cDNA. This is not possible, however, because highly purified cellulose synthase is not presently available. An investigation of translation capacity during encystment led to an interesting result. It could be shown that the incorporation rate of radioactive amino acids into a cell-free protein-synthesizing system, prepared from amebae before and after the beginning of encystment, decreases to nearly one-tenth of the initial value just when the cellulose synthase rises. The polysome-monosome ratio decreases also at this time. The relative inactivation of translation capacity is nonspecific and is not caused by a change in the amount of only that mRNA that is able to translate. All mRNA species prepared from amebae in different encystment stages have been active in translation and show no remarkable differences. It is astonishing that despite a decrease in the translation activity the cellulose-synthase activity

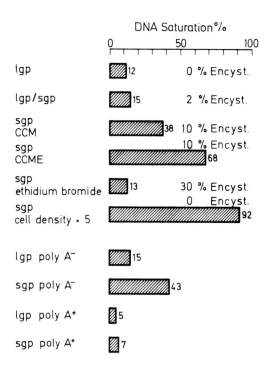

Fig. 16. DNA saturation with RNA synthesized by ameba (*Acanthamoeba castellanii*) under different culture conditions and growth phases of the cells in correlation with and without encystment. *lgp*, logarithmic growth phase; *sgp*, stationary growth phase; *CCM*, complete culture medium; *CCME*, minimal culture medium inducing encystment; poly A[+] or poly A[-], isolated from whole RNA. (From Jantzen, pers. commun.)

steadily increases after the beginning of encystment. In view of these results, it is highly unlikely that encystment is a problem of gene activation. Much more promising is the assumption of a cytoplasmic control mechanism involved in cell determination. Deichmann and Jantzen (1977) showed that at the very moment, that cellulose synthase activity is detectable, cellulase activity decreases to more than 90 % of their initial activity. It is plausible that cellulose synthase is active during the whole life cycle of amebae, but an accumulation of cellulose is possible only from the moment when cellulase activity is reduced at the beginning of encystment. The ameba falls asleep during encystment, but the single functions decrease with different velocities; the cell remains the same. Is encystment a case of differentiation?

G. The Problem of Dedifferentiation

The foregoing discussion has shown that the concept of a quantal cell cycle is not sufficient to explain all the phenomena of differentiation in terms of molecular biology. This situation is not astonishing, because too many different processes fall under the term differentiation and morphogenesis. We speak of differentiation when an ameba is transformed into a cyst or a chondroblast into a chondrocyte. However, there is a considerable difference between these two developmental steps. A cyst reverts to an ameba at once if the culture medium is changed into a growth-promoting medium, but the differentiated state of a myoblast is considerably stable. Possibly the first step, the adjustment of stem cells or areas of organ-anlagen, so far understood as a determination, is in reality the decisive differentiation step; the subsequent steps leading to the visible expression of the differentiated state may be a problem of cell metabolism or regulation of synthesis, especially concerning macromolecules. A specific substance that inter-

feres with a real differentiation step seems to be bromodeoxyuridine
(BUdR). Its depressive action on the expression of terminal developmental steps is reversible, but early developmental steps like gastrulation of the sea urchin egg are irreversibly blocked (Tencer and
Brachet, 1973). Early somites that are grown for two days in a BUdR-
medium and then transferred into normal medium never develop into myogenetic or chondrogenetic cell lineages (Holtzer et al., 1972). In
my opinion, only these fundamental developmental steps depend on activation of different gene sets and only these are irreversibly blocked
by BUdR. The high grade of stability of the determinated state supports
this argument. However, the stability of the determinated state is not
absolute - transdetermination is possible, but only as a mistake after
a prolonged series of mitotic cell divisions. This phenomenon is not
restricted to imaginal disks in Dipters. Metaplasia of human organs
is a well-known phenomenon in pathology.

After certain conditions of culture or regeneration cells can lose
their differentiated phenotypic character and assume a "blast"-like
appearance. This phenomenon is usually called dedifferentiation or
modulation. The modulatory changes can be considerable, but are restricted to the attributes of cell lineages.

A special case of dedifferentiation is cancer. It is commonly considered to be a heritable alternation of cells and is of permanent
and irreversible nature. It is thought that cancer may be a somatic mutation
of nuclear genes in the form of deletions or a stable reorganization
of genetic formation. Recently, however, accumulating evidence indicates that a return to normal growth is possible (Braun, 1968).
There are probably many types of tumors that do not have any similarities other than uncontrolled growth. Pathologists know that the organs
that most often develop cancer normally show periodic mitotic activity, such as mammary or prostate glands, or are defect regenerates, e.
g., the gastric mucosa after multiple healings of ulcer. In this case
the development of the cancer often begins as a metaplastic change of
gastric mucosa into mucosa of duodenum and into mucosa of small intestine with goblet cells and Paneth cells. This metaplasia is finally
the cause of a cancer. There is a striking similarity between metaplasia and transdetermination of imaginal disks in insects. In both
cases reversal to normal tissue is possible under certain conditions.
Following subcutaneous injections of carcinogenic hydrocarbons into
newts, Seilern-Aspang and Kratochwil (1962) observed the growth of
malignant tumors arising from the basal cells of the mucous glands of
skin. Such tumors infiltrated normal tissues, metastasized to other
organs of the newts, and in many cases, finally caused, the death of
the animals. Occasionally, however, a large infiltrating tumor rever-
ted to normal tissue. Such recovery could readily be brought about
when the carcinogen was injected into a region close to the tail base
and a part of the tail was afterward removed. The newts are well known
for their capacity to regenerate. The stimulation of the regenerative
processes apparently caused a reversal of the cancerous state. In this
case the tumor appears as a "modulation" of epidermic cells, induced
by environmental conditions as a regressive but reversible phenotype
in the spectrum of possible forms of expression within a determinated
cell lineage. It seems that the most important object of future research in the field of developmental biology should be the elucidation
of the processes leading to the formation of stem cells from which the
organ-anlagen arise. Let me finish with a sentence from J.W. von Goethe:

"Den Stoff sieht jedermann vor sich,
den Gehalt findet nur der, der etwas dazu zu tun hat
und die Form ist ein Geheimnis den meisten."

(From: Maximen und Reflexionen)

References

Aviv, H., Voloch, Z., Bastos, R., Levy, Sh.: Biosynthesis and stability of globin mRNA in cultured erythroleukemic Friend cells. Cell $\underline{8}$, 495-503 (1976)

Brachet, J., Hubert, E.: Studies on nucleocytoplasmic interaction during early amphibian development. I. Localized destruction of the egg cortex. J. Embryol. Exp. Morph. $\underline{27}$, 121-145 (1972)

Braun, A.C.: The multipotential cell and the tumor problem. In: The Stability of the Differentiated State. Results and Problems in Cell Differentiation. U. Ursprung (ed.). $\underline{1}$, 128-135 (1968)

Briggs, R.: Further studies on the maternal effect of the o-gene in the Mexican axolotl. J. Exp. Zool. $\underline{181}$, 271-280 (1972)

Briggs, R., Justus, J.T.: Partial characterization of the component from normal eggs with corrects the maternal effect of gene o in the Mexican axolotl (*Ambystoma mexicanum*). J. Exp. Zool. $\underline{167}$, 105-116 (1968)

Britten, R.J., Davidson, E.: Genetic regulation in higher cells: A theory. Science $\underline{165}$, 349-357 (1969)

Brothers, A.J.: Stable nuclear activation dependent on a protein synthesized during oogenesis. Nature (London) $\underline{260}$, 112-115 (1976)

Buckingham, M.E., Caput, D., Cohen, A., Whalen, R.G., Gros, R.G.: The synthesis and stability of cytoplasmic messenger RNA during myoblast differentiation in culture. Proc. Natl. Acad. Sci.U.S. $\underline{71}$, 1466-1470 (1974)

Bünning, E., Biegert, F.: Die Bildung der Spaltöffnungsinitialen bei *Allium cepa*. Z. Botanik $\underline{41}$, 17-39 (1953)

Carlson, J.G.: Microdissection studies of the dividing neuroblast of the grasshopper, *Chortophaga viridifasciata* (De Geer). Chromosoma (Berl.) $\underline{5}$, 199-220 (1952)

Cather, J.N.: Cellular interactions in the regulation of development in annelids and mollusca. Advan. Morphogenesis $\underline{9}$, 67-125 (1971)

Crouse, H.V., Keyl, H.-G.: Extra replications in the "DNA-puffs" of *Sciara coprohila*. Chromosoma (Berl.) $\underline{25}$, 357-364 (1968)

Davidson, E.H.: Gene Activity in Early Development. New York: Academic Press 1968

Davidson, E.H., Britten, R.J.: Note on the control of gene expression during development. J. Theoret. Biol. $\underline{32}$, 123-130 (1971)

Deichmann, U., Jantzen, H.: Das Cellulaseenzymsystem während Wachstum und Entwicklung von *Acanthamoeba castellanii*. Arch. Microbiol. (in press)

Duspiva, F., Scheller, K., Weiß, D., Winter, H.: Ribonucleinsäuresynthese in der telotroph-meroistischen Ovariole von *Dysdercus intermedius* Dist. (Heteroptera, Pyrrhoc.). Wilhelm Roux' Archiv $\underline{172}$, 83-130 (1973)

Gehring, W.: The stability of the determined state in cultures of imaginal disks in *Drosophila*. In: The Stability of the Differentiated State. Results and Problems in Cell Differentiation. Ursprung, H. (ed.). $\underline{1}$, 136-154 (1968)

Gross, P.R.: The control of protein synthesis in embryonic development and differentiation. In: Curr. Topics Develop. Biol. New York: Academic Press 1967, Vol. $\underline{2}$, pp. 1-46

Groudine, M., Holtzer, H.: Lineage-dependent transcription of globin genes. Cell $\underline{3}$, 243-247 (1974)

Gurdon, J.B.: Changes in somatic cell nuclei inserted into growing and maturing amphibian oocytes. J. Embryol. Exp. Morphol. $\underline{20}$, 401-414 (1968)

Gurdon, J.B.: Nuclear transplantation and the cyclic reprogramming in gene expression. In: Cell Cycle and Cell Differentiation. Results and Problems in Cell Differentiation. Reinert, J., Holtzer, H. (ed.). $\underline{7}$, 123-131 (1975)

Gurdon, J.B.: Transplanted nuclei and cell differentiation. Scient. American $\underline{219}$, 24-35 (1968)

Gurdon, J.B., Laskey, R.A.: The transplantation of nuclei from single cultured cells into enucleate frogs' eggs. J. Embryol. Exp. Morphol. $\underline{24}$, 227-248 (1970)

Gurdon, J.B., Woodland, H.R.: The cytoplasmic control of nuclear activity in animal development. Biol. Rev. $\underline{43}$, 233-267 (1968)

Hadorn, E.: Differenzierungsleistungen wiederholt fragmentierter Teilstücke männlicher Genitalscheiben von *Drosophila melanogaster* nach Kultur in vivo. Develop. Biol. $\underline{7}$, 617-629 (1963)

Hadorn, E.: Kontanz, Wechsel und Typus der Determination und Differentierung in Zellen aus männlichen Genitalanlagen von *Drosophila melanogaster* nach Dauerkultur in vivo. Develop. Biol. 13, 424-509 (1966)

Holtzer, H., Weintraub, H., Mayne, B., Mochan, B.: The cell cycle, cell lineages and cell differentiation. In: Curr. Topics Develop. Biol. Moscona, A.A., Monroy, A. (eds.) New York: Academic Press 1972, pp. 229-256

Hörstadius, S.: Über die Determination des Keimes bei Echinodermen. Acta Zool. (Stockholm) 9 (1928)

Humphries, S., Windass, J., Williamson, R.: Mouse globin gene expression in erythroid and non-erythroid tissues. Cell 7, 267-277 (1976)

Jantzen, H.: Änderung des Genaktivitätsmusters während der Entwicklung von *Acanthamoeba castellanii*. Arch. Microbiol. 91, 163-178 (1973)

Jantzen, H.: Die Entwicklung von *Acanthamoeba castellanii* zur Cyste mit und ohne Veränderung des Genaktivitätsmusters. J. Protozool. 21, 791-795 (1974)

Jantzen, H.: Über den Informationswert neuer Transkriptionsprodukte während der Entwicklung von *Acanthamoeba castellanii*. Stuttgart: Verh. Dtsch. Zool. Ges. 1974 (1975)

Kühn, A.: Vorlesungen über Entwicklungsphysiologie, 2nd ed. Berlin, Heidelberg, New York: Springer 1965

Lodish, H.F., Small, B.: Different lifetimes of reticulocyte messenger RNA. Cell 7, 59-65 (1976)

Pavan, C., Da Cunha: Chromosomal activities in *Rhynchosciara* and other Sciaridae. Ann. Rev. Genet. 3, 425-450 (1969)

Pflugfelder, O.: Lehrbuch der Entwicklungsgeschichte und Entwicklungsphysiologie der Tiere. Jena: VEB Gustav Fischer 1962

Potter, J.L., Weissman, R.A.: Correlation of cellulose synthesis in vivo and in vitro during the encystment of *Acanthamoeba*. Develop. Biol. 28, 472-479 (1972)

Rappaport, R.: Cytokinesis in animal cell. Intern. Rev. Cytol. 31, 169-211 (1971)

Robinson, K.R., Jaffe, L.F.: Polarizing fucoid eggs drive a calcium current through themselves. Science 187, 70-72 (1975)

Robinson, K.R., Jaffe, L.F.: Calcium gradients and egg polarity. Cell Biol. 70, 37a (1976)

Samarina, O.P., Lucanidin, E.M., Molnar, J., Georgiev, G.P.: Structural organization of nuclear complexes containing DNA-like RNA. J. Mol. Biol. 33, 251-263 (1968)

Seilern-Aspang, F., Kratochwil, K.: Induction and differentiation of an epithelial tumour in the newt (*Triturus christanus*). J. Embryol. Exp. Morphol. 10, 337-356 (1962)

Spek, J.: Das pH in der lebenden Zelle. Kolloid Z. 85, 162-170 (1938)

Spirin, A.S.: On "masked" forms of messenger RNA in early embryogenesis and in other differentiating systems. In: Curr. Topics Develop. Biol. New York: Academic Press 1966, Vol. 1, pp. 2-36

Steward, F.C.: From cultured cells to whole plants: The induction and control of their growth and differentiation. Proc. Roy. Soc. (London) Ser. B 175, 1-30 (1970)

Stossberg, M.: Die Zellvorgänge bei der Entwicklung der Flügelschuppen von *Ephestia kühniella*. Z. Morphol. Ökol. Tiere 34, 173-206 (1938)

Tencer, R., Brachet, J.: Studies of the effect of bromodeoxyuridine (BUdR) on differentiation. Differentiation 1, 51-64 (1973)

Verdonk, N.H.: The relation of the two blastomeres to the polar lobe in *Dentalium*. J. Embryol. Exp. Morphol. 20, 101-105 (1968)

Weiss, P.A.: Differentiation in retrospect. Differentiation 1, 3-10 (1973)

Weissman, R.A.: Differentiation in *Acanthamoeba castellanii*. Ann. Rev. Microbiol. 30, 189-219 (1976)

Winter, H., Wiemann-Weiss, D., Duspiva, F.: Endogene Synthese kurzlebiger Messenger-RNA-Spezies in der Oocyte von *Dysdercus intermedius* Dist. nach Abschluß der Vitellogenese. Roux' Archiv (in press)

Winter, H.: Ribonucleoprotein-Partikel aus dem telotrophmeroistischen Ovar von *Dysdercus intermedius* Dist. (Heteroptera, Pyrrhoc.) und ihr Verhalten im zellfreien Proteinsynthesesystem. Wilhelm Roux' Archiv 175, 103-127 (1974)

Zepf, E.: Über die Differenzierung des Sphagnumblattes. Z. Botanik 40, 87-118 (1952)

Discussion Session VI: Mitosis in Differentiation, Morphogenesis, and Cancer

Chairman: E. SCHNEPF, Heidelberg, F.R.G.

I. Orientation of Spindle Axis and Its Role in Differentiation

J.G. Carlson, Knoxville, Tennessee, USA

I should like to comment on polarity, unequal division, and experimental induction of equal division in the grasshopper neuroblast. This cell maintains a constant, identifiable polarity throughout the cell cycle. It is concavoconvex in form from middle telophase to the time in late prophase when it rounds up just before membrane breakdown. The nucleus is doughnut-shaped with a central cytoplasmic core in which the centrioles are situated. The caps of the prophase spindle coincide with the caps of this core and are in line with the ganglion cells previously formed by that neuroblast. When cleavage occurs, the small daughter ganglion cell (about 3/5 the diameter of the large daughter neuroblast) forms contiguous with its sister ganglion cells, which make up a column extending toward the center of the ventral nerve cord. At anaphase the spindle normally shifts toward these ganglion cells as the neuroblast elongates, so that the cleavage furrow, which forms approximately midway between the sets of kinetochores, divides the cell unequally. It is possible, however, to induce equal divisions experimentally. One way is to prevent the spindle from shifting by holding it away from the ganglion cell side of the neuroblast with a microneedle, so that the cleavage furrow passes through the center of the cell. Another way is to decrease the size of the spindle by treatment with colchicine or 225 nm ultraviolet radiation to a length and diameter approaching half its normal size, after which it will frequently undergo equal division. Further, it was found by Kawamura that if the neuroblast is mechanically separated (by squashing) from its cap cells, which lie against the convex side of the neuroblast opposite the ganglion cells, division is nearly always equal. On the other hand, if the cells are chemically separated by trypsin, division is equal. This is not affected by the addition of trypsin-inhibitor, indicating that the effective part of the trypsin is not the proteolytic portion of the trypsin molecule.

H.W. Sauer, Würzburg, F.R.G.

With respect to experimental disorientation of the mitotic spindle, does this manipulation change the developmental track of the daughter cells?

J.G. Carlson, Knoxville, Tennessee, USA

When division is equal, both daughter cells are morphologically like neuroblasts rather than like ganglion cells, but we have not determined their behavior in subsequent divisions.

B. Beetz, Saarbrücken, F.R.G.

How long before anaphase-movement do you have to separate the cap cell from the neuroblast cell mechanically to get an equal cleavage?

J.G. Carlson, Knoxville, Tennessee, USA

The cell separation you refer to was done many years ago and I do not remember exactly the stages at which it was effective. I believe that equal division could be induced by cell separation at late prophase, metaphase, and possibly early anaphase.

N. Paweletz, Heidelberg, F.R.G.

The beautiful example of the formation of scales in butterflies clearly demonstrates that differentiation is related to mitosis. The direction of the long axis of the spindle of the scale mother cell is turned to a definite degree. This brings us to the question of what is responsible for the direction of the spindle and exactly what is the role of the spindle poles. There must be cooperation between neighboring cells. If the neighborhood plays a role, then there should be a transition from one cell to the next. How can this occur? For instance, through the desmosomes or by some other mechanisms. Do we know enough about the transfer of proteins or large molecules through the cell membrane from one cell to the other?

F. Duspiva, Heidelberg, F.R.G.
It is well known that in multicellular systems the cell membrane is permeable to
molecules of medium molecular weight. So, it is possible that proteins of small
molecular weight can be exchanged between two cells. This process is dependent on
the Ca^{2+}-ion concentration.

P.F. Baker, London, England, U.K.
I know of no experimental evidence for the transfer of molecules of molecular
weight greater than 1000 across low resistance junctions. These experiments have,
in general, used fluorescent molecules . In embryonic material, the work of
Warner suggests that the junctions may restrict the movement of molecules
of even smaller molecular weight.

H.W. Sauer, Würzburg, F.R.G.
There is good evidence that in some cases, where purified inducing molecules have
been immobilized on plastic beads and therefore could not enter the cells, "induction" of competent cells is triggered. (Experiments on embryonic induction in
amphibia, by Tiedemann and others). Therefore, in general terms, embryonic induction may be a *permissive* rather than an *instructive* event.

N. Paweletz, Heidelberg, F.R.G.
There may be a transfer of smaller molecules or ions. Can they act as determinants?
Even if we assume that large molecules cannot enter the neighboring cells, a signal
must be delivered from one cell, must reach the other, and perhaps be transformed
at the cell surface and then be incorporated to affect the neighboring nucleus or
cytoplasm to trigger it to have, e.g., an unequal cell division.

O.G. Meyer, Amsterdam, The Netherlands
Did not the work on gap junctions indicate that circumstantial evidence exists
concerning the role of intercellular communications and the difference between
normal and malignant cells?

W.W. Franke, Heidelberg, F.R.G.
Although there appears to be a tendency to decrease the number of gap junctions in
some tumors this has been shown not to be the case in others. Recent literature
indicates that no generalization as to a decrease of gap junctions in tumor cells
can be made.

II. Mode of Action of Determinants in Morphogenesis and the Morpho-
genetic Capacity of Nuclei During Development

J.H. Frenster, Atherton, California, USA
Dr. Duspiva has mentioned the important opportunity for the reprogramming of gene
expression during mitosis by the removal of epigenetic control molecules, such
as acidic proteins and RNA from the nucleus in metaphase, and the return of such
molecules from the cytoplasm in telophase.
Since different molecules may return to the chromosomes in the daughter cells than
were present in the maternal cells, the opportunity presents itself for reprogramming gene expression. We have shown that DNA helix openings persist through mitosis
in *neoplastic* cells but *not* in *normal* human cells. If such DNA helix openings are
preserved by the binding of control molecules, then the persistence of the DNA
helix openings in neoplastic cells suggests that the neoplastic cells *cannot* reprogram themselves during mitosis.

K.B. Moritz, München, F.R.G.
Dr. Duspiva has not mentioned those cases of chromosomal differentiation during
development, in which qualitative genome DNA alterations take place. Indeed, there
are many different kinds of chromosomal DNA restriction phenomena in normal development, such as elimination of L chromosomes in insects, diminution of germline
limited chromatin in nematodes and lower crustaceans, selective degradations of
single sex chromosomes as well as of whole haploid sets in soma and germline of
different insect species, DNA digestion in the macronucleus anlage of *Stylonychia*,
etc. There is the well-known differential overreplication of rDNA in oogenesis of
many organisms; during polytenization in different cell types satellites are under-
replicated.

During the diminution mitosis in *Ascaris* large amounts of heterochromatic regions are eliminated from the plurivalent germline chromosomes.

J. Bryan, Philadelphia, Pennsylvania, USA
Is there an increase in the number of kinetochores in the cells undergoing chromosome diminution?

K.B. Moritz, München, F.R.G.
During the diminution some 25 tiny single somatic chromosomes arise from the holokinetic germline sammel chromosome in *Parascaris equorum univalens*. It is still open whether these chromosomes have diffuse or localized kinetochores.

H. Sato, Philadelphia, Pennsylvania, USA
Two questions:
1. Are there any structural differences on kinetochores in prometaphasic or metaphasic somatic chromosomes vs. germline chromosomes?
2. Did you ever compare the state and fine structure in the holokinetic region on metaphase Lepidopteran chromosomes vs. *Ascaris* chromosomes?

K.B. Moritz, München, F.R.G.
We have not yet studied the mitosis in germline and presomatic cells by electron microscopy.

N. Paweletz, Heidelberg, F.R.G.
Endomitosis often happens in some small glands. Again, there must be an influence of the surrounding cells because endomitosis is an event restricted to definite areas within the organism. Therefore, it is obviously not, at least, directly, controlled by the whole organism.

M. Bopp, Heidelberg, F.R.G.
Endomitosis or better endopolyploidy in plants can occur spontaneously and can be induced by external factors. For example in *Lobularia maritima* leaf cells that have been grown in a more highly concentrated salt solution, the degree of polyploidy is greater than in those grown with no salt. In many seedlings darkness, which stimulates the growth of the stem axis, at the same time enhances the number of cells with a higher polyploidy. This means that in plants the endomitosis is not only induced by the neighborhood of certain cells, but more so by different external factors.

III. The Quantal Cell Cycle

D. Mazia, Berkely, California, USA
In a broad sense, it is hard to dispute the idea of quantal mitosis. The classical paradox is that mitosis produce daughter cells identical with the mother cell. If there is to be differentiation, there must be mitoses that produce daughters that are different from their mothers: Such mitosis would represent distinct steps in development. We need, then, to take a closer look at such events. Do they follow from a certain number of rounds of chromosome replication or must there be separation of chromosomes and division of the cell at each round?

F. Duspiva, Heidelberg, F.R.G.
Quantal mitosis is not necessary if the daughter cells have the same properties as the mother cell. However, we find a differentiation between the daughter cells if the regulatory proteins of the mother cell are distributed unequally during mitosis.

J.F. López-Sáez, Madrid, Spain
Mitosis must be considered as an event in the differentiation program and not the cause that induced the cell to take this path.

P. Malpoix, Rhode-St-Genese, Belgium
The question we have to answer is whether in mitosis, the cell is programmed simultaneously a) for a finite number of all divisions or b) for specific protein synthesis.
The two events may be dissociable in time since we have shown that when cultures of precursor cells are stimulated by erythropoietin in the presence of bromodeoxyuridine, the earliest uncommitted cells can continue to divide but do not initiate

hemoglobin synthesis. When cytokinesis is blocked in erythropoietin-stimulated cells, hemoglobin synthesis is nevertheless induced; so again, the two events may be partially dissociable, at least in time. However, repeated inhibition of cyto-kinesis finally produces cells in which the capacity to divide has become un-limited, and the control of differentiation altered (see abstract of Poster 33). So somewhere, control mechanisms interfere when one or another process is disturbed.

W.W. Franke, Heidelberg, F.R.G.
Did you look for infection with viruses or induction of production of endogenous C-type viruses in your cells?

P. Malpoix, Rhode-St-Genese, Belgium
No exogenous virus has been added, but the possibility that endogenous virus has been activated cannot be excluded. Temin's hypothesis that RNA viruses evolved from cellular regulatory molecules should also be taken into account.

Session VII

Chromosome Movement: Facts and Hypotheses

R. B. NICKLAS, Department of Zoology, Duke University,
Durham, North Carolina 27706, USA

What follows is a brief and informal summary of my remarks at the
workshop. Fortunately, several recent reviews are available for additional information and references. These include Bajer and Molè-
Bajer, 1972; Forer, 1974; Inoué and Ritter, 1975; McIntosh et al.,
1975; Nicklas, 1971, 1975. Research reports from several laboratories
may be found in Goldman et al., 1976.

The usual focus of discussion is chromosome movement in anaphase, but
here prometaphase movement will be considered as well. At the workshop, recent results from our laboratory on spindle organization were
also presented: experimental studies of the role of chromosomes in
spindle organization (Marek, 1977), and electron microscopic studies
of the origin of chromosome association with the spindle (Brinkley,
Kubai, Nicklas, and Pepper, unpublished). These results will not be
considered further in this paper.

A. Anaphase Chromosome Movement

I. The Current Status of Hypotheses

Hypotheses of three general types are currently receiving the greatest
attention; the postulates and current status of each can be outlined,
on one person's view, as follows:

1. *Governor Plus Separate Motor (Forer, 1974; Nicklas, 1975)*

In this hypothesis, spindle microtubules are linked to one another to
form a skeleton, and the rate of change in microtubule length (the
"governor") limits the velocity of chromosome movement caused by a
mechanically separate force producer (the "motor"). There is good experimental evidence that length changes in birefringent elements,
probably microtubules, are velocity-limiting (see, e.g., Inoué and
Ritter, 1975; Salmon, 1975). An actin/myosin system is a popular candidate for the motor in hypotheses of this sort. The possible presence
of functionally significant actin within the spindle has been much
discussed during the workshop (Sanger and associated Discussion, this volume. The present evidence seems to me just sufficient to motivate further
work, and while others take a more positive view, all jurors return
the Scottish verdict "not proven."

2. *Governor-Motor-Microtubule Length Changes (e.g., Dietz, 1972; Inoué and Ritter, 1975)*

Here, changes in microtubule length not only determine the velocity
of chromosome movement but provide the motive force as well. For example, in chromosome-to-pole movement, the shortening of kinetochore
microtubules would occur by the loss of interstitial tubulin subunits
from previously intact microtubules. This would be followed by reconstitution of the original microtubule lattice through a process
that generates force. The experiments of Inoué and co-workers (e.g.,
Inoué and Ritter, 1975; Salmon, 1975) show that chromosomes can be
moved poleward following treatments known to depolymerize microtubules.

As already noted, these experiments provide clear evidence for microtubules as velocity-limiting elements, but they are unavoidably ambiguous on whether or not depolymerization produces the force. Thus
once the importance of microtubules as possible skeletal elements is
recognized, it is evident that depolymerization could merely *permit* a
separate motor to produce poleward movement (i.e., according to hypotheses of type 1) rather than *cause* the movement itself. The situation is clearer for the chromosome movement during experimentally
induced spindle elongation, which also has been studied by Inoué and
co-workers - here the movement probably is caused by microtubule polymerization (e.g., Inoué and Ritter, 1975; Salmon, 1975). In sum, the
results of these elegant and extensive experiments must be explained
by any satisfactory hypothesis of chromosome movement, but they do
not provide proof that microtubule polymerization/depolymerization
itself is the motor, especially for chromosome-to-pole movement.

3. Sliding (e.g., McIntosh et al., 1969)

Hypotheses of this type are distinguishable from those of type 1 if
the designation "sliding" is limited to systems truly comparable
mechanically to muscle and cilia/flagella: the motor (e.g., dynein-
like or myosin-like) is permanently attached to one microtubule and
is transitorily associated with another microtubule or filament (e.g.,
actin). The observed slow velocity of chromosome movement could result from motors that are less powerful or less numerous than in muscle or cilia/flagella, but it would likely result at least in part
from regulation of the rate of change in microtubule length, as in
hypotheses of the other two types. The current status of such hypotheses may be characterized as "lost innocence." The elegant McIntosh
et al. (1969) hypothesis probably must be discarded - something more
complex than a single motor, a single type of microtubule, and a
single type of motor/microtubule interaction is necessary (McIntosh
et al., 1976). A good example of the contrary evidence comes from the
reconstruction of the microtubule distribution in *Diatoma* spindles
by McDonald and co-workers (see McIntosh et al., 1976). However, the
same study provides, as did some earlier ones, clear, if circumstantial, evidence *favoring* sliding as a mechanism for spindle elongation.
So the sliding hypothesis is far from dead; it merely has lost its
former simplicity and, in some measure, testability - it is now no
better than other hypotheses in these regards. But it is no worse
either!

II. The Future

Some necessary advances are obvious to all of us; the question is
how to achieve them. First, in any hypothesis of chromosome movement
we must understand microtubule dynamics - the control of assembly/
disassembly and the sites of assembly and disassembly within the
spindle. Second, the motors must be identified and their position
and molecular interactions relative to other spindle components must
be determined. Surely the spindle is likely to contain a sample of
cellular contents that includes many components irrelevant to chromosome motion, so we need physiologic or structural (e.g., actin fiber
polarity) evidence for a functional role to recognize relevant components.

For progress toward these goals, we badly need a useful cell or spindle
model, and despite valiant efforts we do not yet have one. The obvious
criteria for utility are (1) the continuation of chromosome movement
despite unambiguous permeability of the model to ions and macromolecules, and (2) the capability to selectively extract and then add back
particular components, without secondary loss of function (e.g.,

through generalized destruction of spindle structure. The nearest approach to these criteria to date was made in the studies of Cande and co-workers (see McIntosh et al., 1975) using cells lysed by detergent in which chromosome movement was stopped by KCl extraction but was reinitiated after the addition of dynein. Unfortunately these results have proved difficult to reproduce and a satisfactory model system using detergent lysis has not yet been achieved (McIntosh, this volume). I have been studying a new model system entirely different from detergent lysis except that its utility is a promise not yet fulfilled. The new system involves mechanical demembranation of grasshopper spermatocytes. Anaphase movement continues and the model is unequivocally permeable, but the second criterion has not been satisfied. The method and the results to date are described in Nicklas, 1977.

B. Prometaphase Chromosome Movement

When the spindle forms, the chromosomes are scattered about, but eventually they come to lie exactly midway between the poles. This "congression" of chromosomes to the equator during prometaphase poses a challenge that seems even greater than movement during anaphase, but it cannot be evaded (and who would want to evade it anyway?). I wish to describe two sets of observations that shed some light on how the equatorial position is reached.

I. Induced Prometaphase Movement

The first set of observations is from micromanipulation experiments. The question asked is whether or not congression can be induced artificially by pushing on a chromosome lying off the equator with a micromanipulation needle. The rationale is that if chromosomes do congress with an artificially applied force, then force, not structure, is the primary determinant of position. More specifically, the alternatives are that the equatorial position is achieved (1) because the forces toward opposite poles on a mitotic chromosome or a bivalent in meiosis are in balance only when an equatorial position is reached (Östergren, 1950) or (2) because lengthening of kinetochore microtubules on the side toward the farther pole is balanced by shortening on the side toward the nearer pole; in this view, whatever the magnitude of the forces, length changes in microtubules are necessary for movement to occur. Put another way, is it primarily the regulation of forces or of microtubule length that produces congression?

Briefly, the experiment is performed as follows: a microneedle is placed against a bivalent in a grasshopper *(Melanoplus differentialis)* spermatocyte in culture, and the needle is then displaced until the bivalent is slightly stretched toward the farther spindle pole. The relative force applied is gauged from the strain (a measure of the amount of stretching induced), and the displacement of the chromosome toward the opposite pole is measured at various strains. The chief results from the eight most informative experiments can be summarized as follows:
1. Artificially applied force invariably and immediately produces chromosome displacement toward the farther pole. For technical reasons, the strain imposed at first is about two times greater than that which occurs during natural congression in these same cells, but is not grossly outside the normal range.
2. Usually, the resistance to displacement increases (the strain must be increased) with increasing distance of the chromosome from the pole it was nearer initially, as the bivalent is displaced from one side

of the equator to the other. Total displacements of 8-12 µm (over half
the spindle length) are easily achieved.
3. The attachment of the bivalent to the spindle (tested by prodding
with the microneedle) is maintained at all times. Most significantly,
the kinetochore fiber toward the (initially) farther pole *shortens* in
exact coordination with the lengthening of the fiber on the (initially)
nearer side (fiber length can be directly determined after appropriate
micromanipulation). Or alternatively, the whole unit - bivalent plus
attached short segments of microtubules toward opposite poles - is
gliding along some other spindle component (e.g., interpolar microtubules).

Conclusion. In these experiments at least, the equilibrium position of
chromosomes in prometaphase is determined by forces, not by the length
of spindle fibers or other structural constraints -structure seems always ready to yield to force, and yet the mechanical attachment of
the chromosome to the spindle is maintained without lapse.

II. Multivalents

Östergren (1945) long ago realized that the final position at metaphase
of multivalents with unequal numbers of chromosomes oriented to opposite poles, constituted a natural experiment to test his view of congression. For instance, trivalents should lie nearer the pole to which
two of the three chromosomes are oriented if poleward forces determine
the final position, whereas if the length of kinetochore microtubules
regulates position, such trivalents would be expected to lie at the
equator. Since Östergren's pioneering thoughts and observations, numerous additional observations have been made, some of which seemingly
support his conception and some of which seemignly do not. I wish to
suggest criteria for the least ambiguous observations on multivalents
and to mention the results to date which meet these criteria.

Criteria. (1) The multivalents should be of recent origin. (2) The multivalents must be studied in living cells or, if in fixed cells, only
where it is certain that the multivalent is observed in its final,
stable position at metaphase. A justification for these criteria will
be offered in a paper now in preparation.

Results. On the most stringent application of the second criterion (the
study of living cells) only five published examples are usable: four
crane fly spermatocytes with a trivalent (Bauer et al., 1961) and one
grasshopper spermatocyte with a mal-oriented quadrivalent (Wise and
Rickards, 1977). In all five cases, the multivalent lies off the
equator at metaphase nearer that pole to which the greater number of
kinetochores is oriented.

III. Implications

Prometaphase chromosome movement intrigues me especially when it is
compared with anaphase movement. Thus in anaphase, a role of kinetochore microtubule length in determining chromosome position, and thus
velocity, seems a natural assumption (e.g., as in hypotheses type
1 and 2 above), while in prometaphase congression, chromosome position
appears to be governed by opposed spindle forces, not the length of
kinetochore microtubules. I certainly do not wish to suggest that the
observations on prometaphase make any present hypotheses for anaphase
unlikely to be true, because the mechanisms may well differ in some
regards at least and also because the observations on prometaphase
need further scrutiny before the interpretation just offered is accepted as the final word.

My current preference among hypotheses to explain prometaphase congression is sliding in the sense used above. In this view, both force production and chromosome association with the spindle arise from active cross-bridges between kinetochore microtubules and other microtubules or filaments; congression occurs because the number of cross-bridges is proportional to the length of kinetochore microtubules (Luykx, 1970), but the velocity is controlled by the rate of kinetochore microtubule disassembly near the poles. Thus the rate of change in microtubule length would determine the *rate* at which the equilibrium position is reached but not the position itself.

C. Conclusion

Let us not be overly preoccupied with discovering the motor that drives anaphase movement, for in many ways this is one of the less interesting questions posed by chromosome movement. I urge an equal effort to understand the *regulation* of the motor(s), especially in the control of chromosome velocity and of congression in prometaphase.

Acknowledgments. I am grateful to Tom Hays for skillful technical assistance. Our studies were supported in part by research grant GM-13745 from the Division of General Medical Sciences, United States Public Health Service.

References

Bajer, A.S., Molè-Bajer, J.: Spindle dynamics and chromosome movements. Int. Rev. Cytol. Suppl. 3, 1-271 (1972)

Bauer, H., Dietz, R., Röbbelen, C.: Die Spermatocytenteilungen der Tipuliden. III. Das Bewegungsverhalten der Chromosomen in Translokationsheterozygoten von *Tipula oleracea*. Chromosoma (Berl.) 12, 116-189 (1961)

Dietz, R.: Die Assembly-Hypothese der Chromosomenbewegung und die Veränderungen der Spindellänge während der Anaphase I in Spermatocyten von *Pales ferruginea* (Tipulidae, Diptera). Chromosoma (Berl.) 38, 11-76 (1972)

Forer, A.: Possible roles of microtubules and actin-like filaments during cell-division. In: Cell Cycle Controls. Padilla, G.M., Cameron, I.L., Zimmerman, A.M. (eds.). New York and London: Academic Press 1974, pp. 319-336

Goldman, R., Pollard, T., Rosenbaum, J. (eds.): Cell Motility. Book C. Microtubules and Related Proteins. Cold Spring Harbor: Cold Spring Harbor Laboratory 1976

Inoué, S., Ritter, H.: Dynamics of mitotic spindle organization and function. In: Molecules and Cell Movement. Inoué, S., Stephens, R. (eds.). New York: Raven Press 1975, pp. 3-30

Luykx, P.: Cellular Mechanisms of Chromosome Distribution. Int. Rev. Cytol. Suppl 2, 1-173 (1970)

Marek, L.M.: Chromosome control of spindle form and function in grasshopper spermatocytes. Dissertation, Duke University, 1977. University Microfilms, Ann Arbor, Michigan

McIntosh, J.R., Cande, W.Z., Lazarides, E., McDonald, K., Snyder, J.A.: Fibrous elements of the mitotic spindle. In: Cell Motility. Book C. Microtubules and Related Proteins. Goldman, R., Pollard, T., Rosenbaum, J. (eds.). Cold Spring Harbor: Cold Spring Harbor Laboratory 1976, pp. 1261-1272

McIntosh, J.R., Cande, W.Z., Snyder, J.A.: Structure and physiology of the mammalian mitotic spindle. In: Molecules and Cell Movement. Inoué, S., Stephens, R. (eds.). New York: Raven Press 1975, pp. 31-76

McIntosh, J.R., Hepler, P.K, Van Wie, D.G.: Model for mitosis. Nature (London) 224, 659-663 (1969)

Nicklas, R.B.: Mitosis. Advances in Cell Biology 2, 225-297 (1971)

Nicklas, R.B.: Chromosome movement: current models and experiments on living cells. In: Molecules and Cell Movement. Inoué, S., Stephens, R. (eds.). New York: Raven Press 1975, pp. 97-117

Nicklas, R.R.: Chromosome distribution: experiments on cell hybrids and in vitro. Phil. Trans. Roy. Soc. Lond. B 277, 267-276 (1977)

Östergren, G.: Equilibrium of trivalents and the mechanism of chromosome movements. Hereditas 31, 498 (abstr.) (1945)

Östergren, G.: Considerations on some elementary features of mitosis. Hereditas 36, 1-18 (1950)

Salmon, E.D.: Spindle microtubules: thermodynamics of in vivo assembly and role in chromosome movement. Ann. N.Y. Acad. Sci. 253, 383-406 (1975)

Wise, D., Rickards, G.K.: A quadrivalent studied in living and fixed grasshopper spermatocytes. Chromosoma (Berl.). In press (1977)

Discussion Session VII: Chromosome Movement: Facts and Hypotheses
Chairman: D. SCHROETER, Heidelberg, F.R.G.

Chairman:
Before the lecture of B. Nicklas, films were shown by
1. R. Wolf, Würzburg, F.R.G.
   Embryonic development of the gall midge *Wachteliella*;
2. H. Sato, Philadelphia, Pennsylvania, USA
   Meiosis I in grasshopper spermatogenesis;
3. U.-P. Roos, Zürich, Switzerland
   Spindle formation and chromosome orientation at early prometaphase in rat kangaroo
   cells;
4. A.S. Bajer, Eugene, Oregon, USA
   Lateral changes in birefringence during mitosis in *Haemanthus* endosperm cells.

These contributions, thematically connected to the lecture, should be discussed
first.

T.E. Schroeder, Friday Harbor, Washington, USA
Dr. Sato, this is perhaps theoretical, but is there any way of explaining wave-
like motion and changes in birefringence other than by explaining it as movement
of parts. Could there be changes in the polarization properties rather than translocation of structures?

H. Sato, Philadelphia, Pennsylvania, USA
Spermatogenic cells of the grasshopper in meiosis I were immersed in FC-47 fluor-
ocarbon oil to prevent possible hypertonicity and were recorded by time-lapse
cinematography. The prometaphase spindles were not stable structures. The birefringent spindle fibers swam and swung like northern light flickers. This could
partly be due to the possible aggregation and disaggregation of spindle microtubules. But knowing the labile but dynamic nature of the mitotic spindle, I am inter-
preting this phenomenon as a process to establish a local and overall equilibrium
in metaphase. Spindle microtubules may be assembled and disassembled locally
during prometaphase. However, the spindle becomes stable at full metaphase and
poles show little oscillatory movement during anaphase.
The behavior of prometaphase spindles in cultured salamander epithelium can be
explained in the same way. However, exaggerated movement such as the migration or
rotation would be assisted by the actin bundles that are distributed between the
outer spindle and the cell membrane.

A.S. Bajer, Eugene, Oregon, USA
The short film in polarized light on *Haemanthus* endosperm (made with Dr. H. Sato
and Dr. Molè-Bajer) demonstrates that lateral changes of birefringence (northern
lights phenomenon - Inoué and Bajer, Chromosoma, 1961) are correlated with lateral
movements of structures inside the spindle, such as mitochondria (exceptionally
long) and long chromosome arms. Lateral change of birefringence is even more
clearly evident in grasshopper spermatocytes, as demonstrated by Dr. Sato. In
plant cells such movements are exceptionally well seen in phragmoplasts. The
simplest explanation of the movements is the assumption that microtubules change
their arrangement laterally. I would like to stress, however, that such an interpretation is in full agreement with Dr. Sato's statement that the movements "reflect continuous adjustments of equilibrium" in the spindle.
I agree with Dr. Sato that the "northern light phenomenon" (lateral change of
birefringence) may reflect "equilibrium adjustments." On the electron microscopic
level this is seen as pronounced changes of microtubule arrangement, especially
clearly seen in the phragmoplast (Lambert and Bajer, Chromosoma, 1972). Progressive change of the shape of microtubules (progressive bending of microtubules) is
well demonstrated at various stages of phragmoplast formation and in various regions of the phragmoplast. Such change of arrangement may be interpreted as (1)
continuous disassembly followed by assembly of microtubules at different positions,
(2) change of microtubule arrangement, or (3) these two processes combined. I feel
that assumption No. 1 has a very low thermodynamic probability and I am inclined

to interpret the changes according to scheme 3, with the restriction that at least in some cases the rearrangement of microtubules is the major factor.

Chairman:
We thus observe a gradual change in spindle motility when the cell passes from prometaphase into metaphase. As the film of Roos demonstrated, the placement of chromosomes in metaphase is dependent upon their orientation to the poles during spindle formation in prometaphase.

U.-P. Roos, Zürich, Switzerland
I would like to add some information about spindle formation and chromosome orientation at the beginning of prometaphase in rat kangaroo cells.
The first slide (Fig. 1) shows some pictures taken from a film sequence of mitosis in living rat kangaroo cells, and I want to draw your attention to the fact that during early prometaphase some or many chromosomes may lie near the spindle poles and that they usually all congress later on.
As Dr. Nicklas mentioned, Östergren presented a hypothesis for chromosome orientation. I should like to introduce a few terms coined by Bauer et al. (Chromosoma 12, 116-189, 1961). A mitotic chromosome normally is in amphitelic orientation, i.e., its two kinetochores are oriented toward opposite poles. In certain cases, however, both kinetochores may be oriented toward the same pole, and this is called *syntelic* orientation. During my investigations of rat kangaroo cells, I have found another type of orientation, where only one kinetochore is oriented and I called this *monotelic* orientation. Eventually, such chromosomes achieve *amphitelic* orientation prior to congression.
In thin sections of cells (Fig. 2) fixed shortly after the nuclear envelope began to disintegrate, one can observe chromosomes whose kinetochore region is about equidistant from the poles. Such chromosomes are amphioriented. Other chromosomes lie close to a pole and they are monooriented, i.e., only the kinetochore facing the nearer pole bears microtubules running toward the pole. My interpretation is that such chromosomes necessarily move toward this pole, because there is no counterforce applied to the unoriented kinetochore.
The second point I consider important is that, in rat kangaroo cells at least, the spindle is poorly organized at the beginning of prometaphase (Fig. 3). In thin sections of cells fixed in very early prometaphase, the asters are very large, there are relatively few microtubules in the area between the poles, and fragments of the nuclear envelope partially envelope the chromosomes. Thus, the situation for the chromosomes is not such that they are "placed" in a well-organized spindle framework. Perhaps I will have the opportunity later to present a model attempting to explain how initial chromosome orientation is achieved and in what kind of chromosome behavior it results.

G.K. Rickards, Durham, North Carolina, USA
I would like to ask Dr. Borisy whether he has any evidence, or would like to speculate on the question, as to whether in his in vitro system in which microtubules are polymerized onto kinetochores, the growing end of the microtubule is proximal to the kinetochore or distal to it. Intuitively one might expect it to be the distal end. The point is, however, that there is some cytologic evidence, and some hypotheses of chromosome movement, that suggest that polymerization might in fact be taking place at the kinetochore. The question is of obvious relevance in relation to the "forced congression" experiments described by Dr. Nicklas. Thus, does the kinetochore fiber that "backs away" from one pole slide through the rest of the spindle, with microtubule growth at the distal end; or does growth proceed by addition of subunits at the kinetochore?

G.G. Borisy, Madison, Wisconsin, USA
This is a very important question. However, we have no answer; we are trying to do the experiment. In principle these experiments are straightforward: It is now possible to obtain tubulin assembly onto chromosomes in vitro, and it is possible to decorate these tubules. In principle, all the ingredients to do the experiment are on hand.

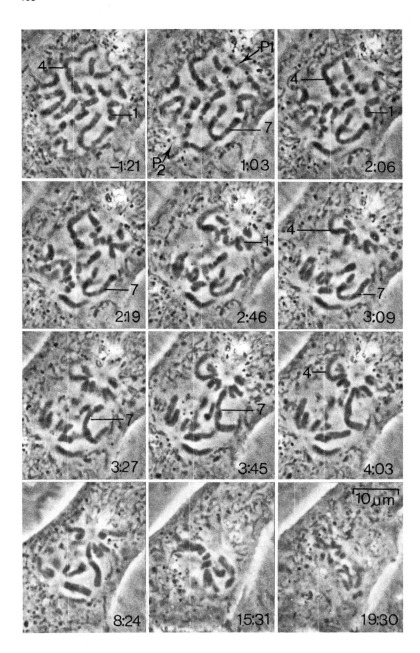

Fig. 1. Late prophase (-1:21), prometaphase (1:03 to 15:31), and anaphase (19:30).
Primary constriction and sister chromatids of several chromosomes in favorable positions are recognizable in the prophase nucleus (-1:21). Chromosomes 1 and 4 move
to and associate with the proximal pole (P_1) (2:06 to 2:46 and 2:19 to 3:45, respectively). Chromosome 7 illustrates movement to and association with the distant pole
(3:09 to 3:45). By late prometaphase all the chromosomes have congressed (15:31)
and normal anaphase ensues (19:30)

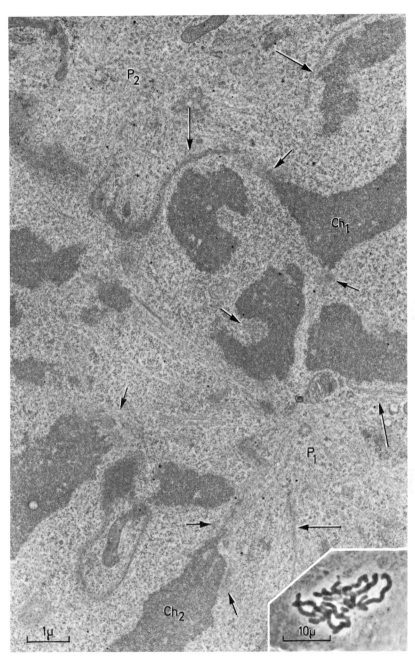

Fig. 2. Very early prometaphase. The NE (*long arrows*) is discontinuous near the two asters, whose centers are marked P_1, P_2 (the centrioles at P_2 were located in other serial
sections). One chromosome (Ch_1) lies approximately midway between the poles and both its
kinetochores (*short arrows*) appear oriented and stretched. Another chromosome (Ch_2) has
one kinetochore stretched and oriented toward pole no. 1 (P_1); the sister kinetochore
is unoriented (*short arrows*). Note the structure and position of other kinetochores
(*short arrows*), distribution and direction of MTs. *Black dots* are stain marks. *Inset:*
Phase contrast micrograph of the same cell in plastic

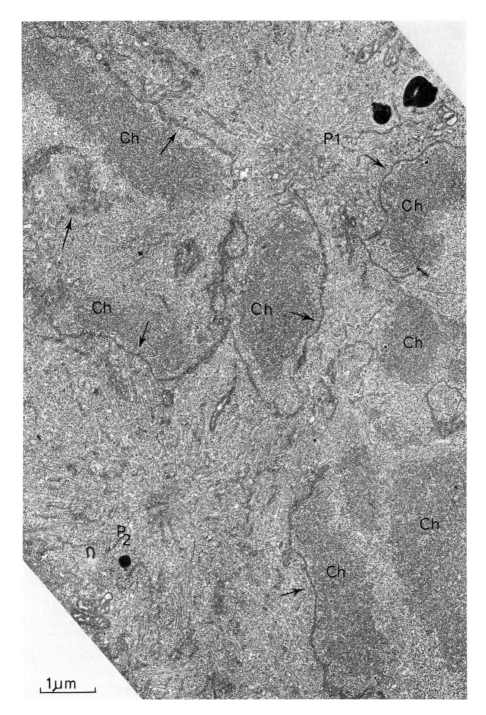

Fig. 3. Approximately median longitudinal section of the mitotic apparatus of a pro-metaphase cell fixed 3 min after the nuclear envelope began to break down. Astral MTs are numerous, but relatively few MTs occur in the central area of the spindle. Large fragments of the nuclear envelope partially wrap the chromosomes (*arrows*)

J.R. McIntosh, Boulder, Colorado, USA
The problem of the polarity of growth is an interesting one, but there are really two problems, and it is important not to confuse them. If a tubulin dimer contains α and β tubulin, we can represent the molecule by an arrow running from α to β. Now the polymerization is diagrammed as $\rightarrowtail\rightarrowtail$. If a kinetochore had a high affinity for β tubulin, then the first subunits to stick would point like this $(K)\overset{\leftarrow}{\rightarrow}$. If polymerization were distal, then $(K)\overset{1\,2\,3}{\underset{\leftarrow\leftarrow\leftarrow}{}}$ would describe the order of the addition of the subunits. If a mitotic center had a high affinity for α tubulin, and a proximal position for subunit addition, then $(MC)\overset{3\,2\,1}{\underset{\rightarrow\rightarrow\rightarrow}{}}$ would describe polymerization from this organelle. Thus opposing kinetochores and poles could have four different polarity relations with respect to intrinsic molecular polarity, none of which is necessarily related to the four possible answers to the problem of the polarity of growth. The questions of the polarity of growth and intrinsic molecular polarity are independent of each other and must be answered by different experiments.

A. Wegner, Basel, Switzerland
An observation which might be relevant to the model of tubulin assembly-disassembly may be that muscle actin is able to shorten at one end and simultaneously to lengthen at the other. The driving force of this translocational polymerization is ATP hydrolysis, which is connected with the assembly of actin. Because tubulin assembly, as well, is associated with a nucleotide hydrolysis, it is to be expected that tubulin also associates and dissociates in a similar manner. One interesting point is that the lengthening end of actin filaments is that end which is bound to Z-lines in muscles, while the shortening end is free. It would be of interest to know if there are receptors for actin filaments or tubulin filaments that permit or even catalyze insertion of subunits.

H. Sato, Philadelphia, Pennsylvania, USA
Electron-microscopic analysis of the D_2O effect on echinoderm spindles revealed that the total number of spindle microtubules is increased 2.5 times within 3-5 min. They also elongate. However, the fact that a fixed number of microtubule clusters always exists suggests that they are kinetochore microtubules. Hence, we can conclude that D_2O greatly increases the number of continuous microtubules while the number of chromosomal microtubules remaines constant. Kinetochore microtubules should show distal growth because the other end is blocked. But continuous microtubules will grow in both directions. However, we have no idea of how to determine the growing point.

R. Tiggemann, Heidelberg, F.R.G.
I would like to ask Dr. Nicklas a question concerning the configuration of the kinetochore fibers mentioned in your lecture. If, in addition to other theories, you believe that tubulin mainly polymerizes and depolymerizes during chromosome movement, how do you explain the mechanism of such rapid reactions? Can an eventual pool for unpolymerized tubulin be localized?

B. Nicklas, Durham, North Carolina, USA
Because chromosomes move so slowly, the required polymerization-depolymerization reactions need not be atypically rapid, and one thing we know about chromosome-to-pole movement in anaphase is that the kinetochore microtubules do shorten. The pool of unpolymerized tubulin has not been localized to any particular region or regions of the cell, and that is certainly an interesting possibility.

R. Dietz, Tübingen, F.R.G.
I am going to object to Dr. Nicklas' statement that assembly hypotheses cannot explain the prometaphase maneuvres of chromosomes that you have analyzed further. I agree with you that trivalents often remain significantly off the equator in metaphase, lying closer to that pole toward which two kinetochores are oriented. However (shows slide), you can see a trivalent of *Pales crocata* that has moved almost all the way back into the equator. This behavior is species-specific. In *P. ferruginea* (slide) autosomal univalents also first approach the poles. However, they do not stay there; they move backward - without reorienting - until they almost reach the equator. Hypotheses must explain why bipolarly oriented chromosomes move into the equator, whereas monopolarly oriented chromosomes remain sitting close to their pole, or approach the equator, depending on the species, and within a species, on the chromosome.

Let me now try to explain these findings. According to my hypothesis, which is presented in more detail in poster 39, there is a disassembly of structural spindle elements throughout mitosis. Assembly occurs only in the vicinity of organizing centers (kinetochores and pole-determining organelles). During mitosis each organizing center surrounds itself with a field in which tubulin polymerizes. The apparent polymerization constant is at maximum next to an organizing center, and decreases with increasing distance. Moreover, the apparent polymerization constant next to kinetochores is high throughout prometa-metaphase, and it is turned down, and finally switched off, during anaphase. The fields of all polymerizing centers within a cell superimpose accumulatively. The smallest structural spindle element may be visualized as a microtubule that is unraveled at both ends into its protofilaments. With the help of one of their protofilamentous ends, some microtubules are anchored within the kinetochores. All other microtubules are also interconnected, since labile bonds exist between adjacent protofilamentous ends throughout the spindle. Thus, the whole spindle forms a mechanical unit. Forces can be transmitted between poles and more so, between kinetochores and poles. Within the spindle, one can distinguish between the tubular phase and the protofilamentous or reticular phase. All spindle segments grow by end addition of protomers to the protofilaments within the reticular phase, and by subsequent phase transition and anisodiametric growth of the tubular phase. The reverse processes happen during depolymerization, provided that the anisodiametric shortening of the spindle segment is not prevented by opposing mechanical forces. If coaxial strain is applied to a spindle segment, it grows longer according to the principle of Le Châtelier and Brown, i.e., without the addition of protomers. Since the tubular phase bridges a longer distance with the same number of tubulin molecules, it is reversibly favored over the reticular phase. A spindle segment that is under depolymerization conditions but which cannot shorten because of external mechanical strain, will stay at the same length during depolymerization. For a while at least it depolymerizes isometrically. It is assumed that microtubules that depolymerize isometrically become more and more labile, until they finally collapse cooperatively. A fraction of the subunits that are released in the course of collapse is incorporated into adjacent spindle elements. Therefore, a spindle segment that is under depolymerization conditions in which, however, the spindle elements cannot shorten, grows longer and thinner. A few microtubules grow longer at the expense of many more microtubules that cooperatively collapse. From these assumptions the following conception of mitotic processes can be developed. Throughout division microtubules continue to grow next to organizing centers, and since new ones are nucleated again and again, the newly assembled material displaces the older assembly products in radial directions. After the assembly products have been shifted out of assembly fields, they are depolymerized. Depolymerization is at maximum in the polar regions and all along the spindle periphery. In metaphase steady state is reached, i.e., assembly equals disassembly. The displacement of polymerized material presumably causes akinetochoric transport and can be directly seen in polarized light movies as lateral displacements. Whereas microtubules all along the spindle periphery finally collapse cooperatively (and this causes the formation of a prop skeleton that holds the poles apart), constituent microtubules of chromosomal spindle fibers can shorten in the polar regions since kinetochore microtubules continue to grow next to kinetochores. This explains the moving of chromosomal spindle fibers with stationary chromosomes during metaphase, which occurs as Forer has shown in experiments. The growth rates of the antagonistic chromosomal spindle fibers of a bipolarly oriented chromosome are the same. The depolymerization rate, on the other hand, is higher in the longer fiber. Equilibrium is reached when the two antagonistic chromosomal spindle fibers have attained the same length. In other words, bipolarly oriented chromosomes are bound to move into the equator. They are displaced poleward again whenever a new kinetochore microtubule is nucleated. The assembly field that surrounds each chromosome grows rapidly weaker if the chromosome approaches one pole, since superimposition with the assembly fields of other chromosomes rapidly decreases. Thus a chromosomal spindle fiber that had been allowed to shorten because no or only very weak antagonistic forces existed, is in an environment in which the assembly rate is greatly reduced. Only in very strong kinetochores are the assembly rates high enough at late prometaphase to cause their chromosomal spindle fibers to grow longer. Only these chromosomes are pushed backward into the equator. Weak kinetochores can maintain very short chromosomal spindle fibers, at best. If they do not reorient rather soon, chromosomes with weak kinetochores stay close to their poles.

A.S. Bajer, Eugene, Oregon, USA
Dr. Nicklas stressed the importance of prometaphase as a stage for "model" studies of chromosome movement. I have been working on prometaphase movements for many years, but I am inclined now to use anaphase as a basic model. It is impossible to predict in practice during prometaphase in what direction the chromosome will move. On the contrary we expect that during anaphase the chromosome gradually approaches the polar region. Therefore, e.g., if chromosomes move backward in anaphase (Lambert and Bajer, Cytobiologie 15, 1-23, 1977) in experimental conditions, it is clear that such a movement is the result of experimental treatment. I would like to stress therefore the practical usefulness of anaphase for experimental studies.

K.B. Moritz, München, F.R.G.
There is an old, genetically defined term - the kinetochore strength. Could Dr. Nicklas and Dr. Dietz comment on the processes with respect to the models discussed here.

R. Dietz, Tübingen, F.R.G.
In certain divisions, one chromosome controlls the behavior of one, two or even several other chromosomes. This is indicative of the existence of strong and weak kinetochores. In my explanation, the strength of a kinetochore is equivalent to the strength of the assembly field which it is forming.

G.G. Borisy, Madison, Wisconsin, USA
Dr. Roos, would you describe your model for chromosome orientation.

U.-P. Roos, Zürich, Switzerland
The model concerns the initial orientation of chromosomes and the resulting movements, but it does not explain congression in the sense of Dr. Nicklas' presentation. Figure 3 illustrated that, at the beginning of prometaphase in rat kangaroo cells, the chromosomes are faced with a rather poorly organized spindle. This is evident from thin sections of cells fixed in very early prometaphase. The nuclear envelope begins to break down in the vicinity of the asters, and from this, as well as other observations, I assumed that there is a gradient of tubulin extending from the center of each aster toward the center of the prometaphase nucleus and that the initial orientation of every chromosome depends on the position of its kinetochores relative to the two poles.
For instance, if we consider a chromosome lying closer to the left pole, as indicated in Figure 4a, the kinetochore facing this pole is exposed to a greater concentration than its sister. I therefore consider it likely that it can organize a functional fiber more rapidly, and as soon as this fiber has attained a certain degree of organization, it will move the chromosome toward the left pole. This can happen even if the other kinetochore facing the more distant pole organizes a fiber simultaneously (which does not seem to be the case), for that fiber will grow toward an as yet poorly organized part of the spindle that does not favor force production.
In fact, those chromosomes whose kinetochore region lies relatively close to one pole at the beginning of prometaphase are the first to move and, with rare exceptions, they move to and/or associate with the nearer pole. At a later stage of prometaphase the situation is more like that depicted in Figure 4b, where the spindle is now well organized, and I have assumed that the concentration of tubulin is more uniform throughout the spindle. Thus, the heretofore unoriented kinetochore can now organize a fiber, which can interact with the by now quite well-organized midregion of the spindle, and the chromosome can congress.
I have not speculated on the forces that move chromosomes. Rather, I think that any of the current hypotheses can account for these movements, although I have assumed that the kinetochore fibers need to interact with other spindle elements (see Roos, Chromosoma, 54, 363-385, 1976).

D. Mazia, Berkeley, California, USA
Metakinesis is not merely an event that happens between prophase and metaphase. It is an activity that the spindle can and will repeat. If the metaphase spindle is

164

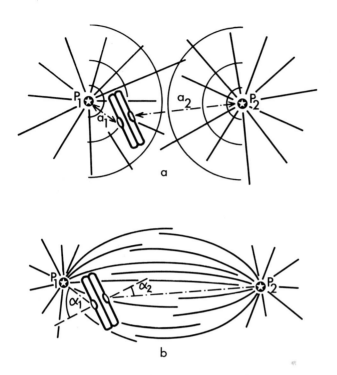

a

b

Fig. 4a and b. Diagrammatic illustration of kinetochore orientation in two different spindle systems. For simplicity the spindles are depicted as two-dimensional
systems, but similar relationships apply in three dimensions. *Solid lines* represent
MTs. (a) An early prometaphase spindle that consists essentially of two large asters.
Semicircles symbolize gradients from the poles to the center of the nuclear area of
subunits available for MT assembly (the tighter the semicircles, the higher the concentration). The kinetochore region of the chromosome illustrated is nearer P_1 than
P_2 (distance $a_1 < a_2$) and because the concentration of subunits is greater and the
structural organization of the spindle is better in the polar areas than in the
equatorial region, the conditions for fiber formation and force-producing interaction with complementary elements are not equal for the sister kinetochores. Each
kinetochore will principally orient to the "most direct" pole [see (b)], but the
one more nearly facing the proximal pole (P_1) will form a fiber sooner and/or
more rapidly than its sister. Furthermore, its fiber will grow into the well-endowed
polar region where the conditions for immediate force-producing interaction are fulfilled. As a result, the chromosome will move to the proximal pole. (b) A later
prometaphase spindle with a well-developed, axially oriented framework of MTs. The
distribution of MT subunits in the spindle body is assumed to be uniform. The angles
α_1 and α_2 indicate the "most direct" pole for each kinetochore (cf. Nicklas, 1967,
for meiotic bivalents). The conditions are equal for both sister kinetochores and
the direction of initial movement is independent of the distance from the kinetochore region to the poles and it is therefore unpredictable

disturbed and the chromosomes and poles are displaced, a good metaphase spindle
will be reestablished when the disturbance is removed. We can ask this question of
the cell: will chromosomes scattered anywhere in the cell be drawn into a metaphase
plate?

N. Paweletz, Heidelberg, F.R.G.
I have the feeling that we are looking for a unique model explaining mitotic events
and valid for all kinds of cells. We have experimental data showing that in one
type we have sliding, in another zipping, in a third assembly-disassembly.

Would it be fair to assume that there is more than one mechanism responsible for
the movement of chromosomes?

B. Nicklas, Durham, North Carolina, USA
I like uniformity, but when I cannot have it, I go with nature.

J.F. Lopéz-Saéz, Madrid, Spain
Do you think, Dr. Nicklas, that the spindle fibers might be pulling continuously
on kinetochores from prometaphase to the end of anaphase?

B. Nicklas, Durham, North Carolina, USA
I agree with Dietz that in many cells there does seem to be a lessening of the
force probably associated with an increased average length of the chromosome fibers,
relative to the interpolar distance during late prometaphase and metaphase, which
probably results in a diminution of the force the chromosomes see at that time.
It is known that chromosomes can be violently pulled to the poles in early prometaphase, just after the spindle forms, yet these forces have diminished by metaphase
(Hughes-Schrader, Biol. Bull. 8S, 265, 1943).

N. Paweletz, Heidelberg, F.R.G.
Dr. Nicklas, how can one explain the different experimental data that favor one or the
other mechanism in terms of a unique mechanism?

B. Nicklas, Durham, North Carolina, USA
You think there are experimental data that clearly speak in favor of one theory or
the other. I do not know of any observations that absolutely demand a particular
interpretation at the present time. For example, the decrease in microtubule overlap in the interzone in *Diatoma* does not mean that sliding must operate there,
since depolymerization at the microtubule ends could produce the observed configuration (McDonald et al., J. Cell Biol. 74, in press, 1977).

R.J. Wang, Columbia, Missouri, USA
We have isolated a collection of temperature-sensitive hamster cell mutants with
defective prophase chromosome condensation, prometaphase congression, metaphase
progression, anaphase movement, and cytokinesis. In at least two of the mutants,
chromosomes move into normally appearing metaphase plates. Subsequent anaphase
motion, however, either fails to occur or is defective. The results suggest, but do
not prove, that prophase and anaphase chromosome movement may involve different
mechanisms.

A.-M. Lambert, Strasbourg, France
(Showed a short 16-mm film about the effect of cold shock on mitosis and some ultrastructural data).
Reversible arrest of chromosomes during mitosis can be obtained in dividing plant
cells (endosperm) by lowering the temperature from 22°C to +3.5°C or below. Normal
movements are reinitiated within 2-5 min by returning the cells to room temperature
- slight backward movements (5-6 μm) of the kinetochores are observed during
cold shock.
Combined light- and electron-microscopic studies of the same cells show that in
cold-treated cells, when chromosomes did stop:
- the total number of microtubules is reduced to around 1/3
- continuous microtubules are no longer present
- some kinetochore microtubules persist and show typical parallel arrangement - their
number is still higher than in late anaphase control cells.
During the reinitiation of movements by return to room temperature, the number of
microtubules increases, continuous microtubules are well developed, and the arrangement of kinetochore microtubules reverses from a parallel to a divergent configuration.
Such data suggest that arrangement of microtubules has an important role in chromosome
dynamics.

I.B. Heath, Downsview, Ontario, Canada
Dr. Lambert showed that at 4°C the spindle totally lost nonchromosomal microtubules,
yet it clearly retained most of its structure. Could you, Dr. Lambert, explain what
it was that retained the structure of the spindle? If, as seems likely, it was some
component of the matrix of the spindle, does this not point out the importance of
considering the properties of the matrix as an important component of the spindle
and its structure.

A.M. Lambert, Strasbourg, France
The role of continuous microtubules has to be further investigated - I think they are involved in keeping polar regions apart; they interact also with kinetochore fibers during chromosome movements - integrity of the spindle during cold shock may be related also to membrane development in the matrix.

A.S. Bajer, Eugene, Oregon, USA
At least in one plant tissue (endosperm in single cell and early syncytium stage) there is, without any doubt, a predetermined direction when the spindle begins to form. In the first, earliest stage of spindle formation, the so-called clear zone (Bajer, 1959) is formed. It is a uniform, "clear" (in the phase microscope, and uniformly birefringent in polarizing microscope) zone around the nucleus. In *Haemanthus* it grows uniformly (increases in thickness) and slowly at the beginning, and only later when nonuniform shape appears does it develop faster and assume a nonsymmetric shape (extranuclear spindle). The shape seems to depend mostly on mechanical factors, i.e., the shape of the spindle. Even during prometaphase the final axis is not yet fully determined and may change considerably. This situation is certainly different in most other plant tissues (meristems), but the factors influencing orientation are not clear. In my opinion, however, the shape of the cell always plays some role in the final orientation, i.e., in the determination of the long axis of the spindle.

H. Ponstingl, Heidelberg, F.R.G.
We have now seen a number of brilliant records of mitosis and listened to proponents of almost all current hypotheses on chromosome movement. Since none of these considerations sufficiently explains what is going on, I wonder whether we could arrive at a list of essentials that should be accounted for by any future interpretation.

J.R. McIntosh, Boulder, Colorado, USA
Part of the problem in talking about models for mitosis at this time is that we are in the midst of a basic change in our understanding of mitotic mechanisms. At the cellular level, we understand mitosis quite well. We know that there are pole-directed forces exerted on the chromosomes by the spindle. We know that chromatids find spindle fibers at sister kinetochores that are back-to-back so the sisters must go to opposite poles. Other aspects of mitosis at the cellular level are clear. The questions we are now trying to answer are basically questions of molecular mechanisms. What molecules are responsible for the pole-directed forces on the chromosomes? What molecular changes initiate the separation of the bound sister kinetochores at the metaphase to anaphase transition? Each of these cellular processes is a complex event, and it is small wonder that we are having difficulty in working them out. Experiments are in progress in many laboratories to obtain molecular information and, as Dr. Nicklas pointed out, we have to some extent lost our naiveté with respect to molecular models. I do not think anyone expects simple answers to questions about molecular mechanisms for chromosome movement, and so I, for one, am no longer able to talk about simple, general models. We try to model individual processes and experiments to see if these "minimodels" can be disproved.

Session VIII

Mitosis in vitro: Isolates and Models of the Mitotic Apparatus

J. R. McINTOSH, Department of Molecular, Cellular, and Developmental Biology,
University of Colorado, Boulder, Colorado 80302, USA

A. Introduction and Glossary of Terms

Mitosis in its broadest sense is the set of processes that occur to
achieve the accurate segregation of the duplicate chromosomes during
division of a eukaryotic cell. The set includes chromosome condensation,
nuclear envelope breakdown, formation of the fibrous spindle with associated organization of the chromosomes into the metaphase configuration, the breaking of that arrangement with concomitant separation
of the chromosomes, shortening of the chromosome-pole fibers and elongation of the spindle, the reformation of the nuclear envelope, decondensation of the chromosomes, and final dissolution of whatever spindle
fibers remain. Because of the limits defined by present knowledge and
my own point of view, this paper is addressed to a subset of the relevant processes: In vitro studies on spindle formation and dissolution
and on chromosome movement.

It must be stated at the outset that our knowledge of mitosis in vitro
is far from perfect. The ideal system for study would be an isolate
containing biochemical quantities of pure spindles that could function
"normally" by moving chromosomes at physiological rates over their natural distances upon the addition of some standard reagent, such as
ATP. No such preparation exists, so we must piece together the fragments
of information as we find them and try to develop as substantial an
understanding as possible in our current state of ignorance. The work
to be discussed may be divided into two rather distinct categories:
(1) the study of isolates containing the relevant cellular components
in a well-defined milieu, and (2) the characterization of cell lysates
or extracts of whole cells, socalled cell models, that retain some
physiological characteristic of interest. Further, the work falls into
two rather rough groups based upon the time at which it was done relative to the development of our knowledge of the chemistry of proteins
relevant to cell movement. For this reason, I have broken my discussion into what are essentially temporal sections: studies not dependent
upon the in vitro polymerization of tubulin and those that take advantage of this advance.

The goals of many of the laboratories working on this subject are similar. We all want to find a molecular explanation for this complex
cellular event. One imagines the goal to be worthwhile not only because
cell division is interesting and important in its own right, both
biologically and medically, but because the structure of many cells
that are undergoing a morphogenetic change suggests that these processes too may be based upon molecular mechanisms similar to those
functional during mitosis. For this reason workers from several rather
distinct fields are converging upon a single field, and there is occasionally semantic confusion. To help avoid that problem here, I include below a glossary of terms used in this paper. The words are organized in a logical sequence derived from the interrelationships between their meanings rather than in alphabetical order, but it is
hoped that the reader who is not familiar with mitosis will find the
list useful.

The Spindle is functionally defined as the fibrous cellular machinery that segregates eukaryotic chromosomes at cell division. Structurally, it is so variable from cell to cell that it is difficult to describe in general. Its significant parts are listed below.

A Chromosome is a naturally occurring information-bearing piece of DNA and associated macromolecules. In eukaryotic cells it is usually sufficiently big and condensed at mitosis to be easily stained and visible in the light microscope. At the beginning of cell division, the chromosomes are double, and each part is called a chromatid.

A Kinetochore is the specialization on a chromatid to which the spindle fibers attach. In some organisms it is an easily identified structure with a trilamellar or tufty appearance in the electron microscope. Each mitotic metaphase chromosome has two kinetochores at the primary constriction, arranged back-to-back so that the microtubule attachment sites face in opposite directions. The term "centromere" is sometimes used to mean a kinetochore and sometimes to refer to the whole primary constriction that contains two kinetochores.

A Pole is the region at each end of the spindle toward which the chromosomes will move at anaphase.

In some animal spindles, the pole contains a pair of centrioles arranged at right angles to one another and surrounded by a halo of amorphous, darkly-staining material where the microtubules end. These poles are derived from the cell center or "cytocentrum" which duplicates during interphase and then changes in an unknown way to initiate the growth of the polar spindle fibers.

Polar structures in some algae and fungi do not include centrioles, but contain some other well-defined structure. In higher plants there is no particular structure visible with light or electron microscopes at the pole, but the region at the ends of the spindle still accomplishes all the relevant functions.

A Spindle Tubule is a microtubule found in the mitotic spindle. There are some spindle tubules that end on a kinetochore and are called kinetochore microtubules or chromosomal microtubules. Some do not attach to a kinetochore, but have one end at a pole and are called polar microtubules. Others have both ends free and are called free or fragment microtubules.

A Spindle Fiber is a fibrous element of the spindle visible in the light microscope. Spindle fibers are bundles of microtubules and the associated matrix material: some ribosome-like particles, some small vesicles, and some ill-defined filamentous material. Immunofluorescent studies and studies using heavy meromyosin as a tracer for actin suggest that both actin and myosin may also be present in spindle fibers, but the evidence is still incomplete. There are "chromosomal fibers" that run from chromosome to pole and "continuous fibers" that run from pole to pole.

An Aster is a radial array of spindle fibers emanating from the poles of some spindles. They are impressive structures in the spindles of marine eggs; they occur in the spindles of many somatic animal cells; they are not found in the spindles of higher plants.

A Mitotic Center is a region capable of organizing an astral array of microtubules in a cell. Two mitotic centers together with the chromosomes can organize a normal spindle.

A Mitotic Apparatus was defined by Mazia in 1960 to include the spindle, the chromosomes, the poles, the asters (if any occur naturally in that cell type), and associated matrix components.

A Centric Spindle is one with structurally defined polar regions where the spindle fibers focus in toward a point at each end of the spindle.

A Half Spindle is the portion of the spindle running from one set of kinetochores to and including the pole they face. A half spindle shortens at anaphase.

The Interzone is the portion of the spindle that lies between the sister sets of kinetochores. The interzone elongates at anaphase and is then seen to contain fibers called "interzone fibers." During anaphase the microtubules of the interzone fibers accumulate an osmiophilic material where the metaphase plate used to be. These regions have been called stem bodies, following the German term meaning "pushing bodies," but their function is not known, so the term is not a good one.

A Midbody is found at telophase in cells that accomplish cytokinesis by means of a cleavage furrow. The furrow bunches the interzone fibers together to form the midbody. The osmiophilic portions of the interzone fibers are comparatively stable and often survive within the bridge into interphase. This spindle remnant is often called the Flemming body.

A Phragmoplast is the set of interzone microtubules that form during telophase in higher plants. It is involved in making the "cell plate," the middle lamella of the wall that forms between the two daughter cells.

B. Studies Not Dependent upon Polymerizable Tubulin

I. Spindle Models

The success of glycerol in supporting muscle contraction after cell lysis inspired a spate of work in which other motile systems were glycerinated to make cell models for these functions too. The leader in this field, Hoffmann-Berling, turned his hand to mitosis. His studies made it clear that chromosome movement did not occur by a process identical to muscle contraction (Hoffmann-Berling, 1954). He was able to obtain chromosome movement in glycerinated fibroblasts, but the conditions resembled those suitable for relaxing muscle more than those for inducing its contraction. These intriguing results went largely unexplored for about 20 years until the advent of polymerizable tubulin made it possible to improve the conditions for lysis. In the interim Goode and Roth (1969) tried to develop a working model for spindles from giant amebae by varying the ionic composition of the medium into which the cell was lysed, but the addition of ATP to the lysate served only to support limited chromosome motion, and even that would occur only as the cell was lysed.

II. Spindle Isolates

The mitotic apparatus (MA) was first isolated by Mazia and Dan in 1952. They identified marine eggs as a suitable material to solve the problem of getting a large number of cells synchronized within their cell cycles to sufficient precision that essentially all of the cells would be in a selected stage of mitosis. The spindles from these and most other cells are labile structures, so the initial problem in spindle isolation was that of preventing the rapid dispersal of the MA upon its release from the cell. The first method for spindle isolation was based upon a treatment of the cells with cold ethanol, which probably resulted in mild fixation by denaturation, followed by a detergent solubilization of as much unwanted cytoplasm as possible. Fixation is an essentially irreversible process and precludes most physiologic study, so later work has pursued methods that would stabilize the MA reversibly in the extracellular medium. Mazia and his co-workers found that sucrose and dithiodiglycol would serve this stabilizing role (Mazia et al. 1961). Kane reported on the stabilizing effects of slightly acidic pH and of several glycols, drawing attention to the correlation between spindle stabilization and poor protein solubilization (Kane, 1962, 1965). Aqueous solutions of 1 M hexylene glycol (2-methyl-2,4-pentanediol) at pH 6.4 will stabilize MAs from sea urchins for hours. Kane's methods have been adapted to clam eggs (Rebhun and Sander, 1967), to starfish eggs (Bryan and Sato, 1970), and mammalian cells (Sisken et al., 1967; Sisken, 1970; Wray and Stubblefield, 1970). MAs prepared by these methods do not move chromosomes and are not so labile as are spindles from living cells. Nonetheless, they have been useful for various kinds of studies: microtubule structure (Kiefer et al., 1966); Cohen and Gottlieb, 1971) microtubule birefringence (Rebhun and Sander, 1967; Sato et al., 1975), microtubule polymerization kinetics (Goode, 1973), the thermodynamics of microtubule assembly (Sato and Bryan, 1968), the effects of D_2O and glycols on microtubules (Marsland and Zimmerman, 1965; Rebhun et al., 1975); the synthesis of MA precursors (Bibring and Cousineau, 1964; Wilt et al., 1967), changes in mass distribution through mitosis, (Rustad, 1959; Forer and Goldman, 1972); the protein composition of the isolated MA (see for example Miki-Nomura, 1965, 1968; Sakai, 1966; Borisy and Taylor, 1967; Bibring and Baxandall, 1968, 1971; Cohen and Rebhun, 1970), and for study of MA-associated ATPases (Mazia et al., 1961; Miki, 1963; Dirksen, 1964; Weisenberg and Taylor, 1968). More recently, a lithium-cold ethanol method has been discovered that preserves a Ca^{2+}-activated, MA-associated ATPase activity (Mazia et al., 1972; Petzelt, 1972a, b). Especially in the areas of protein chemistry and enzymology of the isolates, however, the existence of cytoplasmic contamination has been a persistent problem, honestly acknowledged by the investigators themselves.

The fine-structure studies of most MAs show that they contain substantial amounts of material in addition to microtubules. For this reason they have not generally proved more useful than living cells as a starting material for electron-microscopic investigation of the fibrous component of the spindle. Sisken's preparations of hexylene glycol-isolated HeLa cell spindles, on the other hand, are remarkably clean. We have used a high voltage microscope to obtain stereo electron micrographs of HeLa spindles isolated by Chu and Sisken, glutaraldehyde-osmium fixed, and dried by the critical point method. The details of this work will be presented elsehwere (McIntosh, Chu and Sisken, manuscript in preparation), but I include three pictures here to help illustrate the morphology of the microtubule component of an MA.

Figure 1 is a stereo pair high-voltage electron micrograph of a metaphase spindle. With the low concentrations of divalent cation used, the chromatin has unraveled and broken, so the chromosomes are absent from the metaphase plate. Some of the granular material coating the spindle and the individual microtubule bundles is probably chromatin. The general, three-dimensional architecture of the microtubule component of the spindle can be seen in these micrographs. The overall shape of the spindle is approximately spherical. The spindle poles are evident as regions upon which the spindle fibers focus. The chromo-

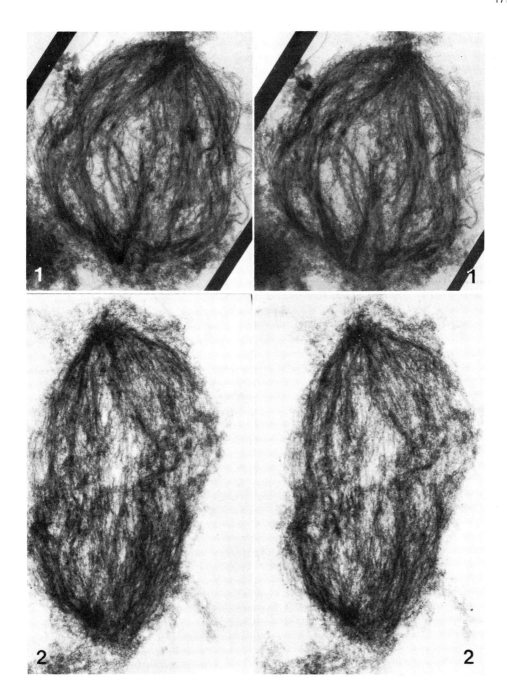

Figs. 1-3. Morphology of the microtubule component of an MA

Fig. 1. Stereo pair high-voltage electron micrograph of a metaphase spindle

Fig. 2. Chromosomal and polar tubules in anaphase spindles; interzonal granules

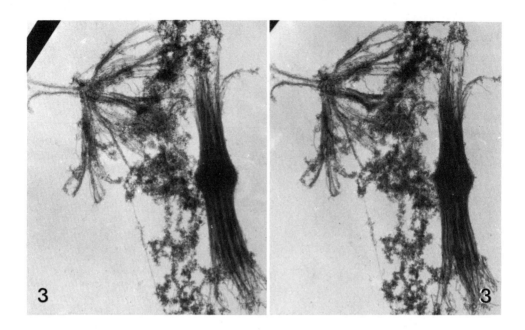

Fig. 3. Stereo pair of a midbody and fragment of a metaphase or anaphase spindle

some fibers are identifiable as bundles of microtubules that extend
from the poles to the midplane of the spindle and sometimes a little
across it. (The isolated spindles are probably damaged by handling
during isolation, and there is some damage from the electron beam
during microscopy.) Each chromosome fiber has an ill-defined tuft of
fibrous material at its end distal to the pole. I interpret this tuft
as the residuum of the kinetochore and the chromosome. There are also
some less well-bunched microtubules in the spindle that are longer
than the chromosome fibers. These are likely to be the polar microtubules of the metaphase spindle. The polar tubules commingle with the
chromosome fibers.

Anaphase spindles are often disrupted by the isolation process, and
only a few specimens have been found intact. In the example shown as
Figure 2, chromosomal and polar tubules are present, and some interzonal tubules can be seen. As anaphase proceeds in HeLa cells, the
interzonal microtubules cluster to form several slender bundles that
run through the interzone (Robbins and Gonatas, 1964; McIntosh and
Landis, 1971). The cleavage furrow then bundles these together to
form the telophase midbody. This structure is very stable and is frequently found in the preparations. Figure 3 is a stereo pair of both
a midbody and a fragment of a metaphase or anaphase spindle. In this
fragment (which reflects a commom image in the preparations), a single
set of chromosome fibers is seen to focus upon a central point in a
fashion reminiscent of the spindle pole. I interpret this image as
a chromosomal half spindle that has broken away from the polar spindle
and yet retained its structural integrity. I infer that the connection between chromosomal fibers and poles is comparatively strong.

This image depicts some aspects of spindle architecture, but it also
serves to illustrate some of the problems one encounters in studying

preparations of MAs. (1) Material has been lost from the MA (chromosomes, vesicles, ribonucleoprotein particles, and probably other unidentified structures). (2) It is likely that there is cytoplasmic material not relevant to spindle structure still present in the preparations. Similar points have been made by Forer and Goldman (1972) on the basis of quantitative optical studies. (3) The isolates are not functional, in the sense of moving chromosomes, so it is not clear how useful a structural or biochemical analysis of the material really is.

It has been obvious to all concerned that the isolation of a lifelike spindle would be a major contribution, since functional criteria could then be applied to determine the importance of any one component to the properties of the MA. Forer and Zimmerman (1974) have made an advance toward this goal by preparing sea-urchin MAs with the glycerol-dimethylsulfoxide medium of Filner and Behnke (1973) to stabilize the structure. These spindles are more lifelike in their behavior than those stabilized in hexylene glycol, and three papers on their properties have now been published (Forer and Zimmerman, 1976a,b; Forer et al., 1976). The isolates are birefringent, contain microtubules, chromosomes, and substantial amounts of the vesicular and granular material seen with the electron microscope in fixed mitotic cells. The birefringence (technically, the retardation) of these MAs is sensitive to pressure and cold, as is that of the living spindle, provided that the isolates are placed in a diluted isolation medium. Some birefringent components of the MAs are soluble in 0.5 M KCl, but there is no corresponding decrease in the number of spindle microtubules seen by these authors, in disagreement with the work on spindles isolated with hexylene glycol by Sato et al. (1975). Forer et al. conclude that in their preparations there is a birefringent component other than the microtubules, and from indirect arguments from the fixation properties of the MAs, they conclude that this substance (or substances) must interact with the microtubules. While the arguments are not compelling, their data reinforce the need for functional criteria to evaluate different components of any isolate.

Sakai and Kuriyama (1974) have discovered a glycerol-ethylene glycol-bis (2-amino-ethylether)-N,N,N',N'-tetraacetic acid (EGTA) method for isolating spindles that not only stabilizes the spindle structure but also stabilizes the colchicine binding of the spindle tubulin. Their preparation is noteworthy in that it has served as a basis for the development of what is probably the most lifelike spindle isolate. This material will be discussed in a later section.

C. Studies Using Polymerizable Tubulin

In 1967 two groups independently succeeded in using the property of colchicine binding to identify a subunit of the microtubule (Borisy and Taylor, 1967; Wilson and Friedkin, 1967). The work based upon this advance led to the discovery by Weisenberg (1972a) of conditions that would bring about the polymerization in vitro of tubulin and several "microtubule-associated proteins" (MAPs). This volume contains a review of this work by Borisy to which the interested reader is referred. The knowledge of tubulin and MAP chemistry that has grown up in the last few years has changed our ways of approaching the spindle in vitro, and the two sections of this paper to follow deal with findings that have depended upon tubulin assembly chemistry.

I. Spindle Models

1. Spindle Formation

Weisenberg was the first to use his conditions for the assembly of microtubule protein to investigate the properties of mitotic centers in cell lysates (Weisenberg, 1972b). Equal volumes of eggs from the surf clam *Spissula solidissima* and a buffer comprised of 0.5 M 2(N-morpholino) ethane sulfonate, pH 6.5 with 1 mM EGTA were mixed and the eggs mechanically ruptured. No microtubule formation occurred at $0^{\circ}C$, but at room temperature and higher, tubules formed in the extracts of either activated or unactivated eggs. If live eggs were allowed to proceed before lysis to the breakdown of the egg nucleus (the end of prophase), then the microtubules that formed would radiate from small, punctate structures in the lysate. These structures were found to correspond to mitotic centers, and the microtubule aggregates formed in vitro bear a significant resemblance to normal asters (Weisenberg and Rosenfeld, 1975).

Mitotic centers from activated eggs can be partially isolated and mixed with tubulin-containg fractions from either activated or inactivated eggs to produce aster-like structures. Similar fractions prepared from inactivated eggs have no such polymerization-inducing effect, suggesting that it is the mitotic center, not the tubulin or ooplasm, that changes when an aster begins to form. I will refer to this acquisition of microtubule organizing capacity as the "maturation" of a cell center into a mitotic center. Weisenberg and Rosenfeld (1975) described a structural change in the centriole located at the middle of the mitotic center as it matures: There are no triplet microtubules visible in the centers of *Spissula* 2.5 min after activation, but they are well defined by 15 min. Treatment with 0.5 mM colchicine does not inhibit this change. Maturation of the mitotic center is clearly a process of great interest, since it may well constitute a significant regulatory mechanism in the control of spindle formation. I will return to the subject several times below.

Tubulin appears to be a well-conserved protein, based upon the data describing its chemistry (for review, see Snyder and McIntosh, 1976), the wide-spread property of colchicine binding (for review, see Wilson and Bryan, 1974), and upon the limited primary structure data currently available (Luduena and Woodward, 1973). Several laboratories have therefore pursued the logic that neurotubulin might be used to do experiments on mitotic spindles. Neurotubulin possesses the significant advantages that it is comparatively easy to prepare in quantity, and it will polymerize efficiently in vitro. Our group has used mammalian cells, strain PtK1 grown on glass coverslips and lysed with Triton X-100 while under observation in the light microscope to study spindle formation in the presence of various concentrations of neurotubulin obtained from porcine brain (Cande et al., 1974). We found that metaphase spindle birefringence and microtubule number could be both increased and decreased by varying the concentration of microtubule protein at lysis. Electron microscopy shows that in this system it is the polar microtubules that respond most rapidly to conditions favoring either increase or decrease of polymer (McIntosh et al., 1975b). An increase in spindle birefringence after lysis can be reversed by cooling the augmented spindle to $0^{\circ}C$ or by treating it with 2 mM Ca^{2+} ions. (Both are conditions that reverse assembly of neurotubulin in vitro.) A spindle made to fade by cold treatment will reappear upon rewarming to $37^{\circ}C$, although the cold lability is itself labile and decreases over tens of minutes after cell lysis. Colchicine will block the increase in spindle birefringence, but nonphysiologically high concentrations are required to solubilize the spindle in vitro.

The spacial distribution of the birefringence changes seen as a function of temperature in these in vitro spindles resembles those seen with rapid cooling and warming of living spindles in marine eggs. The changes are relatively uniform over the body of the spindle. They do not appear as a wave of birefringence starting at the poles or kinetochores and sweeping over the rest of the spindle. It is as if the cooling caused the tubules to fall apart everywhere at once and regrowth were initiated over the entire spindle.

The length of the lysed metaphase spindle does not vary appreciably with changes in temperature or the concentration of free Ca^{2+}, but it does respond in certain ways to changes in the concentration of tubulin in the lysis mixture. In high concentrations of exogenous tubulin, individual polar spindle tubules elongate and will even grow to extend out of the spindle and beyond the opposite pole (McIntosh et al., 1975b). Surprisingly, however, the distance between the spindle poles is almost constant, even with prolonged periods of growth (1 h) and with concentrations of tubulin high enough to induce a fourfold increase in the maximum number of microtubules observed with the electron microscope in serial cross sections of the spindle. At low tubulin concentrations, on the other hand, the mammalian spindle will both fade in birefringence and shrink in length (Cande et al., 1974). Electron microscopy of this material at various stages in the shrinking process shows that the polar spindle tubules disappear more rapidly than the kinetochore tubules, so cells can be prepared in which essentially all of the polar tubules are gone, but 50-75 % of the kinetochore tubules remain. With longer or more severe treatment, the kinetochore tubules too disappear, and as they disassemble, the poles move in toward the metaphase plate. Spindles in living cells also possess the property that the polar tubules are more labile than the kinetochore tubules (Brinkley and Cartwright, 1975).

Inoué et al. (1974) have used oocytes of the marine worm *Chaetopterus* to study the growth and lability of metaphase spindles in varying concentrations of porcine neurotubulin. The MAs from this material will not only increase in spindle birefringence, but also in length after treatment with high concentrations of exogenous tubulin. The lack of electron microscopy in this study precludes a determination of whether the kinetochore tubules elongate as the spindle poles move apart, but it is clear that both polar and astral spindle fibers increase in number and in length. As in the case of the mammalian spindle, the augmented MAs are sensitive to Ca^{2+} ions but rather insensitive to colchicine. Inoué et al. (1974) showed that the in vitro spindles in their system would increase and decrease in birefringence in a fashion that was quantitatively similar to the changes observed with the same spindles in vivo.

With the lysed mammalian cell system we have tried repeatedly to obtain some in vitro analogue of a motile process relevant to spindle formation, such as the separation of the mitotic centers or the congression of the prometaphase chromosomes to the spindle equator. Thus far all our efforts to that effect have failed. Adjacent mitotic centers can initiate microtubules far longer than the distance between the two centers, but the growing microtubules do not push the two centers apart. Prometaphase cells can be lysed in high concentrations of neurotubulin so there is a substantial increase in spindle birefringence, but the chromosomes do not move on the spindle. We infer that the protein mixtures and lysis buffers we are using do not stabilize the relevant functions, and I think it likely that these processes depend upon more elaborate processes than the controlled assembly of microtubule protein.

When spindle tubule number is increased by lysis in high concentrations of exogenous tubulin, some of the increase is probably due simply to the elongation of microtubule fragments that normally reside within the living spindle (see Fuge, this volume), but several systems have been developed that allow one to study microtubule initiation where no tubules previously existed. Snyder and McIntosh (1975) found that PtK1 cells grown for 1 h in 10^{-5} M colcemid will complete prophase in an apparently normal fashion. Following chromosome condensation, the nuclear envelope breaks down, but the chromosomes do not become organized and the spindle fails to form. Electron microscopy of complete serial sections of these cells reveals no recognizable microtubules. When such colcemid-treated cells are lysed with Triton and warmed to 37°C in polymerization buffer without tubulin, no microtubules form. If lysis is done in the presence of tubulin that has been spun at high speed prior to use (Borisy and Olmsted, 1972), numerous microtubules form within 5 min at the mitotic centers, while the same tubulin warmed by itself, i.e., without any structures to serve as polymerization initiator, shows a 15-45 min lag time before polymerization begins (Snyder and McIntosh, 1975). Thus the mitotic centers of these cells are serving to initiate polymerization of tubulin that it itself is incompetent to initiate. The mitotic center is truly working as a "microtubule organizing center" (Pickett-Heaps, 1971). The lysed cells in this preparation can be staged in the mitotic cycle by the appearance of their nuclei. Interphase cells contain a cell center that initiates only a few microtubules with the initiation-incompetent neurotubulin. Prophase cells are similar to the interphase cells, but as soon as the nuclear envelope has broken down, marking the end of prophase, the cell center shows a marked improvement in its capacity to initiate microtubules. Thus the maturation of the cell center into a mitotic center as identified by Weisenberg and Rosenfeld (1975) in *Spissula* eggs is also found in mammalian cells. In the mammalian material there is no visible change in the centriole as the cell proceeds from interphase to metaphase, but there is an increase in the amount of amorphous, osmiophilic material associated with and surrounding the centriole pairs (Robbins et al., 1968). It seems likely that this change in pericentriolar material is a morphologic manifestation of the maturation of the cell center, and our attention is thus drawn to this amorphous material as a possible active principle of microtubule initiation in vivo.

Kinetochores too can be identified as microtubule-organizing centers in lysed cells and cell extracts. McGill and Brinkley (1975) showed that the chromosomes in colcemid-arrested HeLa cells will initiate the assembly of neurotubulin. Telser and Rosenbaum (1975) isolated chromosomes from the same cell type, attached them to a carbon specimen support for electron microscopy, floated them on tubulin that has been centrifuged hard, and observed substantial microtubule polymerization initiated at the kinetochores. These studies do not include a quantitation of the number of tubules per kinetochore, but the data presented suggest that the number of neurotubules initiated is similar to the number of kinetochore tubules found in vivo, suggesting that the microtubule initiating system at the kinetochore can count. Snyder and McIntosh (1975) also observed some initiation of microtubule assembly at the kinetochores of the chromosomes of their lysed cells, but the growth was not impressive. Since they did find better tubulin initiation from metaphase kinetochores than from those of early prometaphase, it is probable that the kinetochore, like the cell center, goes through a time-dependent maturation process. The use of rather extensive colcemid blocks in both of the other studies of microtubule initiation at kinetochores may have allowed sufficient time for further kinetochore maturation and thus account for their greater degree of success.

Borisy et al. (1975) showed that structures capable of initiating growth of neurotubulin can also be identified in lysates of the yeast *Saccharomyces*. In this organism the mitotic spindle forms within the nuclear envelope, and there are two specializations called spindle plaques or spindle pole bodies attached to the nuclear envelope that seem to organize the polar spindle. Borisy et al. show that these structures will initiate the assembly of neurotubulin in vitro. This system is particularly attractive because mutants of yeasts that affect the normal duplication of the spindle pole bodies (Byers and Goetsch, 1975) are already known, and it is reasonable to hope that genetic dissection may be used here to help in the analysis of spindle pole maturation.

The process of maturation of a cell center is not yet understood in any detail, but there are two complementary studies that indicate strongly that an RNA specific to the centriole is in some way involved. Zackroff et al. (1975) have shown that mitotic centers from *Spissula* eggs treated with RNase initiate smaller asters than untreated centers. Weisenberg (1976) has run numerous controls on this observation, showing that either basic or acidic RNases will have the effect, that poisoning the enzymatic activity of the RNase stops the effect, and that DNases have no effect.

There are several cytochemical studies of structures similar to centrioles called basal bodies, showing that these organelles contain RNA (Hartman et al., 1974; Dipple, 1976). A basal body looks essentially the same as a centriole, and several organisms exist in which one is converted into the other, so it seems possible that Weisenberg's results stem from a direct effect of the RNase on the centrioles. This possibility has recently been given considerable support by the work of Heidemann and Kirschner (1975, 1977). They have purified basal bodies from *Chlamydomonas* and *Tetrahymena* and then injected them into the cytoplasm of eggs from the frog *Xenopus laevis*. In activated eggs where the egg nucleus has already broken down, a large number of asters form about 1 h after the injection of the basal body preparation. A substantial number of control injections have no aster-inducing effect: e.g.., soluble tubulin, microtubules, fragments of flagella, or fractions from bacteria. Since the number of asters that form is directly related to the number of basal bodies injected, these workers have concluded that the egg can induce the maturation of the basal bodies, converting a substantial, but unknown fraction of them into active mitotic centers that proceed to organize the egg tubulin into asters. Just as in *Spissula*, aster formation is not seen in experiments on eggs in which the nuclear envelope has not yet broken down.

Heidemann and Kirschner (1977) recently found that treatment of the basal bodies after isolation but before injection with RNase will completely block the capacity of the egg to form asters after basal body injection. The controls they have run exclude the possibility that the small amount of RNase injected with the basal bodies is responsible for the effect. Again, RNase of several types has been used with good effect. DNases have no effect, and while protease treatment of the basal bodies reduces the subsequent astral growth upon injection, the effect requires extensive proteolysis, and the electron microscope shows that these treated basal bodies have been morphologically damaged by the enzyme. With the RNase effect, on the other hand, only small amounts of enzyme are required, the effect is rapid, and there is no obvious damage to the general structure of the basal body. (Detailed fine-structure study of treated basal bodies has not yet been done.) Large numbers of RNase-treated basal bodies have been injected into eggs at the proper stage with no subsequent formation of asters et all, indicating strongly that an RNase-sensitive component in the basal body fraction is required for the egg to be able to make mature mitotic centers from the injected material. The connection between these results and the observation that centriole maturation in vivo is associated with an accumulation of pericentriolar material is not yet clear. Dipple (1975) finds RNase-sensitive material inside a basal body in *Paramecium*, and there is as yet no information concerning the chemical composition of the pericentriolar material. The closing of the gap between these different findings is a field of active research in several labs.

In summary, the in vitro studies with spindle models have provided direct evidence that in at least some senses the spindle is legitimately regarded as a dynamic equilibrium between ordered and disordered states of some subunit, as postulated by Inoué and Sato (1967) and by Dietz (1969) on the basis of optical studies of the living spindles. The fact that in vitro the spindle loses its lability with time after cell lysis may mean that spindle assembly and disassembly normally occur by different paths and that the physiologic pathway for disassembly fails to survive in the conditions used for lysis. Certainly microtubule structures are generally disassembled by different routes than those by which they are assembled (Bloodgood, 1974), and the spindle may be no exception. The "head-to-tail" polymerization of polymers like actin and tubulin, identified by Wegner (1976), may be important in this asymmetry of assembly and disassembly. Further, if chromosome movement is actually caused by controlled disassembly of spindle tubules, as suggested by Inoué and Sato (1967) and Dietz (1969, 1972), then the loss of spindle lability would account for the difficulties experienced in making chromosomes move on isolated spindles.

The chemical nature of the maturation process by which a cell center is changed into a mitotic center stands out as one of the central problems in this field. The identification of centriole-associated RNA as a likely candidate for some role in the process is an important step, and future work will probably clarify the role of this RNA and other macromolecular components.

The insufficiency of heterologous microtubule protein to support a transition from prometaphase to metaphase in vitro, or even bring about some separation of adjacent mitotic centers (except when they are already serving as poles of a spindle formed in vivo), underscores how far we really are at the present time from understanding the chemistry of the mitotic spindle. There may not only be a difference between neural and mitotic tubulin, there may also be a complete set of microtubule-associated, spindle-specific proteins whose identity is not yet known. Such proteins may well be required before we can realize a phenomenon truly analogous to spindle formation in vitro.

2. *Chromosome Movement*

Cande et al. (1974) found that mammalian cells lysed at anaphase with low concentrations of Triton X-100 into the right tubulin-containing buffers would continue to move chromosomes after lysis. The rate and extent of chromosome movement are approximately normal when the lysis mixture contains 2 mM ATP, 4 mM $MgCl_2$, and 1 mM dithiothreitol in addition to 1 mM GTP, about 2 mg/ml of microtubule protein and the buffers normally used in tubulin assembly studies (McIntosh et al., 1975a). Low detergent concentrations are required to preserve chromosome motion, but there is good evidence that the cells are well lysed by the treatment. In the light microscope, the cell surface is seen to bleb and vesiculate with the arrival of the detergent. The mitochondria pale and disappear. The spindle poles become visible with differential interference contrast optics. The phase contrast of the chromosomes substantially increases. In the electron microscope, cells fixed after detergent treatment and during chromosome motion show damaged mitochondria and extracted cytoplasm, and the texture of the chromosomes is abnormal. If tubulin is replaced during lysis by equal concentrations of bovine serum albumin, the chromosomes stop moving within a few seconds and the spindle fades away, usually within 90 s. We conclude that under our conditions for lysis, the chromosomes continue to move after the cell's permeability barrier has been broken.

The useful lysis conditions are rather sensitive to Triton concentration. There is only about a factor of three between the minimum concentration that will give rapid and effective lysis and the maximum that the motile properties of the resulting cell model will tolerate. The capacity of tubulin to polymerize and to add to microtubule organizing centers in vitro is much less sensitive to detergent, suggesting that some other spindle component is being affected by the treatment.

We have tried to use this model system to identify the enzymes important for chromosome movement, but have experienced little success so far. Antiserum directed against actin (Lazarides and Weber, 1974) or DNase 1, which binds strongly to actin (Lazarides and Lindberg, 1974), or phalloidin, which also binds to actin and makes it very stable (Wieland and Govindan, 1974) has no effect on the rate of chromosome movement. Since we find that the same reagents do not interfere with the contraction of glycerinated muscle fibers upon the addition of ATP, it is difficult to conclude anything from the observations (McIntosh et al., 1975a). We have occasionally obtained a dramatic restart of salt-stopped chromosomes by washing out the excess salt with tubulin plus a dynein-containing extract of starfish sperm tails (McIntosh et al., 1975a), but this result has been so difficult to repeat that the six convincing positives obtained with two batches of sperm are impossible to interpret clearly.

The most reasonable inference concerning the lysed mammalian cell model seems to be that it is presently too complicated and too fragile to be very useful. If we can find conditions that allows more extreme lysis, yet preserve chromosome motility, those would help us to extract the cell more thoroughly and simplify the chemistry of the system. A better knowledge of general spindle chemistry might give the necessary insights into the proteinaceous components that we should add to the preparations of microtubule protein to provide a proper milieu for stability of function after lysis. Probably there are several spindle proteins in labile equilibrium between assembled and disassembled forms, so it is not really surprising that brain microtubule protein cannot do the whole job. The composition studies that could be done on spindle isolates would be very helpful to further work on the lysed cell models.

3. Spindle Composition

Spindle models have been used for certain cytochemical examinations of the protein composition of the MA. Sanger (1975) has used glycerinated PtK cells as specimens for study of the binding of fluorescein-labeled heavy meromyosin to nonmuscle cells. His work on mitotic cells is described elsewhere in this volume and need not be treated here. Cande et al. (1977) used the Triton-extracted models decribed above to study the relation between the distribution of actin and tubulin in the mammalian spindle. Their results confirm Sanger's findings in most ways and exclude certain possible sources of artifact present in the glycerinated material. It therefore seems likely that there is actin in spindle models, but there is, as yet, no functional evidence to allow any conclusions about the significance of this actin for chromosome movement.

II. Spindle Isolates

Rebhun et al. (1974) used a modification of Weisenberg's method for assembling neurotubulin to isolate MAs from several marine eggs. *Spissula* was the material of choice, and its MA was found to be stable in Weisenberg's buffer supplemented with additional EGTA, a proteolysis inhibitor, and 0.2-1 % Triton. No exogenous tubulin is necessary in this procedure to preserve a birefringence spindle in vitro. The MAs retain their birefringence and lifelike appearance through partial purification by differential centrifugation. They are cold labile, although they lose their cold lability rather rapidly. MAs cooled to 0°C, 5-6 min after isolation, will lose only 1/2 to 3/4 of their birefringence over about 15 min. MAs isolated in buffer supplemented with polymerization-competent neurotubulin are cold labile for a longer period. The MAs will lose birefringence upon cooling and regain it when the preparation is rewarmed. If the tubulin concentration is high, the reformed spindles will have more birefringence than the original isolate, and over a 2 h period, the pole-to-pole distance can increase in length by a factor of 3 or 4. The regrown spindles are still cold labile, but like the original isolates, they are insensitive to tubulin dilution. No conditions have been found in which the spindle will shorten. Birefringence loss and gain occur essentially uniformly over the entire spindle body in a fashion resembling both the in vitro spindles described above and the living marine egg spindle as it responds to rapid cooling and warming (Inoué et al., 1975).

These spindle isolates are similar in all properties described to the *Chaetopterus* spindles reviewed above. They are much like the mammalian spindles studied in our lab, but differ from them in three ways: 1) The marine spindles will elongate but the mammalian will not. 2) The mammalian spindles will shorten in low concentrations of tubulin, but the marine spindles will not. 3) The spindles of gently lysed mammalian cells will move chromosomes. These latter two properties may be functionally related.

Sakai et al. (1975) used the procedure of Sakai and Kuriyama (1974) mentioned above, a protease inhibitor, and polymerizable brain tubulin to isolate spindles that may be the most lifelike isolates yet obtained. Their spindles are cold labile, solubilized by a sulfhydryl reagent and by Ca^{2+} ions, but insensitive to 10^{-4} M colchicine. The addition of 3 mg/ml of polymerizable tubulin to the isolates will induce them to increase in both birefringence and length.

When anaphase spindles are treated with these high concentrations of tubulin, the chromosomes and spindle poles move, but in a nonphysiologic way: Accurate chromosome segregation does not occur, and although the pole-to-pole distance increases, the chromosome-to-pole distance does too, suggesting a general microtubule eleongation rather than a model of anaphase. In 0.5 mg/ml of tubulin, however, the pole-to-pole distance still increase, and the chromosome-to-pole distance decreases by a small but measurable amount. The motion is reproducible, but not dramatic. The chromosomes move about half the way to the poles (between 2 and 5 μm) during 1 h incubation. In vivo they would have gone more than twice as far in less than 5 min. The in vitro motion, though small, requires ATP; GTP will not serve. The Japanese group is certainly pursuing this promising start, and a published abstract suggests exciting things to come. Mohri et al. (1976) used an antiserum prepared against dynein from sea urchin sperm tails and indirect immunofluorescence to examine both fixed eggs and isolated MAs. Dynein antigenicity is associated with the spindle in both cases. Further, they found that anti-dynein inhibits the limited chromosome movement in the isolated spindles, while antiserum prepared against sea urchin myosin does not. These suggestive findings should soon appear in a more complete puslished form.

WHAT IS ITS MOLECULE . .

D. Summary and Conclusions

Several experimental systems have now been developed for studying aspects of mitosis in vitro. As one might expect, the gently lysed cell models generally preserve physiologic function better than the isolates, but these systems are still sufficiently complex biochemically that straightforward answers to simple questions are not easy to find. Nonetheless, the studies on both the models and the isolates have provided some new evidence for old ideas, suggested a few new thoughts, and provided a wealth of opportunity for future experimental work.

The concept of the spindle as a labile equilibrium between monomeric and polymeric states of subunits, as developed by Inoué and Sato (1967) and by Dietz (1969, 1972), has been given some direct support by the observation that low temperature and Ca^{2+} ions solubilize the in vitro MAs and that high concentrations of exogenous tubulin induce an increase in spindle birefringence for all MAs and an increase of length for all but the mammalian MA. It is noteworthy, however, that under conditions in which the in vitro MA is thermolabile, it is not labile to subunit dilution, as one might except for a system in equilibrium. The only in vitro spindle that will shorten is the mammalian model system. This observation suggests that spindle tubules can fall apart when the physiochemical environment no longer favors assembly, but that under "polymerization conditions" where subunits in the microtubule wall can be expected to be comparatively strongly bound to one another, the rate of tubule disassembly is low. Perhaps there is some hindrance to the loss of subunits from the microtubule ends that prevents their shortening. Since the isolated spindle can grow but not shrink, it is not in a simple equilibrium under conditions of low monomer concentration. The shrinking of the mammalian MA upon lysis in low tubulin is reminiscent of the response of the marine egg spindles in vivo to slow cooling or moderate hydrostatic pressure treatment (Inoué et al., 1975). Perhaps both phenomena are related to the close coupling between disassembly of the kinetochore tubule and the motion of the chromosomes to the poles. Inoué and Dietz have both suggested that the disassembly is causal for chromosome movement. An equally possible alternative is that there is a labile machinery that pulls the chromosomes toward the poles, but is prevented from moving them in part by the intervening microtubules. Disassembly of some spindle tubules is an essential prerequisite for chromosome-to-pole movement, but need not cause it. The presence of actin in this region of the spindle, as judged both by fluorescent heavy meromyosin binding and by indirect immunofluorescence, suggests that this protein is likely to contribute to the mechanical system. Perhaps the myosin seen with immunofluorescence by Fujiwara and Pollard (1976) in the region between chromosome and pole is another part of it. The lack of spindle shortening in the isolated MA would then be seen as a different manifestation of the same factors that make chromosome movement small and slow or nonexistent in the MA isolates. Presumably the functioning of this motor is labile to the current methods of isolation.

Two in vitro systems provide limited but direct evidence for the importance of dynein in chromosome movement. The data from both systems are subject to criticism, but hopefully more time will clarify the situation. It is important to recognize that the evidence for dynein is not contradictory to the evidence for actin and myosin or to the evidence showing the importance of microtubule disassembly for chromosome movement, because these systems are not mutually exclusive. There is even a certain elegance to the idea that a spindle may have evolved with a "fail-safe" design incorporating mechanochemically independent systems in the same structure to assure the equipartition of the chromosomes (McIntosh et al., 1976). It seems likely to me that controlled assembly and disassembly of microtubules are regulators for rate and extent of the mechanical action of an actomyosin motor that pulls the chromosomes to the poles and a dynein motor that slides the two polar spindles apart.

Controlled microtubule assembly is of course essential for normal spindle formation. The in vitro systems provide solid evidence for the idea that mitotic centers and kinetochores are cellular sites for the initiation of microtubule polymerization. They further implicate these loci as sites of the temporal control for spindle formation. This problem of time-dependent maturation of the poles and kinetochores is now sharply defined and will be an active area of work in coming years. The discovery that RNA has something to do with the maturation of cell centers is one that stimulates the imagination, but I will leave the pleasure of responding to that stimulus to the reader.

The field of molecular biology of mitosis is by now established as a worthy area for research, and it is reasonable to hope that the future efforts of the chemists on purified material and of cytologists on complex mixtures will nurture each other, promoting the development of a substantial body of hard facts about the mechanisms of mitosis, both in vitro and in vivo.

References

Bibring, T., Baxandall, J.: Science 161, 377 (1968)
Bibring, T., Baxandall, J.: J. Cell Biol. 48, 324 (1971)
Bibring, T., Cousineau, G.H.: Nature (London) 204, 805 (1964)
Bloodgood, R.A.: Cytobiol. 9, 143, (1974)
Borisy, G.G., Olmsted, J.B.: Science 177, 1196 (1972)
Borisy, G.G., Peterson, J., Hyams, L., Ris, H.: J. Cell Biol. 67, 38a (1975)
Borisy, G.G., Taylor, E.W.: J. Cell Biol. 34, 535 (1967)
Brinkley, B.R., Cartwright, J.: Ann. N.Y. Acad. Sci. 253, 428 (1975)
Bryan, J., Sato, H.: Exp. Cell Res. 59, 371 (1970)
Byers, B., Goetsch, L.: J. Bacteriol. 124, 511 (1975)
Cande, W.Z., Lazarides, E., McIntosh, J.R.: J. Cell Biol. 72, 552 (1977)
Cande, W.Z., Snyder, J., Smith, D., Summers, K., McIntosh, J.R.: Proc. Natl. Acad.
   Sci. U.S. 71, 1559 (1974)
Cohen, W.D., Gottlieb, T.: J. Cell Sci. 9, 603 (1971)
Cohen, W.D., Rebhun, L.I.: J. Cell Sci. 6, 159 (1970)
Dietz, R.: Naturwissenschaften 56, 237 (1969)
Dietz, R.: Chromosoma (Berl.) 38, 11 (1972)
Dipple, R.V.: J. Cell Biol. 69, 622 (1976)
Dirksen, E.R.: Exptl. Cell Res. 36, 256 (1964)
Filner, P., Behnke, O.: J. Cell Biol. 59, 99a (1973)
Forer, A., Goldman, R.D.: J. Cell Sci. 10, 387 (1972)
Forer, A., Kalnins, V.I., Zimmerman, A.M.: J. Cell Sci. 22, 115 (1976)
Forer, A., Zimmerman, A.M.: J. Cell Sci. 16, 481 (1974)

PARTICULE PHYSICS
BIG BANG CHEMISTRY

Forer, A., Zimmerman, A.M.: J. Cell Sci. 20, 309 (1976a)
Forer, A., Zimmerman, A.M.: J. Cell Sci. 20, 329 (1976b)
Fujiwara, K., Pollard, T.D.: J. Cell Biol. 71, 848 (1976)
Goode, D.: J. Mol. Biol. 80, 531-538 (1973)
Goode, D., Roth, L.E.: Exp. Cell Res. 58, 343 (1969)
Hartman, H., Puma, J.P., Gurney, T.: J. Cell Sci. 16, 241 (1974)
Heidemann, S., Kirschner, M.S.: J. Cell Biol. 67, 105 (1975)
Heidemann, S., Sander, G., Kirschner, M.S.: Cell 10, 337-350 (1977)
Hoffman-Berling, H.: Biochim. Biophys. Acta 15, 226 (1954)
Inoué, S., Borisy, G.G., Kiehart, D.P.: J. Cell Biol. 62, 175 (1974)
Inoué, S., Fuseler, J., Salmon, E.D., Ellis, G.W.: Biophys. J. 15, 725 (1975)
Inoué, S., Sato, H.: J. Gen. Physiol. 50 (Suppl.), 259 (1967)
Kane, R.E.: J. Cell Biol. 12, 47 (1962)
Kane, R.E.: J. Cell Biol. 25, 137 (1965)
Kiefer, B., Sakai, H., Solari, A.J., Mazia, D.: J. Mol. Biol. 20, 75 (1966)
Lazarides, E., Weber, K.: Proc. Natl. Acad. Sci. U.S. 71, 2268 (1974)
Lazarides, E., Lindberg, U.: Proc. Natl. Acad. Sci. U.S. 71, 4742 (1974)
Luduena, R.F., Woodward, D.O.: Proc. Natl. Acad. Sci. U.S. 70, 3594 (1973)
Marsland, D., Zimmerman, A.M.: Exptl. Cell Res. 38, 306 (1965)
Mazia, D., Dan, K.: Proc. Natl. Acad. Sci. U.S. 38, 826 (1952)
Mazia, D., Mitchison, J.M., Medina, H., Harris, P.: J. Biochem. Biophys. Cytol.
    10, 467 (1961)
Mazia, D., Petzelt, Ch., Williams, R.O., Meza, I.: Exptl. Cell Res. 70, 325 (1972)
McGill, M., Brinkley, B.R.: J. Cell Biol. 67, 189 (1975)
McIntosh, J.R., Cande, W.Z., Lazarides, E., McDonald, K., Snyder, J.A.: In: Cell
    Motility. Goldman, R., Pollard, T.D., Rosenbaum, J. (eds.). New York: Cold Spring
    Harbor Press 1976, Vol. 2, p. 1028
McIntosh, J.R., Cande, W.Z., Snyder, J.A.: In: Molecules in Cell Movement. Inoué,
    S., Stephens, R.E. (eds.). New York: Raven Press 1975a, p. 31
McIntosh, J.R., Cande, W.Z., Snyder, J.A., Vanderslice, K.: Ann. N.Y. Acad. Sci.
    253, 407 (1975b)
McIntosh, J.R., Landis, S.C.: J. Cell Biol. 49, 468 (1971)
Miki, T.: Exptl. Cell Res. 29, 92 (1963)
Miki-Noumura, T.: Embryologia 9, 98 (1965)
Miki-Noumura, T.: Exptl. Cell Res. 50, 54 (1968)
Mohri, M., Mabuchi, I., Ozawa, K., Kuriyama, R., Sakai, H.: In: Contractile Sys-
    tems in Non-muscle Tissues. Perry, M., Adelstein (eds.). Amsterdam: North-
    Holland 1977, p. 336
Petzelt, C.: Exptl. Cell Res. 70, 333 (1972a)
Petzelt, C.: Exptl. Cell Res. 74, 156 (1972b)
Pickett-Heaps, J.: Cytobios. 3, 205 (1971)
Rebhun, L.I., Mellon, M., Jemiolo, J., Nath, J., Ivy, N.: In: The Biology of Cyto-
    plasmic Microtubules. Soifer, D., (ed.). Ann. N.Y. Acad. Sci. 253   (1975)
Rebhun, L.I., Rosenbaum, J., Lefebvre, P., Smith, G.: Nature (London) 249, 113
    (1974)
Rebhun, L.I., Sander, G.: J. Cell Biol. 34, 859 (1967)
Robbins, W., Gonatas, N.K.: J. Cell Biol. 21, 429 (1964)
Robbins, E., Jentzsch, G., Micali, A.: J. Cell Biol. 36, 329 (1968)
Rustad, R.C.: Exptl. Cell Res. 16, 575 (1959)
Sakai, H.: Biochim. Biophys. Acta 112, 132 (1966)
Sakai, H., Hiramoto, Y., Kuriyama, R.: Develop. Growth and Differ. 17, 265 (1975)
Sakai, H., Kuriyama, R.: Develop. Growth and Differ. 16, 123 (1974)
Sanger, J.: Proc. Natl. Acad. Sci. U.S. 72, 2451 (1975)
Sato, H., Bryan, J.: J. Cell Biol. 39, 118a (1968)
Sato, H., Ellis, G.W., Inoué, S.: J. Cell Biol. 67, 501 (1975)
Sisken, J.E.: In: Methods in Cell Physiology. Prescott, D.M. (ed.). New York: Aca-
    demic Press 1970, Vol. 4, p. 71
Sisken, J.E., Wilkes, E., Donnelly, G.M., Kakefuda, T.: J. Cell Biol. 32, 212 (1967)
Snyder, J.A., McIntosh, J.R.: J. Cell Biol. 67, 144 (1975)
Snyder, J.A., McIntosh, J.R.: Ann. Rev. Biochem. 45, 699 (1976)
Telser, B., Rosenbaum, J.: Proc. Natl. Acad. Sci. U.S. 72, 4023 (1975

Wegner, A.: J. Mol. Biol. <u>108</u>, 139 (1976)

Weisenberg, R.C.: Science <u>177</u>, 1104 (1972a)

Weisenberg, R.C.: J. Cell Biol. <u>55</u>, 277a (1972b)

Weisenberg, R.C., Rosenfeld, A.C.: J. Cell Biol. <u>64</u>, 146 (1975)

Weisenberg, R.C., Taylor, E.W.: Exptl. Cell Res. <u>53</u>, 372 (1968)

Wieland, T.H., Govindan, V.M., FEBS Letters <u>46</u>, 351 (1974)

Wilson, L., Bryan, J.: Adv. Cell Molec. Biol. <u>3</u>, 21 (1974)

Wilson, L., Friedkin, M.: Biochemistry <u>6</u>, 3126 (1967)

Wilt, F.H., Sakai, H., Mazia, D.: J. Mol. Biol. <u>27</u>, 1 (1967)

Wray, W., Stubblefield, E.: Exptl. Cell Res. <u>59</u>, 469 (1970)

Zackroff, R., Rosenfeld, A., Weisenberg, R.: J. Cell Biol. <u>67</u>, 469a (1975)

Discussion Session VIII: Mitosis in vitro
Chairman: C. PETZELT, Heidelberg, F.R.G.

J.W. Sanger, Philadelphia, Pennsylvania, USA
In your experiments where you used the myosin antibodies to see if they could stop chromosome movement, do you know if the antibody was made against the rod portion or the head region of the myosin molecule?

J.R. McIntosh, Boulder, Colorado, USA
I do know that it was made against myosin from the same organism, and it is not stated in the published abstract whether it is rod or head. As you well know, almost all antibodies against myosin are to the rod portion and therefore, may legitimately not have an effect upon myosin ATPase activity and function.

H. Ponstingl, Heidelberg, F.R.G.
We have now so many muscle-like molecules in the cell that may be slightly different from muscle but may have the same function. Is work proceeding to isolate these molecules by affinity columns and antibody columns, mix them with the tubulin broth, add it to the lysed cell, and check for effects?

J.R. McIntosh, Boulder, Colorado, USA
We have tried such experiments in two ways. One way is to use platelet myosin, skeleton muscle α-actinin, and skeleton muscle actin together with the tubulin lysis mixture to see if we could make the lysed cell system more resistant to detergent destruction. These experiments have had no positive results. We have also just used crude cell extracts. Based on the premise that we did not know what we should add, we added everything, and those have given us no positive results.

H. Ponstingl, Heidelberg, F.R.G.
Here in the Deutsches Krebsforschungszentrum another group found that detergents irreversibly bind to tubulin. Apparently there is a large hydrophobic site, which should be able to bind Triton. This may have an effect, too.

J.R. McIntosh, Boulder, Colorado, USA
In our hands the Triton effect upon tubulin assembly properties of the models is mild. All of the tubulin lysis experiments that I described first, which had to do with spindle assembly and birefringence, were done in 0.1 % Triton. We found that in 0.2 % Triton we get the same critical concentration for tubulin assembly mixture as without the Triton. So, although it may bind, it does not seem to affect that. However, your point is applicable to the things we do not understand in the spindle - the molecules we do not yet know how to add. They may have a detergent effect. What are the detergents that bind? I am not familiar with this work.

H. Ponstingl, Heidelberg, F.R.G.
There is a published paper on charge shift electrophoresis by Helenius and Simons (Proc. Natl. Acad. Sci., U.S. 74, 529-532, 1977). They think that they can determine whether a protein is a membrane protein by binding positively or negatively charged detergents, cetyltrimethylammoniumbromide and sodium desoxycholate, respectively, to the molecule, performing electrophoresis, and seeing whether there is a shift in mobility. This has also been done for tubulin.

W.W. Franke, Heidelberg, F.R.G.
You are giving percent figures of the detergent. Could you indicate the amount of detergent per milligram protein, in particular, of course, cell protein?

J.R. McIntosh, Boulder, Colorado, USA
When microtubule protein is 1 mg/ml, it is a 0.1 % solution. The percentage of detergent to tubulin is the same on a mass basis, and I do not know the relative molarities. I do not know the relationship between detergent and the amount of macromolecular material in the cells on a coverslip.

W.W. Franke, Heidelberg, F.R.G.
Under your conditions of detergent-induced cell lysis how much of the cellular lipids, in particular phospholipids, were extracted?

J.R. McIntosh, Boulder, Colorado, USA
We have avoided investigating the chemistry of the extracts because we do not have a synchronous cell population. We are working with log-phase cells, and we isolated the cells only in the sense that we look at one with the microscope. It is a property of our cells that the mitotic cells are far more resistant to detergents than the interphase cells and, therefore, to try to make a claim on chemistry would be foolish.

Chairman:
What proof do we have that a cell is lysed or that its membrane has become permeable to the larger molecules outside the cell?

J.R. McIntosh, Boulder, Colorado, USA
In my mind I have broken the lysis phenomenon into two parts. The first is lysis and by that I mean a change in the permeability properties in the cell membrane. Following lysis there is extraction. Lysis is difficult to measure because one would be asking the question: In what way is the permeability of the membrane changed? One could do transport studies, but the only studies we have done are the ones I have described. If tubulin can get in and out or if the absence of tubulin on the outside is important, then I presume there must have been a change in the permeability barrier. Following that change, one sees extraction. Extraction is much easier to measure than lysis because one can see changes in cytoplasm. What we are now set up to do but have not yet done, and I would recommend this as an approach to any of you who would like to do lysed cell studies, is to obtain a quantitative interference microscope. With this tool one can measure the changes in refractive index of the cytoplasm during the extraction process and one can make time-dependent plots of dry mass per unit area in the cell as it is extracted. These observations - I cannot say plot, because we have not yet quantitated these changes - show us that you can see what is going on in the way of extraction in a quantitative way. We hope to use this as a tool to analyze detergents because, as Dr. Ponstingl implied, the particular detergent you use may make a lot of difference. I just add one more fact as a warning: We have used the detergent series Brij, Nonidet as well as Triton. In our hands Nonidet is very similar to Triton and we have no reason to choose between them. The Brij detergents looked extremely promising for a while because we got beautifully reversible stop-starts in this system. Things seemed much more stable. Chromosome movement would survive at much higher concentration in Brij 32 than it would in Triton X-100. When we examined the cells under the electron microscope, we found that the cell membrane was utterly gone from these cells, unlike the Triton picture that I showed you. There was just no plasmalemma left. However, the cytoplasm is not at all extracted. In other words, with Brij the cytoplasm simply does not disperse. When we finally got around to doing the tubulin controls, we found that the Brij models went perfectly well in the absence of tubulin despite the fact that the membrane was gone. So one must be very careful to study lysis both in the sense of permeability barrier changes and of extraction properties, and be on the lookout for very confusing surprises.
We tried in several cases to obtain some analogue for real spindle growth and I want to ask the assembled multitude if they have ever had success. When we take a cell in which the two mitotic centers are very close, i.e., within 2 μm of another, as a result of the condition when a cell is lysed, we can grow microtubules that are 10 and 20 μm long, and they do not push the mitotic centers apart. It has always been my feeling, and it is certainly part of the *gestalt* of spindle formation in the literature, that tubule growth pushes the mitotic centers apart, and yet, we failed utterly to see this.

D. Mazia, Berkeley, California, USA
The evidence for the idea that the separation of the centers is an action of microtubules is not very strong. It is an attractive idea that we derive from observation of cases where we can distinguish a central spindle, the pole-to-pole fibers elongating as the poles move apart. Serious research will have to be directed toward the polarization of the spindle as a problem in its own right.

M. Girbardt, Jena, G.D.R.
When treated by cold shock, globular entities of the fungal nucleus-associated organelle will move a short distance without microtubules.

N. Paweletz, Heidelberg, F.R.G.
Dr. McIntosh, can you state briefly the conclusions concerning ultrastructure of the spindle that you can make from comparing lysed and unlysed cells?

J.R. McIntosh, Boulder, Colorado, USA
No conclusions can be made. One advantage is that, as a result of lysis, one has extracted some material from the spindle. What remains is easier to see. That was the key factor in the high-voltage microscope thick sections that I presented. However, if fixation artifacts present difficulties for accurate fine-structure observations, lysis fixation artifacts are a worse problem. Therefore, I feel that, until I can say that this spindle works, detailed fine-structure analysis is not particularly worthwhile.

A.M. Lambert, Strasbourg, France
Did you find any microfilaments in your lysed cells?

J.R. McIntosh, Boulder, Colorado, USA
If we lyse in regular tubulin we find very few microfilaments. If we lyse with phalloidin we find a rat's nest.

U.-P. Roos, Zürich, Switzerland
Has not Dr. Cande in your laboratory studied, by indirect immunofluorescence, the occurrence and distribution of actin in lysed cell?

J.R. McIntosh, Boulder, Colorado, USA
Almost all of the immunofluorescent pictures in a recently published report are of lysed cells. To my mind, it is more than of passing interest that the electron microscope reveals very few microfilaments and yet the immunofluorescence reveals substantial but unquantified amount of actin. There are so many possible explanations for this, the two most obvious being that not all the actin is in microfilamentous form and that we are not preserving the actin in glutaraldehyde fixation for electron microscopy. I do not think we can interpret the observation.

M. Osborn, Göttingen, F.R.G.
We have shown (Osborn and Weber, Exp. Cell Res. 106, 339, 1977) that treatment with nonionic detergents of cells such as 3T3 and chick embryo fibroblasts leaves the nucleus and their microfilament bundles behind. Gels of the corresponding fraction support this idea. Work from Spudich's laboratory by gels and by electron microscopy also support the idea that the microfilament bundles are not extracted by these treatments.
A comment to Dr. McIntosh. There is one control that might be done on your cells to monitor permeability. Treatment of interphase cells with formalin alone is not sufficient to allow the antibodies to penetrate inside the cell. However, after detergent treatment the antibodies do penetrate without the usual organic solvent treatment. Therefore it might be interesting to find out in your system at what time the cells become permeable to antibodies, e.g., by seeing at what time the microfilament bundles can be stained with actin antibody.

J.R. McIntosh, Boulder, Colorado, USA
This brings up an interesting point with respect ot actin preservation. One knows from the electrofocusing result that there are multiple actins, and it is clear from the studies on the assembly chemistry of nonmuscle versus muscle actin that different actins have different assembly properties and that one must not generalize from the fibrous actin that one sees where it is copolymerized with a α-actinin, tropomyosin and myosin into well-defined bundles to the obviously ephemeral actin of the spindle. F-Actin could perhaps be renamed artifactin for the spindle because one has a clear sense that you do not know what is going on. I would not feel that the preservation of a fiber in a cell would be in any sense a control for the preservation of spindle actin.

J.W. Sanger, Philadelphia, Pennsylvania, USA
When you stain the lysed cell with the antibody against actin, have you looked
in the electron microscope to see if you have filaments under those cases?

J.R. McIntosh, Boulder, Colorado, USA
When we look in the electron microscope after antiactin staining, which is a
reasonable thing to do because you figure that perhaps the antibody is a fixative
by itself, one sees filaments. The filaments are, in the words of Ray Stephens,
"ugly fibers." The spindle is poorly preserved by the formaldehyde treatment. The
number of microtubules in the spindle is less than the physiologic number. I
have no way of knowing whether a particular ugly fiber is an actin fiber thickended
by the antibody or a microtubule protofilament bundle which resulted from the fixation breakdown of the microtubules. If this would be done with good electron
microscope immunofluorescence, one might learn something interesting.

J.W. Sanger, Philadelphia, Pennsylvania, USA
Work of Dr. L. Lewis shows that in the micronucleus of a paramecium during mitosis
microfilaments are very abundant. They interact laterally with microtubules. Moreover, these microfilaments attach to the kinetochore of the chromosomes (this
material was obtained by using standard fixation of glutaraldehyde and osmium
fixation). The filaments are found in the micronucleus only during mitosis. The
chemical identity of these filaments has not yet been determined.

N. Paweletz, Heidelberg, F.R.G.
Dr. McIntosh, I have observed large amounts of vesicles in your electron micrographs. In relation to interactions between membranes and tubulin or microtubules,
would lysed cells be a proper model to study these problems or is the larger
number only due to thick sectioning and extraction of obscuring material?

J.R. McIntosh, Boulder, Colorado, USA
I have no numbers, but it is my impression that the vesicles are there in the PtK-
spindle when one fixes directly.

Dr. Mazia, Berkeley, California, USA
Dr. McIntosh has defined "lysis" in a commendably generous way, to include various
means of applying known reagents to the mitotic apparatus. I wonder if the future
of this kind of work includes the use of various strategems that have been developed by the membrane biologists. One would be the use of osmotic changes in the
way they are employed in the study of reversible hemolysis. Another would be the
"injection" of various media by fusion of cells with liposomes or ghosts containing these media.

W.W. Franke, Heidelberg, F.R.G.
As to the identification of such fibrils and filaments associated with the spindle,
do you observe any filaments that resist further treatment with detergent and
with very high and low salt concentrations? In other words, do you find residual
material with preparative characteristics similar to the matrix polypeptides?

J.R. McIntosh, Boulder, Colorado, USA
In the spindle the answer for our material is clearly no. In the cytoplasm that is
not the case. PtK cells have a substantial number of intermediate-sized filaments.
These are extremely stable and survive any treatment to which we have subjected
the cells. May I say a word to Dr. Mazia's comment. You have pointed out that one
can generalize the idea of lysis away from the sense in which I have used it. In
my opinion there is a caution that a cell is so complex that models can be useful
only when you can ask some straightforward questions and get straightforward answers. The glycerinated model is so useful for muscle because it is the contractile machine with most other things removed. One of the difficulties we face in
the spindle is that the cell seems to be able to react with the spindle in a number
of ways. It is my opinion that the cell has a specific method for disassembling
the spindle. That opinion is based upon the fact that a spindle in a cell lysed
slowly is much more labile than a spindle in a cell lysed fast. Everything else is
the same. If you put in 0.2 % Triton you can isolate a mammalian spindle without tubulin, whereas in 0.02 % Triton the spindle will just shrink down to nothingness,

suggesting that there is a detergent-labile disassembling function in the cell. One of the difficulties with sticking close to the cell is that one has all the magic one does not understand. I would argue that models should strive to get as far away from the cell as possible.

M. Bardele, Tübingen, F.R.G.
I have a question concerning the RNA in the mitotic apparatus or in the nucleation centers. We know that the RNA inhibits the in vitro polymerization of tubulin in certain cases. I would like to know whether the RNA functions as a messenger. This could perhaps be tested if you use basal bodies from *Tetrahymena* and inject them into the eggs of *Xenopus* at the same time at which protein synthesis is inhibited and see if it works. Is there any relation to the function of RNA as a mediator of microtubule initiation as, e.g., we know that RNA in tobacco mosaic virus can mediate the assembly of that protein.

J.R. McIntosh, Boulder, Colorado, USA
You raised several interesting points at once. My recollection is that Heideman and Kirschner have done the experiments you suggested. They injected protein synthesis inhibitors with the basal bodies, and it does not effect the result. They tried to isolate the RNA from the basal bodies and to determine if it could be translated in vitro, and they were not able to prepare basal bodies in such a way that the RNA is a big piece - it comes out as little fragments which are heterogeneous. They believe it is denatured during the isolation process. You pointed out that there is an interesting duality in the literature because Bryan has this elegant system for using RNA to suppress spontaneous initiation in vitro, which apparently concerns the ion exchange properties of RNA, whereas here we are talking about an utterly different function for RNA. It is not that the RNA itself directly initiate tubules, because, as Borisy has shown, tubules will grow off the end of a basal body, indicating that it is a centriole, and Snell and Rosenbaum and Heideman and Kirschner have shown the same thing. There you do not see the RNA that I presume is present in all the preparations having anything to do with microtubule initiation. The RNase-sensitive material is only useful to the egg at this point. It knows how to do something with it that we do not know how to do. The question of what the egg does with that RNA is a very interesting one, and I think it is probably the secret of understanding this maturation process that turns the cell center into a mitotic center.

J. Bryan, Philadelphia, Pennsylvania, USA
We can perhaps reconcile the dualism described by Dr. McIntosh, that RNA inhibits neurotubulin assembly and that it is also present in a microtubule organizing center in the following fashion. We can show that RNA binds microtubule accessory proteins (MAPS) and that these lower the critical concentration of tubulin assembly. If this RNA is located in the pericentriolar region, as suggested by the recent work of Berns and others, then this region may also be a zone of high MAP concentration, which should function as a nucleation center. The RNA could function to organize some specific configuration of proteins, such as occurs in a ribosome, or simply provide a high local concentration by a simple equilibrium mechanism. Similar ideas have been put forward by Ruth Dipell.

G.G. Borisy, Madison, Wisconsin, USA
To return to the questions of cell lysis and model systems, I would like to suggest a principle of optimal dirtiness. One wants a system that is not so clean that it is devoid of the interesting molecules and not so dirty as to be hopelessly complex. Two other considerations that might be entertained are (1) to use endogenous tubulin to stabilize the spindle in vitro rather than brain tubulin with its associated proteins; and (2) not to overstabilize the spindle since, after all, in anaphase the spindle depolymerizes.

Chairman:
Let us turn now to the phenomenon of the maturation of the mitotic centers.

D.R. Wheatley, Aberdeen, Scotland, U.K.
I think the point that Dr. McIntosh made was that a maturation process exists from a structure we may call the cell center into one which may be called the

mitotic center, although very little evidence has been presented to support this idea. Might I add that some further evidence may be seen in the behavior of "primary cilia" - cilia originating from the diplosomal centrioles and which are common features of many mammalian and other cell lines in culture - since these structures disappear concomitantly with the development of the mitotic properties of the center. There does seem to be a definite change in character from a cell center to a mitotic center.

J.R. McIntosh, Boulder, Colorado, USA
Your work and that of others has also shown that one can inhibit protein synthesis for a substantial period of time in late interphase and cells will apparently form normal spindles. As one is looking at aspects of maturation, it seems a poor bet that it is a de novo synthesized protein which adds to the mitotic center and thereby gives a new property. I find that an a priori surprising result and difficult to understand because from what we know about phage biology the programmed sequence of assembly of macromolecules would seem an intelligent way to think about mitotic centers. Yet it does not seem to hold up. Can you criticize that view?

D.N. Wheatley, Aberdeen, Scotland, U.K.
In the cells in which we have inhibited protein synthesis in G_2 (HeLa) we have found no primary cilia, and therefore it is not possible to correlate, as you suggest, the change from cell center to mitotic center with protein synthesis. My considered opinion now, however, is that we see reorganization in this process, and therefore de novo synthesis may not be mandatory.

Chairman:
You may remember those old results that one can induce some kind of mitotic centers, namely cytasters, in sea urchin eggs prematurely. Can we compare cytasters with the mitotic centers occurring at mitosis?

D. Mazia, Berkeley, California, USA
The cytasters carry out the same functions as normal mitotic centers. They form microtubules, they engage the chromosomes, and set up a mitotic apparatus that moves chromosomes.

Chairman:
That means we can accelerate the maturation of mitotic centers, at least in the egg.

N. Paweletz, Heidelberg, F.R.G.
Is there any evidence that there is a causal relation between the number of centers and the number of chromosomes or sets of chromosomes, since we mostly observe multipolar mitosis in highly polyploid cells?

J.R. McIntosh, Boulder, Colorado, USA
It seem to me an interesting system in which to observe a process similar to that is the zygote of *Drosophila*, in which a substantial number of rounds of divisions, 11-12, occur before membranes are formed. The cell cycle time is between 7 and 15 min; apparently, protein synthesis is not required at any time. A large number of mitotic centers are formed within the cytoplasm, which had only two to start with, and one mitotic center behaves as if it had the capacity to potentiate the development of another mitotic center, which is of course primarily fancy words used to describe the situation observed in the electron microscope, i.e., one center is usually hooked up to another. Perhaps with the cytaster one simply lowers the concentration in the same way that one can lower the critical concentration for tubulin assembly and get spontaneous nucleation under some circumstances. Under other circumstances it will only assemble if you need it.

P. Harris, Eugene, Oregon, USA
With regard to the maturation of centrioles and their ability to form poles, I would like to recall some work on mercaptoethanol blocked cells. When sea urchin eggs are blocked for the length of time it takes the controls to go through two divisions, on recovery they divide directly from one to four cells. However, the existing centrioles in the cell show a differential competence to form centers,

depending on the time at which the cells are treated with mercaptoethanol. In an early blocked cell, before any visible indication of division, the cell will proceed in the block as far as nuclear membrane breakdown, but only two centrioles are seen located close together and surrounded by a mass of condensed chromosomes. Cells treated at early prophase, before nuclear membrane breakdown, proceed somewhat further. A spindle is formed, about half normal size, with a massive accumulation of dense amorphous material at the poles, and occasionally extra chromosomes not on the spindle are seen associated with one or the other pole with microtubules associated with only one of the kinetochores. If cells are blocked just before metaphase, the centrioles at each pole of the spindle are capable of separating into four individual centers and the cell can divide into four daughters.

J.M. Mitchison, Edinburgh, U.K.
Two questions for the experts: What are the facts and the questions about the chemical composition of the mitotic centers? How do the cells of higher plants manage to organize spindles without centrioles?

B.M. Jockusch, Basel, Switzerland
In answer to Dr. Mitchison's question about the composition of spindle polar area: It has been demonstrated by Haugli and colleagues that in plasmodial nuclei of *Physarum*, the MTOC contain RNA, as seen in cytochemical studies, although no centrioles are developed in the intranuclear spindle.

G. Wiche, Wien, Austria
Concerning the interaction of microtubule-associated protein factors (MAPs) with other cell components, I would like to suggest that their interaction with RNA should not be overemphasized, since very recent studies of Dr. Jesus Avila and coworkers, in Madrid, have shown, that MAPs also interact with DNA.

Chairman:
Is there any indication for DNA in the mitotic centers?

M. Girbardt, Jena, G.D.R.
We have some evidence, that the fungal NAO which acts as an MTOC before and during nuclear division contains DNA. This is proved by cytochemical reaction with the EDTA-method of Bernhard.

J.J. Counce, Durham, USA
Is the replication of MA-associated centrioles tied to the division cycle? In *Drosophila*, mutants blocked in metaphase of cleavage division (16 *n* stage), centrioles apparently continue to multiply at the poles. In several species mutants occur sporadically in which achromatic but otherwise well-organized spindles with well-defined centrioles and polar regions are observed.

D.N. Wheatley, Aberdeen, U.K.
Most of my results on centrosomal studies in many cell lines suggest that the "life history" of the centrioles consists of replication occurring in S phase of the cell cycle (although not necessarily correlated with it) such that by G2, cells possess two diplosomes. The diplosome pair does not begin to move rapidly apart until prophase has begun. I think it is noteworthy that the replicating diplosome retains it primary cilium and that this persists until late G2 or early prophase. It is then resorbed and cells have never been found to retain such structures during mitosis. Immediately after telophase the diplosome in each daughter cell can develop cilia again. This cycle of events again suggests that the centrosomal region of a cell acts in broad terms either as a cell center or a mitotic center, and that it cannot do both jobs at the same time.

P. Dustin, Brussels, Belgium
I wish to recall that centrioles are more complex structures: microtubules organized as triplets link central structures, probably RNA, pericentriolar structures and dense material surrounding the microtubules.
The principal role of centrioles appears to be to provide the cells with organelles which will (or may) be used for the formation of basal bodies and cilia. In non-dividing cells, centrioles may often be seen to multiply considerably, leading to ciliogenesis.

Last, one should not forget that centrioles may be formed de novo without any morphologic precursor, as observed in activated eggs and also in the ameboflagellates, such as *Naegleria gruberi*.

W.W. Franke, Heidelberg, F.R.G.
Just a word of caution as to the findings of RNA associated with centrioles and basal bodies. There is almost no cellular structure including membranes of the ER and the plasma membrane that have not been described in the past two decades as RNA-containing: However, very often detailed studies have shown that such observations based on ribonuclease treatment and cytochemistry were due to methodologic pitfalls or, in the cases of fractionation studies, to cross-contamination, often by degraded RNAs. What would be conclusive would be the demonstration of a special characterizable RNA species.

J.R. McIntosh, Boulder, Colorado, USA
Therefore I draw particular attention to the work of Heideman and Kirschner, which I find compelling because there is a function associated with RNase-sensitive material. One can imagine artifacts, I agree. However, I find this the most interesting evidence because of the function. In regard to the point made by Dr. Dustin, in my understanding all you say is true and yet here comes the Heideman and Kirschner paper. What they are isolating are basal bodies, and, thanks to Pickett-Heaps and others, I have tended to think of a centriole at mitosis as something which is on the spindle for a ride. Just as you want to segregate the chromosomes you want to segregate the basal bodies, so each daughter will have one. But now you are taking basal bodies, putting them into an egg, and they appear to grow into mitotic centers. It seems to me, then, that what one would be obliged to do is to go beyond the statements that have been made in describing mitosis in the plants that do have motile sperm, like *Massilia*, in which it seems that when a basal body is present, it tends to organize the spindle-organizing material. That is probably what is going on in the case of Heideman and Kirschner's results. But it draws attention to the multiple ways in which the cell can approach the problem and why it is so difficult.

M. DeBrabander, Beerse, Belgium
I would like to suggest an additional experimental model for studying the function of centrioles in spindle formation. These are the spontaneous and colchicine-induced multinucleated cells in many types of cell cultures. These often contain a multitude of centrioles and give rise to multipolar mitosis that can be studied, e.g., with immunocytochemical techniques.

G.B. Borisy, Madison, Wisconsin, USA
We have results that show that the nucleation capacity of spindle pole bodies from yeast is not sensitive to treatment with DNase or RNase. So this system appears different from the basal body system described previously. Here, perhaps the organizing centers in hypha and lower eukaryotes will have interesting differences. Would Drs. Osborn and McIntosh comment on why antitubulin fluorescent images show larger numbers of microtubules arising from the interphase centrosome whereas the electron-microscopic observations on cells lysed in the presence of tubulin show the initiation activity of the interphase centrosome to be weaker than that of the mitotic centrosome.

M. Osborn, Göttingen, F.R.G.
I want to describe our experiment briefly because I think even though it is published (Osborn and Weber, Proc. Natl. Acad. Sci. U.S. 73, 867, 1976) it is important in terms of thinking about nucleation models, as opposed to self-assembly models. The experiment showed that if 3T3 cells were first treated with colcemid to destroy all the microtubules and the colcemid was then removed, repolymerization of microtubules was first observed around the centrospheric regions. At later times the microtubules elongate and cells can be found where most of the microtubules in the cell terminate at a point intermediate between the nuclear and the plasma membranes. At 75 min, recovery is complete and the microtubules again seem to reach all the way to the plasma membrane. Thus the apparent direction of growth of microtubules in the cell is from the centrosphere toward the plasma membrane.

One further point. In most of our micrographs we do not see pieces of microtubules. This suggests that in the cell, nucleation from a fixed point may be highly preferred over self-assembly.

M. DeBrabander, Beerse, Belgium
We did similar experiments. However, we used both immunocytochemistry and electron microscopy since I am still not convinced that we can see individual microtubules with immunofluorescence in these experimental conditions. Besides colchicine we also used nocodazole which, unlike the former compound, can easily be washed out. When this is done we see microtubules reforming at the centrioles (nucleation) and all over the cytoplasm in an orderless fashion. With colchicine we can expect that a certain amount will remain bound to tubulin. Due to the lower concentration of polymerizable tubulin, nucleated assembly is probably favored.

Chairman:
Is it true that in cells containing centrioles one always finds the centrioles in the mitotic centers?

H. Fuge, Kaiserslautern, F.R.G.
I am aware of at least one example where basal bodies are present, but do not participate in spindle formation, i.e., do not function as mitotic centers. This is *Trypanosoma*.

M. Hauser, Bochum, F.R.G.
This is not so unusual. In many protozoa, e.g., in all ciliates as far as it is known, basal bodies never are engaged in building up the mitotic apparatus.

Chairman:
What do we know about mitotic centers in plants?

A.M. Lambert, Strasbourg, France
In general in the normal higher plant cells that I studied, I never found any characteristic structures at the polar regions. Microtubules (MTs) end close to the plasma membrane (very often in dense material). Sometimes close contacts are seen between MTs and the plasmalemma. But when motile cells are formed, such as antherozoids, e.g., in gymnosperms or mosses, then centrioles or centriolar structures are present.

T. Schroeder, Friday Harbour, Washington, USA
The asters of mitotic apparatus in eggs appear to communicate with the cell cortex for the purpose of localizing the mitotic apparatus in various ways, and, perhaps, for stimulating cell cleavage contraction. The asters exhibit well-known enlargement, which is particularly pronounced during metaphase and anaphase. In some cases, the component of the asters that enlarge appears to be the centrosphere - not the hollow shell of radiating microtubules. It is interesting to compare the interzone (which appears at anaphase) and the centrosphere. Both of the zones are remarkably devoid of microtubules and vesicles. Do these kinds of zones exhibit their own intrinsic mechanisms of enlargement?

T. Bibring, Nashville, Tennessee, USA
Was this all done on isolated mitotic apparatus , and what species was it on?

T. Schroeder, Friday Harbour, Washington, USA
The species of egg is *Strongylocentrotus droebachiensis* and the basic observations in isolated MAs have been confirmed from sections of whole, fixed eggs.

P. Harris, Eugene, Oregon, USA
Perhaps a different fixative should be tried. Different species of sea urchins often respond in quite different ways and lack of structure as you find in *Strongylocentrotus droebachiensis* may simply indicate a greater lability of the centrosome structures than those found in *S. purpuratus*.

H. Sato, Philadelphia, Pennsylvania, USA
The isolated sea urchin anaphase spindle certainly does possess enhanced centrospheres compared with metaphase spindles. The protein concentration of these centrospheres is estimated by interferometry as about 18 % and is definitely higher

194

than that of the interzone. Tubulin immunofluorescence also revealed intense
fluorescence in the centrospheric area, but weak in the interzone. These observations suggest that not only the fine structure, but also the initial density
in these two regions may now resemble each other.

H. Ponstingl, Heidelberg, F.R.G.
I would like to remind you of the problem of chromosome splitting. There are some
observations which may allow us to tackle the problem experimentally: 1) Dr. Mazia
observed that if he induces chromosome condensation by ammonia, chromosomes will
split without microtubules being present. 2) Nonhistone proteins are enriched in
the centromeric region of the chromosome (Matsukuma and Utakoji, Exp. Cell Res.
105, 217, 1977). 3) All chromosomes split simultaneously. That would suggest a
general signal, not only one from the kinetochore.

J.R. McIntosh, Boulder, Colorado, USA
Bajer showed this beautifully in some of the early movies on *Haemanthus* with a
colcemid block. We have tried to get this process to go in vitro utterly without
success. If one could resolve the problem one would make a major technical break-
through with respect to the study of motile systems, i.e., chromosome movements.
We must lyse our cells after the cell has begun anaphase, since we cannot start
anaphase. It would be a tremendous help if one could add some magic reagent at
metaphase and get the chromosomes to move.

B.R. Nicklas, Durham, North Carolina, USA
This falling apart of the chromosomes at the beginning of anaphase does occur in
our model system, i.e., cells can enter anaphase after lysis or after demembranation, and the chromosomes do fall apart. Therefore, because this can happen in
that model system and because we have access to chromosomes that have not yet
fallen apart we might try the experiments Dr. McIntosh has just referred too, i.e.,
what chemicals are normally involved in holding them together. They will naturally fall apart in our permeable cell model system.

A.S. Bajer, Eugene, Oregon, USA
I have two questions for Dr. Nicklas: 1) What are the criteria you use to determine that the cells are demembranated? 2) Was electron-microscopic control made?

R.B. Nicklas, Durham, North Carolina, USA
I do not know whether there are tiny fragments of the membrane, but there is
nothing obstructing permeability because we can come in with a micropipette or
needle and move mitochondria or chromosomes from one spindle to an adjacent one:
There is no mechanical or chemical barrier between the cells at all. In the electron microscope, we simply do not even see fragments of the membrane.

L. Hens, Brussels, Belgium
Interrelationships between chromosomes may possibly complicate the models on
sister chromatid separation. At this moment three types of observations are relevant with respect to the nature of these interchromosomal links:
1. In human cells rDNA strands were demonstrated between the p-arms of the rDNA
bearing acrocentric chromosomes.
2. Microtubules were demonstrated between chromosomes.
3. Analysis of the nonrandom distribution of association patterns of human acrocentric chromosomes with the longer chromosomes provides some evidence that constitutive heterochromatin proteins might also be responsible for interchromosomal
links.
These three types of observations provide the basis for interpretation of the nonrandom distribution of chromosomes in metaphase plates.

G.K. Rickards, Durham, North Carolina, USA
I remind you of the temperature-sensitive mutants described by Dr. Wang yesterday,
in which a normal metaphase is reached, but in which the cells do not proceed into
anaphase. I suspect, or suggest to you, that perhaps these represent mutations in
this process of initial separation of sister chromatids.

R.J. Wang, Columbia, N.Y., USA

One of our temperature-sensitive mammalian mutants shows chromosome movement without chromatid separation. Intact, normally appearing chromosomes move poleward from a metaphase plate. Even though chromosome movement is abnormal, cytokinesis is still initiated but frequently fails to proceed to completion due to chromosome material remaining in the anaphase bridge.

An aspect of mitosis to which we have not addressed ourselves during this workshop is regulation or the initiation and progression of cytokinesis. I would like to thank Dr. Schroeder for placing the following results in their proper perspective. In some cells no anaphase movement is visible, but cytokinesis is still initiated but fails to be completed. In other cells with varying amounts of chromosome movements, cytokinesis almost always begins, and the extent of cytokinetic progress is generally positively correlated with the amount of chromosome movement.

B.M. Jockusch, Basel, Switzerland

Dr. Wang, I would like to clarify one point: Are these conditional mutants you described?

R.J. Wang, Columbia, N.Y., USA

These cells show a normal mitosis at 33°C. When they are shifted to the nonpermissive temperature of 39°C, they show up with defects. They are very stable mutants with a low reversion frequency, but they do revert. This implies that they are single-point mutations affecting single proteins.

Session IX

Future Research on Mitosis

D. MAZIA, Department of Zoology, University of California, Berkeley, California 94720, USA

If we are to contemplate the future of research on mitosis, we might first reflect on the past and the present. The beginnings of this field were stunning; they belong among the greatest events in the history of biology. By pure observation that developed sight into insight, our predecessors were able to discover an unimaginable process - one that could never have come out of anyone's head - which explained the accurate reproduction of cells and was the basis for the immediate explanation of basic problems of sexual reproduction, fertilization, development, and heredity. One sometimes thinks it might have been interesting to be around at that time, but the price of being dead now would be too great.

What about the present? To be frank, the study of mitosis is not a very large field; it fits easily into this room. Presently, we are indeed enjoying a few large steps forward, which are being recorded here, and did so in the past, as I have witnessed, but we have not witnessed breakthroughs and revolutions on a scale that appeals to historians, journalists, and exploiters of scientific trendiness. Research on mitosis itself is not one of those industries that gives rise to a huge literature, although the broader problems of cell proliferation provide the justification for a number of prosperous and prolific industries. Everyone admits the importance of understanding mitosis, but somehow it does not lend itself to preconceptions of wath constitutes a "nice" field of research. The difficulties are real and forbidding: the dimensions of the problems are formidable; it is hard to isolate attackable questions from the complex web of events.

I will try to outline some structure for future work on mitosis and to seek in our present knowledge the means of isolating some of the major mitotic events for future experimental study.

A. Mitosis and the Cell Cycle

To work on mitosis as such, we must extract it from the cell cycle as a whole, taking liberties with the poetry of the interdependence of all the events in the cell to find the prose of events immediately relevant to the mitotic processes. For our purposes, we can divide the mitotic cycle into two phases; (1) a replicative phase and (2) a distributive phase. The replicative phase we know as the reproduction of chromosomes and mitotic centers that takes place within what we normally call interphase. The distributive phase is what we usually call the M phase. The representation in Table 1 may seem a bit banal, but my purpose is to identify a minimum number of basic questions of mitosis (Table 1).

The true mitotic events that take place during interphase are of no interest to the cell in which they are taking place; they are only of interest to the progeny of that cell. They would not be found in cells that are growing but will not divide. While the duplication of the chromosomes and the centers are the defining events of the repli-

Table 1. Events of the whole mitotic cycle

Replicative Phase (Chromosomes and centers)

    Turn on (command to replicate)

    Replication (initiation: program of replicons)

    Shutdown (only two copies of each replicating unit)

    Synthesis of other molecules required for normal
    distributive phase

Distributive Phase

    Chromosome condensation (prophase to anaphase)

    Polarization (separation of poles; establishment
    of mitotic axis)

    Breakdown of nuclear envelope (not universal)

    Prometaphase and metaphase (not universal)

    Anaphase separation (sister chromosomes to
    different poles)

    Reconstitution (chromosomes decondense; nuclei
    reform)

    Cytokinesis

cative phase of the mitotic cycle, they need not be the only ones
directly related to mitosis cognizance of evidence for a fraction of
protein synthesis and RNA. We define the distributive phase (not a
felicitous term, but let us use it) by the engagement and deployment
of chromosomes, but these may not be the only significant events.

We can resolve the cell cycle a little better by viewing it as a bicycle, with a growth wheel and a reproductive wheel. In the "normal"
cycle the wheels are geared together in one way that gives rise to
the conventional phases - G1, S, G2, M. But the "normal" cycle is only
incidental, a plot that is encountered commonly but has no deeper
necessity. Cases of cells without G1 abound (Prescott, 1976). There
are cases where we see no G2. Recently (Liskay, 1977), there has been
a description of lines of Chinese hamster cells in which G2 is absent;
one of them shows neither G1 nor G2. None of this is to say that the
events of the true mitotic cycle are that different in different kinds
of cells. It only means that the necessary events of the replicative
phase can take place anywhere in interphase and are not defined by
the common way in which the two wheels of the bicycle are geared.

Those who use eggs for the examination of the mitotic cycle think we
are studying perfectly normal mitosis in a situation where only the
mitotic wheel is turning. By the same token, we deprive ourselves of
the possibility of investigating the relations between the growth
wheel and the mitotic wheel of the cycle.

B. The Replicative Phase

I. Turning On

A massive wave of modern research has given us what seems to be an indisputable fact: that the impulsion of a nondividing cell into the processes leading to division begins with changes at the cell surface. Gerald Edelmann's essay this volume not only makes the point, but gives us the climate of current research in the field: the systems used for study and the way questions are asked. Accepting this hard-won insight, the future asks us to define the path from the initial events on the outer surface to the initiation of events we recognize as mitotic events. The student of mitosis requires somewhat sharper definitions of the problems than are given by the loose criteria of "mitogenesis," which merely say that something is done and some time later we see cells in mitosis. The lymphocyte folk speak of one thing: a cell not equipped to do much of anything, which, when turned on, first makes the machinery of a normal cell, then uses it to enter the events of mitosis proper some time later. The ovists have quite another problem: the egg is equipped with an apparent excess of all the products of growth - it has turned its growth wheel many times during oogenesis - and is tripped immediately (within minutes) into the replicative events leading to mitosis when it is turned on. There has long been reason to believe that the turning on of the egg is another case where the cell is turned on by surface changes, and there are more recent studies bearing on that point (Mazia et al., 1975). Will the surface changes at these two extremes turn out to be similar when they are analyzed more deeply? Then there are the cases of free-running cell cycles such as we see in exponentially growing cell cultures; does the initial "turn on" suffice for all the successive cycles or shall we look for a turn on event at each cycle?

Edelman's paper led us toward the idea that the signals by which surface changes bring about the turning on of the cell would be macromolecular. We can think of alternatives. Cyclic nucleotides provide one model- small regulatory models that can be produced rapidly when the cell surface is modified. Current research also calls attention to ions and especially Ca^{2+} as mediators of change starting at the surface. In the case of the egg, Heilbrunn's hypothesis (1952) that activation (a surface event) brought about a liberation of Ca ions in the cell interior has now received confirmation (Ridgeway et al., 1977; Steinhardt et al., 1977).

Consideration of mitosis cannot tell us the nature of the regulatory event that triggers the turning on of the replicative phase, but it puts some demands on its character. It is not enough to say that DNA replication is turned on. Turning on chromosome replication will also involve the turning on of histone synthesis and perhaps that of other chromosomal proteins. When we say "DNA replication" or "histone synthesis" we may be speaking of the start up of numerous biochemical pathways, discrete in themselves, whose end result is the macromolecular synthesis we measure. The duplication of centrioles is turned on at the same time; we know so little about this essential event that we cannot guess what processes are initiated. Thus, whatever the regulatory mechanism is, the principle should be that of a master switch, a common regulator for processes that are intrinsically different but must start moving at the same time. Whatever the first event brought about by the change of the cell surface may be, it must bring such a master switch into action - and, in the case of eggs - within minutes.

II. Turning on the First Steps of Chromosome Replication

In ordinary studies of the cell cycle, we mark the beginning of chromosome replication by the beginning of DNA replication, conventionally the G1/S transition. That assumes a cell in which all the molecular ingredients for the replication are already present; enzymes, substrates, cofactors. A different plot for the cell cycle invokes a GO/G1 transition even in continuously cycling cells; in my present terms it would assign the turning on of the mitotic cycle at a time earlier than the beginning of DNA synthesis; the overture must be played before the curtain is raised.

Recent work (Nishioka and Mazia, 1977) deals with two questions about the turning on of the first steps of chromosome replication in the sea urchin egg: 1) Does the initiating event turn on processes necessary for chromosome replication in anticipation of the replication itself, and 2) is the signal for such initiation of chromosome replication communicated by way of the nucleus? Essentially, the experimental design is to give the command for chromosome replication in the absence of chromosomes and to observe how the cell responds to the command.

The sea urchin egg permits such an experiment. Methods for splitting the eggs into nucleated and enucleated halves by centrifugation have long been available. In a series of recent reports, we described how the replication of chromosomes and the chromosome cycles may be turned on in unfertilized eggs by treatments with NH_3 and by various other means. The eggs are still regarded as unfertilized eggs, but they now respond to a command to replicate their chromosomes.

One consequence of fertilization (and of the treatment with ammonia) is the immediate phosphorylation of thymidine. The unfertilized eggs were preloaded with labeled thymidine that accumulated as the nucleoside. When they were fertilized or treated with ammonia, the thymidine was phosphorylated immediately, going through TMP and TTP into DNA without significant accumulation of TMP or TTP. The same was true of half-eggs with nuclei when they were treated with ammonia. However, in the half eggs *without* nuclei, the phosphorylation went to TTP and the TTP accumulated for the obvious reason that there was no replicating DNA to incorporate it. Clearly, the command to replicate DNA had been received and the response was the same in enucleated cells as in cells with nuclei; it was received by the systems that make nucleoside triphosphates in anticipation of DNA synthesis; it did not have to be relayed through the nucleus.

This result is not to be generalized to imply that the production of nucleoside triphosphates is the system that regulates the onset of DNA synthesis. That hypothesis has been around for a long time. It may apply in some cases, but it is not believed to stand up to generalization. It may apply to the sea urchin egg, but here I use it only to point out that the signal to start DNA synthesis may be received at steps antecedent to the actual start of replication and may not involve the chromosomes at all.

In the same study, we showed that the enucleated half-eggs responded to the command to start chromosome replication by synthesizing histones. Histone synthesis, of course, is an essential part of chromosome replication. The experiment shows that the command which starts nucleoside phosphorylation also starts histone synthesis - and neither of these events essential for chromosome replication requires the intervention of chromosomes. It may seem disturbing that histone synthesis is turned on in the absence of DNA synthesis; there is so much evidence in other kinds of cells that suggests that the two are tightly linked. That evidence is not questioned; what is put into question is the assumption that the tight coupling of DNA and histone synthesis is general and essential. There is other evidence that histone synthesis in eggs is not closely tied to DNA synthesis.

Now that there are good methods for enucleating other kinds of cells using cytochalasin (Ringertz and Savage, 1976), comparable experiments need not be limited to eggs. There is a future for efforts to search for the events before the beginning of S in which we will find the first movement of the mitotic wheel.

If there is further evidence that some events of the mitotic cycle are turned on in the absence of a nucleus, that will influence our interpretation of the evidence for genes that determine steps of the passage through the cell cycle. Will we relate the effects of cell-cycle mutants directly to the mitotic events that are impaired by the mutations or indirectly through the synthesis of molecules that might be made at any time but become limiting at a particular point in the cycle of a particular kind of cell? In the former case we would have a mitotic clock in the gene-regulating system; in the latter we would have a powerful tool for identifying limiting molecules in the passage of cells through the mitotic cycle.

III. Chromosomes in Interphase

Whether or not the initial events of the mitotic cycle precede the initiation of chromosome replication, the initiation itself is a distinctive event. The best proof is the fact (Graves, 1972) that the chromosome sets from different species start their replication at the same time in hybrid cells. It also follows from the fact that fusion of a cell in its G1 phase with a cell in its S phase promptly starts DNA replication in the G1 nucleus (Rao and Johnson, 1970).

A speculation made some years ago (Mazia, 1963) postulated that the chromosome cycle of condensation and decondensation, so conspicuous during mitosis, might extend throughout the cell cycle. The telophase decondensation would continue into interphase to a point where the chromatin was sufficiently opened up to replicate; in conjunction with replication, condensation would begin although it only becomes visible at prophase. The general idea that chromatin replicates while in an expanded state, finds ready support in circumstantial evidence, cytologic and biochemical, although I am not sure it can be defended rigorously. For the student of mitosis, the more immediate question is whether the chromosome condensation-decondensation cycle does run through the whole cell cycle.

It does. Thanks to the powerful PCC (premature chromosome condensation) technique, it is possible to impose condensation on chromosomes in interphase and make them visible. The technique calls for virus-induced fusion of a cell in mitosis (it can be held there with colchicine) with a cell in interphase; the interphase chromosomes condense; the "condensation factor" is not species specific or even order specific. (For a general review of the use of the method in cell-cycle studies, see Ringertz and Savage (1976)). Recently, two laboratories to which we owe a great deal of this information, that of Rao in Houston and that of Johnson in Cambridge, have considered the states of condensation of chromosomes during interphase. One report (Schor et al., 1975) shows that the chromosomes in late G1 are much less condensed than those at the beginning of G1 - even when forced to condense by the PCC technique. A still more recent report (Rao et al., 1977) des-

cribes an "ultimate state of decondensation at the end of G1" in which
the prematurely condensed chromosomes show a "diffused morphology ...
instead of discrete single-stranded structures characteristic of early
G1-PCC." In these later experiments, the mitotic cells were fused with
a random population, thus permitting the detection of the degree of
chromosome condensation at all stages of the cycle.

Here we have another handle for future work on the whole mitotic
cycle. One problem is to confirm the earlier hypothesis and the more
recent affirmation that "the decondensation of chromatin during G1
appears to be a prerequisite for the subsequent initiation of DNA synthesis" (Rao et al., 1977). The further work will include a combination
of studies on cells and studies on isolated nuclei and chromatin. For
the student of mitosis as such, the problem will be to extend the investigation of the chromosome cycle (so long a recalcitrant problem
even when studied during the mitotic phases) through interphase.

C. Distribution Phase

In that phase of the whole mitotic cycle of the cell in which the
sister chromosomes are distributed to sister nuclei - which we can
continue to call mitosis in the classical sense - a mitotic apparatus
is made and put to work. The older cytologists perceived the mitotic
apparatus in terms of two components: a "chromatic figure" by which
the behavior of the chromosomes themselves was described, and
"achromatic figure," which was the spindle apparatus that determined
the movements of the chromosomes. The distinction is sound enough for
our present purpose, which is to survey future requirements of research on mitosis. However inseparable the two components may be in
terms of the plot of mitosis, we have to separate them in terms of
realistic research. Research on the chromosome cycle is not the same
kind of problem in the laboratory as research on the spindle fibers.
We do not have to be reminded that the two must ultimately be united,
and I will come to some leads for research on their coordination.

I. Transitional Events

The experience of the recent past does provide some points of de-
parture for future studies of the transition from the replicative
phase into the distributive phase. The metaphor of a "trigger" will
have less appeal simply because we do agree that the mitotic cycle
is fired off early in interphase and that it is not so common for
the cycle to be interrupted between the end of the replicative phase
and the beginnings of visible mitosis.

The idea of syntheses of proteins that take place at the very end of
interphase and are required for the transition into prophase has a
long history (Taylor, 1960) and is reinforced by more recent work
on terminal RNA and protein synthesis (Tobey et al., 1971). The hypothesis that the terminal synthesis of protein is the making of tubulin for the future mitotic apparatus has not been strengthened by
the finding that tubulin is synthesized throughout interphase,
although the rate of production may increase as the cell moves toward
mitosis. In the sea urchin egg, where early tubulin synthesis is not
important (Bibring and Baxandall, 1977), the evidence that early inhibition of protein synthesis prevents entry into prophase (Wilt et
al., 1967) can be confirmed. Unpublished experiments by Wagenaar show
that the prophase condensation of chromatin can be inhibited when
protein synthesis is completely inhibited by emetine early in the
cycle, although DNA replication proceeds. As the evidence now stands,
we are allowed to conclude that there are macromolecules that are

directly concerned with the transition into visible mitosis; here,
clearly, is an attractive problem for the immediate future.

An older clue to the initiation of prophase may have been forgotten.
The effect of low doses of irradiation (UV or ionizing) in delaying
but not stopping the immediately following mitosis was a well-known
phenomenon, carefully studied (Puck, 1961). In the case of the sea
urchin egg, there is convincing evidence (Rao, 1963) that it is the
beginning of chromosome condensation that is delayed. Does the radiation sensitivity of factors responsible for the beginning of visible mitosis offer a handle for future research?

II. Chromosome Cycle

1. Interphase-Prophase Transition

I have already mentioned the evidence for the continuing chromosome
cycle in interphase, as studied by the PCC method. However, the evidence does not preclude the opinion that the further condensation
into prophase may be a sudden event, conventionally describable as
the G2/M transition. Older interpretations of the structure of the
condensed chromosome called for several orders of packing of the fundamental thread - as in the image of a coiled-coiled coil. A recent
version of such a model (Bate et al., 1977) updates it in terms of
present knowledge of chromatin and introduces four orders of coiling
between the primary chain of nucleosomes and the metaphase chromosome.
We can imagine the transition into prophase as the imposition of one
higher order of coiling or packing of the chromosome than that pre-
vailing in interphase: a single step to a microscopically visible
fiber which then is folded to a still higher order of condensation.
The transition is not gradual; when we see chromosomes in prophase,
they are all condensed to about the same degree.

We also note that the technique of premature chromosome condensation
by fusion of a cell containing metaphase chromosomes with a cell in
interphase does not allow complete condensation to the metaphase condition. Even in the fusion of G2 cells with mitotic cells, the chromosomes of the former are much less condensed than metaphase chromosomes. A step is missing. We can anticipate more studies by the PCC
technique that will help to identify the interphase-prophase transition.

III. Chromosome Cycle Without Spindles

It is now possible to isolate the formation and history of the
"achromatic figure" - which I shall term the "nuclear mitotic apparatus" - in one kind of cell, the sea urchin egg. Unfertilized eggs
are turned on by treatments with NH_3 or in other ways - recently I
have been turning on the eggs by a brief exposure to isotonic glu-
cose containing 0. 1 mM CaCl, adjusted to pH 9. The chromosomes replicate and then condense. No mitotic poles are present, no spindle
is formed, the eggs do not divide. The cycles repeat; the unfertilized egg is studied as a machine for making chromosomes and going
through successive chromosome cycles. Since much of this work has
been published (Mazia, 1974), I only draw on it here to make some
points related to our present discussions.

In the first cycle after the treatment, prophase condensation is
seen soon after the completion of DNA replication. At most it requires 10 min, and it will be worthwhile to establish the time more
accurately. The condensation then proceeds through the chromosomal
equivalents of prometaphase, metaphase, anaphase, and telophase, on
a fairly normal schedule, even though there is no spindle to orient

or transport the chromosomes. Thus we can study a free-running chromosome cycle apart from the complete process of mitosis.

First, the situation lends itself to the study of an all-important step in mitosis, the *splitting apart* of sister chromosomes, which normally announces the beginning of anaphase. We confirm directly the older results of Gaulden and Carlson, Bajer, and others who showed that the splitting of chromosomes did not depend on traction of kinetochores to poles. Here there are no poles nor have there been; we are dealing with what Nicklas calls "virgin chromosomes." We are also dealing with a virgin problem. We do not even have a virgin problem. We do not even have a sensible speculation as to the mechanism of the splitting of chromosomes.

These procedures for observing the complete splitting of chromosomes in sea urchin eggs may provide an object for further experimentation, but what questions can we ask of this or any other experimental system? First, the problem is, as has long been recognized by cytologists, the splitting of the kinetochore regions; impediments to splitting of other parts of the chromosomes ("stickiness" in cytologic descriptions) produce anaphase bridges but do not prevent anaphase movement. The contrary phenomenon is seen in effects of colchicine; the arms split apart very well but not the kinetochore regions. Is this telling us that tubulin or microtubules - not microtubules connected to poles - play some role in the splitting apart of the kinetochores? The cycle I have described in the sea urchin egg is stopped by colchicine, even though it does not require a proper spindle. On the other hand, there are microtubules in the "nuclear mitotic apparatus", as described at this workshop (Paweletz and Mazia, this volume. Are we sensing another role for tubulin or microtubules in mitosis?

There is an even deeper question about the importance of the splitting apart of the kinetochores for the whole mitotic cycle. It does not merely determine progress into anaphase, but all progress beyond. We commonly use the colchicine blockade to prevent the entry of the cell into the next cycle. There are experiments in which marginal concentrations of colchicine are used, and in these the spindles may be minute, the chromosomes may barely move apart, a single nucleus may form, but the cell moves ahead. In the experiments on the sea urchin egg where there is no anaphase movement at all, the complete splitting of the chromosome allows the continuation of the cycle. How are we to understand how a single event, seen as a slight separation of two sister kinetochores in space, is so absolute a condition for the progress of the cell through the whole mitotic cycle?

There are other points to be made about the "nuclear mitosis" I have been describing. Even in the absence of centers, the chromosome cycle brings about some of the events of the formation of a mitotic apparatus. In the living cell, the condensed chromosomes are seen to lie in an aster-like clear zone; clear because mitochondria and other large particles are excluded as they are in a normal mitotic apparatus. In the poster offered at this workshop, Paweletz and I this volume describe the fine structure of this clear zone. It consists mainly of a dense mass of membraneous vesicles, perhaps a reticulum, such as is seen in the normal mitotic apparatus of the sea urchin egg. There are microtubules and signs that there are foci for the organization of microtubules, but these foci are not centrioles nor, obviously, can they form mitotic poles. The question we would like most to answer still remains undecided: whether the kinetochores generate microtubules in this situation in vivo where there is no spindle and there has been none. This kind of experimental design is competent to answer a gen-

eral question: What features of the mitotic apparatus are generated
by the chromosome cycle in the absence of poles and a spindle? It may
be difficult to set up comparable systems in material other than eggs,
but the questions remind us that there are natural endomitotic cells
that have not received deeper study.

IV. Coordination of the Chromosome Cycle and the Spindle Cycle

In normal mitosis, the chromosome cycle and the history of the build
up and dismantling of the spindle are closely interlocked. The mitotic
apparatus forms when the chromosomes are condensed to a certain degree;
it breaks down when the chromosomes decondense. We know that the chromosome cycle can proceed without a spindle; we still have to establish
whether spindles can form without chromosomes. How are the spindle
and the chromosomes coordinated in normal mitosis? A simple hypothesis
is that the same factors that regulate chromosome condensation regulate the formation of the spindle. That there are factors that command
the condensation of chromosomes we know from experiments on premature
chromosome condensation. These factors are pervasive - all the chromosomes in the cell condense. The factors are not highly specific; the
same results are obtained when we fuse very different kinds of cells.
If the preceding hypothesis is correct, premature condensation should
be accompanied by premature spindle formation. As far as I know, the
question has not been asked in the work on fused mammalian cells; the
methods used would not necessarily detect the formation of spindles.
However, I have done a comparable experiment on sea urchin eggs. Using
the technique of turning on the unfertilized egg with ammonia, the
eggs are allowed to proceed through the replication of the maternal
chromosomes and toward mitosis. At a certain time, the eggs were fertilized; they had remained unfertilized eggs and would fuse with sperm.
The fertilization amounts to the injection of a set of paternal chromosomes and a pair of centrioles. Given the centrioles, the eggs were
capable of making a mitotic apparatus - but when? The observation was
that the mitotic apparatus appeared at the time when the maternal
chromosomes condensed - quite prematurely, if one considered the
normal time between fertilization and the time of mitosis. Otherwise
stated, the command for the condensation of the chromosomes (the
paternal chromosomes condensed prematurely as expected) was simultaneous with the command for the formation of the mitotic apparatus.
The results conform to the hypothesis that the command is the same
for both.

The future of research on mitosis certainly requires an explanation
of the coordination of chromosome condensation and decondensation
and of spindle assembly and disassembly. I have not really offered
a hypothesis but only the form of a hypothesis: that the two are regulated by a common variable. The form has one advantage for the
future: If we make progress on the problem of the assembly of the
spindle (even of microtubules), we can try out the regulatory principle on the problem of condensation of chromosomes. If the study of
"chromosome-condensing factors" moves ahead, we can see whether these
factors are of value in explaining spindle assembly. For example,
we have discussed at this Workshop the proposal that Ca^{2+} ions regulate the assembly of microtubules. The hypothesis would imply that
chromosome condensation takes place in the cell when the Ca^{2+} levels
decrease. Should we not then ask whether lowered Ca^{2+} concentrations
are factors in chromosome condensation (Mazia, 1974b)?

V. Polarization of the Mitotic Apparatus

We cannot even talk about mitosis without referring to poles. When I reviewed the field some years ago (Mazia, 1961) the identity of the poles in cells that did not have typical animal centrioles was the problem. That problem persists for the higher plants, but meanwhile a splendid literature on mitotic centers in lower eukaryotes has revealed the variety of particles that identify the poles in different organisms. There has been less progress, indeed little discussion, on two other complex problems of the operations of the poles: 1) the separation of centers to form poles; 2) the orientation of the poles, which can be so important in development in animals and especially in plants.

The conception of mitotic centers as microtubule-orienting centers (MTOC) gave us some molecular foundation for thinking about how the poles worked, but it also creates some awkwardness for thought about mitosis. The predicted action of an MTOC would be the production of an astral arrangement of microtubules, and, of course, the formation of asters is pertinent to mitosis. It does not, however, provide an obvious basis for polarity. The work reported by Borisy this volume at this Workshop liberates us from that paradox. He demonstrates a distinction between MTOCs that nucleate a radial arrangement of microtubules in vitro and those that send microtubules in a particular direction; that is just what we need if we are to explain the formation of a mitotic apparatus containing chromosome-to-pole fibers, pole-to-pole-fibers, and asters.

It is a good time to be considering the future of these problems, now that we are catching our breath after the recent explosion of research on tubulin and microtubules. The definition of a mitotic center as a microtubule-organizing center is true, but the truth does not have sufficient content for those who worry about mitosis. We now see evidence that the assembly of microtubules in vitro does not require any nucleating centers, given appropriate conditions (e.g., Himes et al., 1976). The poster by Paweletz and myself in the Workshop (this volume) describes the formation of microtubules - and even a suggestion of osmiophilic foci for their assembly - in sea urchin eggs going through the chromosome cycle after being turned on with ammonia. No centrioles are seen. A bipolar mitotic apparatus is not formed. Microtubule-organizing centers are one thing; mitotic centers functioning as poles are something else.

There have been efforts to examine the production and reproduction of mitotic centers, following the classical work that introduced the concept of centrioles as self-duplicating entities. Much of what we know about the production of centrioles can be discussed in terms of the generative model of their reproduction (Mazia et al., 1960). Here one considers the centriole we recognize by electron microscopy to be phenotypic expression of a centriole that includes a self-replicating genomic component, which can be called the germ. The reproduction of a centriole involves: 1) the replication of the germ; 2) the separation of one copy from the parent centriole; 3) the generation of a procentriole by the daughter copy of the germ; and 4) the growth of the procentriole into a mature centriole. We would not expect to see centrioles dividing and we do not. We can imagine that the germ of a parent centriole could replicate many times, after which the development of numerous descendent centrioles would be possible. Such a generation of numerous centrioles from one is described in the work of Dirksen (1971) on cells of the mouse oviduct. We can also imagine that the microtubular soma of the parent centriole might disappear as the germs replicate. If these germs then generate observable centrioles we will have an apparent production of centrioles de novo, yet the observation would not violate the principle that centrioles are fundamentally self-reproducing. I will discuss these problems in the classical case of cytasters in eggs. The general scheme of generative reproduction is by no means a fantasy; it would describe the production of bacteriophages, for example.

The century-old idea of the centriole as a self-reproducing entity leads to a search for nucleic acids. That difficult problem has been pursued recently by various means. Attempts have been made to inquire into the presence of DNA (Rattner and Phillips, 1973) with results less encouraging than those that look for RNA as the hypothetical genetic material (Dippel, 1976; Hartman, 1975). One thing we now know from studies on several kinds of cells is that the production of new centrioles takes place during the S period; that might seem to point to DNA as the elusive genetic material of centrioles. An interesting hypothesis is that of Went (1977) who proposes a centriolar RNA that reproduces by way of DNA, using an RNA-dependent DNA-synthesizing system. In such a scheme, patterned after the reproduction of some RNA viruses, the genome of the centriole might be RNA and yet could reproduce only under conditions appropriate for DNA synthesis.

Whether we want to look into the mechanism of production of centrioles or whether we are interested in examining the way they work in polarizing the mitotic apparatus, we need to lay our hands on a supply of mitotically competent centrioles. We can look to eggs for that supply. The classical work on artificial parthenogenesis described ways of inducing mitotic cytasters in unfertilized eggs; these could function as mitotic poles. The experience with cytasters was often discussed as a possible disproof of the classical theory of the self-duplication of centrioles or even as a trivialization of the centriole: How could one know that cytasters contained real centrioles? With electron microscopy, such real centrioles could be found in cytasters.
More recently, Weisenberg and Rosenfeld (1975) described a vivid case of the rapid generation of centrioles in material isolated from a clam egg (*Spisula*) after activation. These centrioles did nucleate microtubules to produce asters.
Recent experience in our laboratory may help to clarify the generation of centrioles and to make centrioles more available for closer study. The egg of the sea urchin *Lytechinus pictus* is used. (The method has

---

Fig. 1. Large numbers of asters produced in unfertilized eggs of the sea urchin *Lytechinus pictus*. Chromosome replication was turned on by treatment with ammonia-sea water (sea water brought to pH 9.2 with NH_4OH); exposure was 20 min. The eggs were then kept in D_2O-sea water (50 % D_2O) for 40 min. The asters appeared at about 100 min. The clear zone within the egg is a multipolar mitotic apparatus; some of the asters had engaged the chromosomes. Living egg, compressed; phase contrast

Fig. 2. Multipolar mitotic apparatus in egg treated as described in legend to Figure 1. Eggs were fixed in 3 : 1 ethanol-acetic acid and stained with orcein in 75 % acetic acid

Fig. 3. Asters isolated from eggs induced to form cytasters as shown in Figure 1

Fig. 4. Centrioles in cytasters; thin sections of whole eggs fixed in OsO_4 (1 %; sodium acetate, pH 6.1). Varying numbers of centrioles may be seen in a single section; the total number per aster is not yet known. The time when morphologically normal centrioles appear is not yet known; the centrioles shown were fixed at 150 min from start of the treatment of the eggs

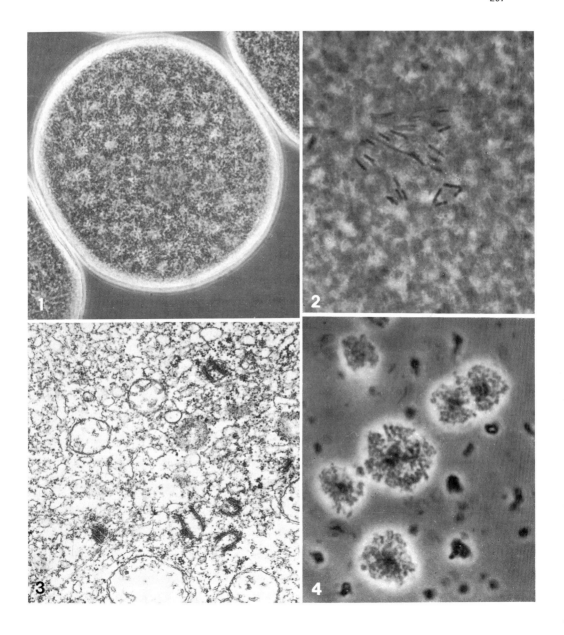

worked well with a Mediterranean sea urchin, *Paracentrotus lividus*, but
not with another American Pacific coast sea urchin, *Strongylocentrotus
purpuratus*.) Unfertilized eggs are turned on by treatment with NH_3 and
then exposed to sea water containing 50 % D_2O. After some time in this
D_2O-sea water, they are restored to normal sea water. At the time when
the maternal chromosomes condense, very large numbers of asters are
seen (Fig. 1). These asters do work as mitotic poles; they engage the
chromosomes, and a very multipolar mitotic figure is formed (Fig. 2);
it appears as though any of these centers is capable of making an

attachment to a nearby chromosome. Moreover, we do find centrioles in these cytasters, as shown in current work by Kallenbach. In fact, we find many centrioles in each aster (Figs. 3 and 4), which suggests that the yield of centrioles may be a good deal higher than is indicated by the number of asters. We have not yet established the time when complete centrioles appear, and of course we hope to learn something about the stages in the genesis of the centrioles. The experimental system may be a starting product for the isolation of reasonable quantities of centrioles, but even before problems of purification are solved the isolated asters - and we know they can be isolated - may be helpful for studies in vitro of how mitotic centers work.

There are more elementary questions to be answered by observing a cell containing many centers that are not near chromosomes and therefore are not engaged in a complete mitotic apparatus. The first is: Do neighboring centers make connections analogous to pole-to-pole fibers in a mitotic apparatus; are we allowed to think of the formation of a spindle without chromosomes?

If we knew all about the mitotic centers themselves, major problems of the polarization of the mitotic apparatus would remain. The first, raised by McIntosh in his lecture (this volume), is the mechanism of separation of the centers to form a bipolar mitotic apparatus. An obvious hypothesis is that the centers are pushed apart by the elongation of microtubules, but I would be hard-pressed to give evidence except by referring to cases of extranuclear spindles in termite flagellates. The problem is important. To begin the polarization of the mitotic apparatus might be examined more thoroughly by electron-microscopic procedures now available. And, when we know more about the gadgets used to separate the mitotic centers, we will still have to inquire into the compass that indicates the direction of separation. The spindle axes are important in determining the cleavages of eggs and dramatically important in accounting for the differentiation of tissues in higher plants.

My guess about the prospects of this field is that major questions about mitotic centers themselves will be answered in the relatively near future. Some of the problems can be solved by progress in fields that are active apart from any focus on mitosis: yeast genetics, nucleation of microtubule assembly; microtubule organization for other aspects of cell morphology; and detection and amplification of nucleic acid molecules. For the specialist on mitosis, the problems of the polarization of the mitotic apparatus remain at the top of the list of those requiring more complete descriptive formation than we now have.

We have enough facts about chromosome replication, chromosome condensation, the mitotic apparatus, and chromosome movement to discuss them seriously - and enough hypotheses to fuel disagreements. We cannot even disagree about the laws of engagement of sister chromosomes to different poles; the basic facts are indisputable and we have to seriously hypothesize about how it works.

We do have some basis for future work on kinetochore-to-center connections in studies of the generation of microtubules, which Borisy (this volume) has discussed. We have an ample platform for further examination of the generation of microtubules by mitotic centers, which has been considered above and in Borisy's lecture. Are kinetochores also generators of microtubules? Certainly there is powerful affirmative evidence for this idea, which was incorporated into theories of mitosis before there was any evidence at all. The picture of microtubules grown from kinetochores of isolated chromosomes are beautiful in themselves (Borisy, this volume; Telzer et al., 1975). The acknowledged criticism of this work - that the microtubules may be nucleated by remnants of microtubules remaining after the isolation of the chromosomes - will surely be probed in the near future. This can be done by studies of cells in which the chromosome cycle is going on in the absence of mitotic centers.

Considering the logic of the situation, the hypothesis of the generation of microtubules by both the centers and the kinetochores does not simplify the situation; it may be true but it creates more problems. We know (Dentler et al., 1974) that the growth of microtubules is polarized; we are not free to propose that growing ends of microtubules will connect when they meet. An alternative idea is that microtubules from kinetochores and those from centers do not connect end-on but side-by-side; that would allow the prediction that chromosomal fibers would be composed of paired microtubules somewhere in the spindle. It would be more attractive to think of pole-to-kinetochore fibers as consisting of microtubules that originate at the centers and screw into the kinetochores - the latter would then function (as is implicit in their definition) as points where microtubules make firm attachments. For purposes of explaining mitosis, the mechanism of attachment of microtubules to kinetochores - and the attachment of the functional kinetochores to the chromatin - is the problem. To explain the law of exclusive attachment, which forbids sister kinetochores to engage to the same pole, is an even more important problem.

It would be helpful to have more complete electron-microscopic descriptions of the moment of engagement of chromosomes to pole. I am aware only of the work of Luykx (1965) on the formation of meiotic spindles in the eggs of the echuroid worm *Urechis*. His observations of what happens at the time of engagement would seem to compound the difficulty of the problem. Serial sections show that at first, both sister kinetochores connect to poles. The law-abiding decision on which chromosome goes to which pole would seem to come as a second step.

What hypotheses can we make to explain the law of engagement of sister chromosomes by different poles? One, for which we can find arguments in the micromanipulation studies by Nicklas (1971), views each kinetochore of a sister pair as pointing toward a different pole at the moment of engagement; each kinetochore would connect to the pole it is facing. That would be hard to fit into Luykx's observations, if those apply generally. The principle would have to work as infallibly as does the phenomenon itself; the facts do not allow for even occasional connections of sister chromosomes to the same pole.

Let me venture a speculation that would lead to a molecular interpretation of the exclusive engagements of sister chromosomes to poles. The speculation calls for an asymmetry in the replication of both kinetochores and centers; the basic proposal is that this replication produces what I will call right-handed and left-handed kinetochores and right- and left-handed centers. We can then imagine that right-handed kinetochores could connect only with right-handed centers, left-handed kinetochores with left-handed centers. (Or the reverse - the only point is that the handedness would exclude attachment of sister kinetochores to the same center.)

Where could we find the handedness? A first place to look is in the DNA itself. The reproducibility of kinetochores assures us that there is a kinetochore-determining DNA; this is proved classically by the fact that kinetochores are never replaced if the kinetochore regions of chromosomes are deleted. The functional kinetochore - the body on

the mitotic chromosome to which microtubules are attached - is the
phenotypic expression of the kinetochore DNA. It is an unusual phenotypic manifestation since we see it at the site of the mitotic chromosome where the corresponding genetic material is located. However,
when we see it with the electron microscope we can observe a definite
structure, sometimes a three-layered plaque (Luykx, 1970), which is
distinguishable from chromatin.

Let us draw on the asymmetry of DNA itself for the source of the asymmetry of kinetochores and propose that the left- and right-handedness
ultimately rest in the differences between the complementary strands
of the kinetochore DNA. A rather extreme version of the speculation
proposes that: 1) Both strands of the kinetochore DNA are transcribed
before replication; 2) the transcripts make different kinetochore
proteins, which ultimately combine with the respective DNA strands to
establish a left-handed and a right-handed functional kinetochore;
3) when the strands have replicated, one daughter chromosome carries
a left-handed kinetochore, the other a right-handed one. The same
supposition would be applied to the mitotic centers. If their replicating material were DNA, the procedure would be the same as just
described. If the genome of the mitotic center were an RNA, the known
mechanisms of replication would produce a complementary strand of RNA
and thus make possible the production of left- and right-handed centrioles.

If there is any substance to this extreme speculation, it is in the
notion of left- and right-handed kinetochores and centers. A simpler
statement would merely call for the association of different molecules
(say proteins) to the left and right strands of the kinetochore DNA.
At least we do know that there is kinetochore DNA and that the functional kinetochores are molecules that somehow recognize that DNA.
A similar conception could be applied to the mitotic centers. We would
have to retain the hypothesis that these centers contain replicating
material but would not have to specify DNA or RNA.

The notion of handedness of kinetochores is a difficult one in that
it violates some assumptions and facts about the randomness of chromosome segregation. (I leave it in this text, even though weighty criticisms were presented at the Workshop, if only to prove that workshops
do some work.)

What means do we have for future experimental work on the engagement
of chromosomes to poles? I have already mentioned some kinds of experiments that surely will be pursued: studies of centers and kinetochores as microtubule-organizing centers. More observational work
on the formation of kinetochore-to-pole fibers would be welcome. Is
it generally true that connections are first made to both poles,
then become exclusive connections of each kinetochore to a different
pole? One useful fact needs to be mentioned, i.e., that the engagement
of kinetochore to poles is not an irreversible event confined to one
moment of mitotic history. If the spindle structure is destroyed, the
connections form again and a normal separation of the chromosomes
follows. There might be some useful information in careful studies
of the recovery of spindles from the action of the many antimitotic
drugs we now have at our disposal.

One is also able to amplify the process of the engagement of kinetochores to centers, showing that any number of chromosomes can engage with two poles. In a version of experiments mentioned earlier,
I have turned on the chromosome cycles in unfertilized sea urchin
eggs, allowed the chromosomes to go through a number of cycles, then

introduced centrioles by fertilizing the eggs. One observes that these
cells now form mitotic apparatus and complete the mitotic cycle. In
cases where there is only one pair of centrioles (sometimes there are
more because more than one sperm penetrates), one can see metaphase plates
and anaphase figures with very large numbers of chromosomes. Obviously
these chromosomes, which have replicated many times without seeing a
mitotic center, can still make the proper engagement to mitotic centers.

D. Perspectives

I have attempted to isolate a number of definite problems of mitosis,
defining them in terms of what is known and of existing means for future attack. (If excessive attention has been given to work with which
I have experience, it has only been to express one observer's optimism
about the accessibility of some difficult and neglected problems.) The
number of questions I have discussed seems large compared to the number of those that now receive concerted effort, but it is not large
compared to what we will have to know before understanding mitosis. To
take one very large example, we have not discussed in depth what we
might call the physiology of the mitotic cycle, that is, the changes
in the cell as a whole as it goes through the whole cycle. The history
of the field has given us a number of biochemical cycles whose status
and future we might consider: the thiol cycle, the Ca-activated ATPase
cycle. It is remarkable that growth and protein synthesis shut down in
most cells during the period of distribution of chromosomes. There is
the well-established fact that the mitotic apparatus sends some kind of
signal to the cell surface, establishing the plane of cell division.
There are all the problems of changes in the cell surface in various
phases, to which recent work has awarded some glances but hardly more.

At the beginning of this talk, I suggested that we extricate the mitotic cycle from the overall growth-division bicycle. The mitotic cycle
defined in its larger sense can be divided into the replicative phase
and the phase during which the chromosomes are distributed to sister
nuclei. Now I propose that these two major phases be viewed as two
distinct states of the whole cell. (To a very small cell biologist,
swimming inside the cell, the scene would appear to be very different
in one phase and in the other.) In a structural sense, the two phases
represent large transformations of the organization of the cell. It is
not trivial that most animal cells lose their morphologic features
and become round during mitosis, nor is it trivial that the formation
of a mitotic apparatus (or even a partial mitotic apparatus in cells
going through the chromosome cycle without centers) involves a gross
redistribution of the major components of cell structure. What will
we see when our vision of cell structure moves a step deeper? From a
biochemical standpoint the fact that all replicative events and syntheses of macromolecules take place during interphase and are largely
shut down during mitosis must be trying to tell us something about two
states of the cell.

Considered more minutely, these two states of the cell in the two major
phases of the whole mitotic cycle may explain - and be explained by -
the regulatory master switches that move the cell around the cycle.
As has been pointed out, all the events of chromosome replication,
whether in the replication of DNA or the synthesis of chromosomal proteins, must be turned on at the beginning of the replicative phase
and turned off at the end. Any case in which we observe one of these
events without observing all of them is likely to be considered excep-
tional. The transition into the active mitotic phase will introduce
simultaneously the changes required for chromosome condensation and

for the assembly of the spindle. Of course the master switches need not act directly; they may well set into motion various specific regulatory events, which we will want to know. If we can subsume the whole cycle broadly into two major states of the cell, it would of course help us to deal with the individual processes of the cycle, especially if it helps us to define proper conditions for experimentation at the subcellular level. This view might also assist with the problem of placing the cell into the organism in which some cells divide and some do not divide.

Acknowledgements. Work from the author's laboratory has been supported by grant GM-13882 from the National Institutes of Health. He also offers his cordial thanks to the Director and staff of the Institut für Zellforschung for valued opportunities to collaborate with them. The force of the late Hans Lettré brought this Institute and ultimately this Workshop into being; may his shade enjoy our record of the present and the future.

References

This list of references is not an adequate bibliography of the topics discussed. It is intended only to provide citations for some of the points raised in the text. The references tend to cite later work, which would give access to the literature, and do not give proper credit to original discoverers.

Bak, A.L., Zeuthen, J., Crick, F.H.C.: Proc. Natl. Acad. Sci. U.S. 74, 1595 (1977)
Bibring, T., Baxandall, J.: Develop. Biol. 55, 191 (1977)
Dentler, W.L., Granett, D., Witman, G.B., Rosenbaum, J.L.: Proc. Natl. Acad. Sci. U.S. 71, 1710 (1974)
Dipple, R.V.: J. Cell Biol. 69, 622 (1976)
Dirksen, E.R.: J. Biophys. Biochem. Cytol. 11, 244 (1961)
Dirksen, E.R.: J. Cell Biol. 51, 286 (1971)
Graves, J.A.M.: Exptl. Cell Res. 73, 81 (1972)
Hartman, H.: J. Theoret. Biol. 51, 501 (1975)
Heilbrunn, L.V.: An Outline of General Physiology, 3rd ed. Philadelphia: Saunders 1952, pp. 716-743
Himes, R.H., Burton, P.R., Kersey, R.N., Pierson, G.B.: Proc. Natl. Acad. Sci. U.S. 73, 4397 (1976)
Liskay, R.M.: Proc. Natl. Acad. Sci. U.S. 74, 1662 (1977)
Luykx, P.: Exptl. Cell Res. 39, 658 (1965)
Luykx, P.: Cellular Mechanisms of Chromosome Distribution (Suppl. 2 of Int. Rev. Cytol.) New York: Academic Press 1970, pp. 47-94
Mazia, D.: In: The Cell. Brachet. J. Mirsky (ed.). New York: Academic Press 1961, Vol. 3, pp. 116-139
Mazia, D.: J. Cell Comp. Physiol. 62, suppl. 1, 123 (1963)
Mazia, D.: Proc. Natl. Acad. Sci. U.S. 71, 690 (1974a)
Mazia, D.: In: Cell Cycle Controls. Padilla, G.M., Cameron, I.L., Zimmerman, A. (eds.). New York: Academic Press 1974b, pp. 265-272
Mazia, D., Harris, P.G., Bibring, T.: J. Biophys. Biochem. Cytol. 7, 1 (1960)
Mazia, D., Schatten, G., Steinhardt, R.: Proc. Natl. Acad. Sci. U.S. 72, 4469 (1975)
Nicklas, R.B.: In: Adv. Cell Biol. Prescott, D.M., Goldstein, L., McConkey, E. (eds.). New York: Appleton-Century-Crofts 1971, Vol. 2, pp. 225-297
Nishioka, D., Mazia, D.: Cell Biology International Reports 1, 23 (1977)
Prescott, D.M.: Reproduction of Eucaryotic Cells. New York: Academic Press 1976
Puck, T.T.: Proc. Natl. Acad. Sci. U.S. 47, 1181 (1961)
Rai, B.: UCRL Reports, No. 10979. Berkeley: E.O. Lawrence Radiation Laboratory 1963, pp. 1-54
Rao, P.N., Johnson, R.T.: Nature (London) 225, 159 (1970)
Rao, P.N., Wilson, B., Puck, T.T.: J. Cell Physiol. 91, 131 (1977)

Rattner, J.B., Phillips, S.G.: J. Cell Biol. <u>57</u>, 1973

Ridgway, E.B., Gilkey, J.C., Jaffe, L.F.: Proc. Natl. Acad. Sci. U.S. <u>74</u>, 623 (1977)

Ringertz, N.R., Savage, R.E.: Cell Hybrids. New York: Academic Press 1976

Schor, S.L., Johnson, R.T., Waldren, C.A.: J. Cell Sci. <u>17</u>, 539 (1975)

Steinhardt, R., Zucker, R., Schatten, G.: Develop. Biol., in press (1977)

Taylor, E.W.: Ann. N.Y. Acad. Sci. <u>90</u>, 430 (1960)

Telzer, B.R., Moses, M.J., Rosenbaum, J.L.: Proc. Natl. Acad. Sci. U.S. <u>72</u>, 4023 (1975)

Tobey, R.A., Petersen, D.F., Anderson, E.C.: In: The Cell Cycle and Cancer. Baserga, R. (ed.). New York: Dekker 1971, pp. 309-353

Weisenberg, R.C., Rosenfield, A.C.: J. Cell Biol. <u>64</u>, 146 (1975)

Went, H.A.: J. Theoret. Biol. in press (1977)

Wilt, F., Sakai, H., Mazia, D.: J. Mol. Biol. <u>27</u>, 1 (1967)

Discussion Session IX: Future Research on Mitosis
Chairman: R. DIETZ, Tübingen, F.R.G.

Chairman:
The lecture by Dr. Mazia has opened a wide field of new aspects for research on mitosis. The first part of the discussion will deal with questions that arose directly after the lecture.

P. Harris, Eugene, Oregon, USA
You said that a centriole injected into an activated egg will produce a spindle, but the injection was by means of a fertilizing sperm, which also brings about the respiratory burst and Ca^{2+} release. Is the resulting spindle due to the centriole, or to the other activation events associated with fertilization?

D. Mazia, Berkeley, California, USA
There is not yet clear evidence as to whether the spindle is due only to the injection of the centriole or to some other events. If one has to sacrifice some results of ambiguity about respiration, etc., for the sake of a natural kind of injection, one chooses the latter.

J.R. McIntosh, Boulder, Colorado, USA
Dr. Paweletz, in your study of the fine structure of ammonia-activated sea urchin eggs with Dr. Mazia, were you able to tell whether the kinetochores were initiating microtubules?

N. Paweletz, Heidelberg, F.R.G.
When the chromosomes are condensed in the ammonia-treated sea urchin egg we can find 1) chromosomes with kinetochores but without any attached microtubules. 2) There are also chromosomes with kinetochores with some microtubules attached to them; and 3) microtubules can be attached to chromosomes at undefined sites. In addition there are many microtubules in the cytoplasm around the chromosomes. We can even find aster-like structures but without centrioles. We do not yet know, of course, whether they grow from the chromosomes or come from outside and then touch the chromosomes. Probably both types of microtubule formation take place.

D. Mazia, Berkeley, California, USA
We have also done the experiment of fertilizing these activated eggs; therefore we have the material for observing the making of the connections after a centriole has been introduced. The material has not yet been fully investigated. This is a collaboration in which the experiments have been done in Berkeley and the electron microscopy here in Heidelberg.

S. Ghosh, Calcutta, India
Dr. Mazia referred to work on premature chromosome condensation by the Johnson and Rao group. Fusing mitotic cells with cells belonging to different stages of interphase, they could demonstrate premature condensation of chromosomes with single or double chromatids depending on their origin from G1 or G2 nuclei. They have not referred to the spindle behavior of prematurely condensed chromosomes in any of their works known to me. Both kinetochores and centrioles are present there. I would like to know what happens in such cells.

D. Mazia, Berkeley, California, USA
That work has not been done.

H. Sato, Philadelphia, Pennsylvania, USA
Ammonia treatment of unfertilized sea urchin eggs can be considered as partial activation, and the effect would be comparable to the butyric acid-sea water treatment. In the latter case, successful activation of female centriole followed the treatment, but why did ammonia fail?

D. Mazia, Berkeley, California, USA
This is "sea-urchinology." This is not the same as butyric activation. There is no cortical reaction. The female centriole has not been found. One can fertilize

the egg many hours later, so we continue to interpret this as an experiment of releasing the repressions in an unfertilized egg, which is, nevertheless, still an
unfertilized egg.

J.G. Carlson, Knoxville, Tennessee, USA
The issue of the importance of metaphase in mitosis has been raised by Dr. Mazia,
and I want to submit some evidence to support the idea that the attainment of a
certain metaphase kinetochore configuration may be the trigger for the start of
anaphase. In the normal, untreated grasshopper neuroblast the prometaphase kinetochores, as seen in side view, are at first slightly staggered in their distance
from one pole or the other. As metaphase develops and progresses, however, they
gradually move closer to a single plane at right angles to the spindle axis. At
the end of about 12 min (at 38°C) after nuclear membrane breakdown, when they form
a straight line in side view, the sister kinetochores begin simultaneously the
movement toward opposite poles, as if the attainment of a one-plane orientation
was the signal for the start of anaphase separation. The same is true of other
nonmeiotic cells I have observed, but it is less striking than in the larger neuroblast with its uncrowded and fairly evenly spaced kinetochores. If a neuroblast
in late prophase is exposed to a rather large dose of 225 nm ultraviolet radiation,
a chromosome frequently attaches to the damaged spindle near one pole. Then, during
a period of 45 min or an hour, or even more, it will move slowly toward the equatorial plane in which the other chromosomes have been waiting long beyond the normal prometaphase-metaphase interval. As soon as it has lined up with the other
chromosomes, anaphase separation begins. A similar event in untreated cells has
been described in other nonmeiotic cells by other investigations. It seems to me
remarkable that kinetochores, which appear to act more or less independently during
metaphase development of microtubules and kinetochore alignment to the spindle,
act as a unit at the start and during the continuation of anaphase separation.
Since, at least in the neuroblast, poleward movement of kinetochores and spindle
elongation begin simultaneously, the triggering action that starts anaphase would
seem to involve the spindle as a single unit.

D. Mazia, Berkeley, California, USA
If I had intended to minimize the metaphase phenomenon I would not have described
the last experiments, which show the tendency of hundreds of chromosomes to go
into a metaphase plate that does not have room for them. The intent of my remarks
was to point out that the metaphase plate is not universal, and as long as we are
gratefully accepting mitosis in some of the lower eukaryotes as mitosis, we had
better not insist on the universality of the metaphase plate.

W. Sachsenmaier, Innsbruck, Austria
Concerning the activation by ammonia of the production of TTP in the absence of the
nucleus, what you see is an activation of thymidine kinase and perhaps also of thy-
midilic kinase, etc.: Do you think that this is activation of a preexisting enzyme
under these conditions or is it new synthesis of the enzymes in the absence of the
nucleus, which would mean that you have a silent messenger present in the active
form?

D. Mazia, Berkeley, California, USA
Based on work done in Japan, showing that in homogenates the enzyme was there in active form, it is certain that it is activation. Also we observed these events within
minutes

S. Ghosh, Calcutta, India
Dr. Mazia's experiment with sea urchin eggs reminds me of the phenomenon of
parthenogenesis in the plant kingdom. In the latter case, however, we get a regular
mitotic division with normal spindle. In sea urchin eggs you do not get centriole
formation, although theoretically you have all the prerequisites, because the latent
centrioles, etc., are present in the cytoplasm. I would like to know your opinion
as to why the centrioles failed to develop in these cells.

D. Mazia, Berkeley, California, USA
The problem about the parthenogenetic activation of the sea urchin eggs has required
a second step of treatment, usually involving hypertonic solution, for which a

superior equivalent is the treatment with heavy water. The interpretation of the phenomenon of parthenogenesis in these organisms is that it takes a second treatment to activate the centriole, so that these treatments are intended to avoid the activation of the egg. The closer they come through a mere derepression of certain events that are all repressed in the unfertilized egg, the better it works. The experiments would not be any good if the eggs develop.

H. Fuge, Kaiserslautern, F.R.G.
I have a question concerning "right"- and "left"-handed kinetochores and centrioles: If such predetermination exists, I cannot imagine how the random distribution of homologues in the first meiotic division, leading to a new combination of the genome, can be achieved?

D. Mazia, Berkeley, California, USA
I have not thought about that particular point, but it seems to me that a mechanism that would randomize the right and left at every replication would produce ultimate randomness, no matter how many replications have taken place.

J.R. McIntosh, Boulder, Colorado, USA
Dr. Mazia has raised the point of specific attachments between chromosomes and poles, and I am reminded of the reports in the older literature that particular chromosomes occupy particular places within the metaphase plate. Does anyone know of a well-documented case for this sort of higher order of organization within the mitotic spindle?

N. Paweletz, Heidelberg, F.R.G.
It is difficult to find a pattern of chromosome distribution during mitosis. Therefore, we looked for special markers. We used a silver impregnation technique which is said to impregnate nucleolar organizing regions. With these markers we could demonstrate a nonrandom arrangement of at least the marked chromosomes. This distribution remains the same in anaphase and telophase, and even during reconstruction of the nucleolus, the arrangement of prenucleoli is nonrandom.

L. Hens, Brussels, Belgium
It was clearly shown that the organization of chromosomes in the human lymphocyte metaphase plate is not random (Warburton et al., Humangenetik 18, 297, 1973; Hens et al., Humangenetik 28, 303, 1975; Hens, Chromosoma 57, 205, 1976; Poster 17). The nonrandom distribution involves, for an important part, the acrocentric chromosomes, but also a lot of combinations between other nonacrocentric chromosomes. There are three types of observations which can provide a functional base for these nonrandom arrangements:
1. DNA strands exist between the p-arms of the human acrocentric chromosomes (Henolerson et al., Nature 243, 95, 1973).
2. There are microtubules between human chromosomes, not only in the centromeric region but also along the whole length of the chromosome arms.
3. There is indirect evidence that some proteins of the constitutive heterochromatin are partly responsible for the nonrandom distribution of chromosomes in the metaphase plate.

N. Paweletz, Heidelberg, F.R.G.
In relation to nonrandom arrangement of chromosomes during mitosis, Rabl described this phenomenon in the 1880s. He reported that the arrangement of chromosomes in prophase is exactly the same as that seen in telophase and reconstruction. This can only be explained by the maintenance of the arrangement during the whole course of mitosis.
This led Professors Lettré to assume that there are persistent spindle fibers. Today we know that they do not persist, but nevertheless there must be a system guaranteeing the maintenance of order even after the breakdown of the nuclear envelope.

J.M. Mitchison, Edinburgh, Scotland, U.K.
If we look into the intermediate future of research into the control of cell division and its relation to growth and size, two types of experiments are likely to be profitable.

The first is cell fusion. This has already been used with great success in *Physarum* and in mammalian cells, but it could be extended to other situations and cell types. For example, it is clear that there is an inducer for DNA synthesis in the cytoplasm of S-phase cells. But is it also being accumulated in G1 cells? If it is, and if it acts as the main initiator of DNA synthesis, then fusion of a late G1 and an early G1 cell should accelerate the early cell into S and, more important, delay the late cell. Again, it would be interesting to see whether a cell running on its "minimum cycle time" could be accelerated into division by fusion with a late G2 cell.

The other type of experiment is a deliberate alteration of the nucleocytoplasmic ratio. Several methods are available: 1) Isolating cell size mutants of *Schizosaccharomyces pombe* - this could be difficult, but it may allow us to detect the gene products and so an important part of the control machinery; 2) using alterations of growth rate by varying the medium - if cell size varies with growth rate, as it does in bacteria and *S. pombe*, this alters the nucleocytoplasmic ratio; 3) removing cytoplasm or nuclei, or adding them - this has already yielded important information in *Amoeba*, *Physarum*, and *Stentor*. It could be tried with other cells, and the ameba experiments could be extended and followed in greater detail than in the earlier work. For example, the expected effect of adding a second nucleus to an ameba would be different in several of the models from fusing two amebae (if this proves possible). The nucleocytoplasmic ratio would be altered in the first case, but not in the second.

G.G. Borisy, Madison, Wisconsin, USA

In this Workshop I think our discussion served to generate questions rather than to establish hard facts. However, for the sake of variety and also to be heretical, I should like to suggest a different viewpoint, namely, that we have more facts than we know what to do with. Indeed, I submit that in principle the problem of mitosis is solved. I suggest that we now understand the basic outlines of the organization, formation, and bipolarity of the spindle in terms of the properties of microtubules and the activity of organizing centers. The major problem remaining is to understand the movement of chromosomes themselves, and for this we do not have to search for a force as yet unknown to us. We are in possession of only two tenable paradigms. The first paradigm is that of muscle, and the active molecules are actin and myosin. The second paradigm is that of flagella, and the active molecules are tubulin and dynein. Therefore we have before us only the relatively trivial problem of identifying which of these two paradigms is indeed the correct one.

J.W. Sanger, Philadelphia, Pennsylvania, USA

On a more serious level I am sure both systems (actin-myosin; tubulin-dynein) are involved in the process of chromosome separation. I think actin-myosin provides the motor force for moving the chromosomes to the poles and that further chromosome separation is achieved by pushing the poles apart using a tubulin-dynein system. Both contractile systems can be assembled anew from their monomers and both systems can work for short periods. Nature has been very innovative in using actin-myosin in such processes as cytokinesis, ameboid movement, and muscle contraction. However, microtubules appear also to be related to these processes. Microtubules may be involved in defining the region (via astral rays) where the actin-myosin filaments should form a belt for cytokinesis. Up to a certain point in mitosis, the drug colcemid will block the formation of a functional cleavage ring. We know that cardiac fibroblasts are active locomoting cells in culture. Yet in the presence of colcemid, while there is ruffling at the perimeter of the cell, there is no longer translational movement. Microtubules appear then to direct the ruffling in the direction the cell "wants" to move: destroy the microtubules, and ruffling then occurs all around the cell with no change in position. Microtubules are involved also in setting up the tubular shape of the myotube. In tissue culture, exposure of myotubes to colcemid leads to a rounding up of the elongated myotube. Remove the drug and the cell will regain its normal asymmetric shape. The clear problem to me in cell motility is to study microtubular microfilament interactions.

A.S. Bajer, Eugene, Oregon, USA

I disagree basically with Dr. Borisy's statement that there are two hypotheses to choose from for interpretation of chromosome movement. At present we have at least three and any fact in principle can be interpreted in an unlimited number of ways, which are on different levels of complexity and are more or less consistent with the facts. I would like at the same time to answer a question posed by Dr. Dietz: Do existing hypotheses help in planning experiments? My answer is yes. We have been working for a few years in my laboratory on the basis of a few assumptions (latest theoretical treatment: Novitski and Bajer, Cytobios, 1977, in press), which permit us to predict the change of spindle structure and to relate the changes to the spindle function. We have heard very little during this meeting about change of spindle structure, understood as tubule arrangement. In any type of mitosis and any type of spindle, there is a pronounced rearrangement of microtubules. Microtubules are not formed (assembled) in a precisely determined position but continuously rearrange, as has been clearly demonstrated, e.g., in the film shown by Dr. Sato. It is unthinkable, for me at least, that such rearrangement is not functionally related to chromosome movement. I would like to stress again that this aspect of the spindle structure has not been discussed during this meeting. I feel, however, that an understanding of "functional" rearrangement of microtubules is a prerequisite for any explanation on the molecular level as to how microtubules may function during mitosis. I favor at present "simplified practical and limited" models that meet some limited criteria, rather than more general, universal models applicable to all types of spindles and all types of processes or movements observed.

J.R. McIntosh, Boulder, Colorado, USA

While the paradigms suggested by Dr. Borisy are highly plausible, we must keep an open mind about a mitotic mechanism. Two other molecular motors are known: the spasmin system of the vorticellids and the rotary motor of bacterial flagella. Others may be found. I think that the best approach for future work is to combine straightforward analytic studies aimed at learning what is in a spindle, with the model system work, which may allow us to identify the functions of the various known parts.

N. Paweletz, Heidelberg, F.R.G.

For the future research on mitosis one would wish to separate the different parts of cells in mitosis and the processes taking place. To clarify the function of the different parts they should then be put together like a "puzzle" to determine the essential structures and processes. Perhaps experiments can be designed for this "puzzle-game."

W.W. Franke, Heidelberg, F.R.G.

The experiment as to what is important in mitosis has been done by evolution. If tubulin or actin, for example, were not essential to the basic process, one should expect to find mitotic systems that do not contain these components. I find it remarkable that all existing mitotic apparatus involve microtubules - in one way or the other.

A.S. Bajer, Eugene, Oregon, USA

I would like to continue and make some more remarks concerning evolution and mitosis. Evolution is not good or bad, it is characterized by the absence of motive and purpose, and most importantly (in relation to the spindle), is disorderly and complex. It works sometimes in the direction that leads to the disappearance of the species. It makes, however, the structures which "somehow" function - probably often not in the simplest way. The spindle, at least of higher organisms, is rather disorderly, judging from fine structural and structural studies. It is in a constant state of change - to maintain its functionality. I am inclined to think that we have many random reactions in the spindle. Mitosis and distribution of chromosomes may be in fact an "avalanche" or chain reaction based on random and not directional interactions between spindle elements. This approach, however, how unappealing it may be, should in my opinion, be investigated.

N. Paweletz, Heidelberg, F.R.G.
What do we really know about evolution of the spindle? May I ask a provocative question: Can it be possible that the spindles of the so-called primitive organisms are in fact not primitive but highly specialized structures? Is it already an established fact that, e.g., the spindle of *Barbulanympha* is the ancestor of the HeLa spindle?

M. Girbardt, Jena, G.D.R.
More information about fundamental processes may be learned from analysis of "premature mitosis." In future workshops like this, they might be discussed in more detail.

Poster Abstracts

1. DNA Synthesis and Differentiation in Imaginal Wing Disks of *Calliphora*
   D.J. EGBERTS

2. Ultrastructural Probes of DNA Helix Openings During in Vivo Mitoses of Normal and Neoplastic Human Cells
   J.H. FRENSTER, S.R. LANDRUM, M.A. MASEK, S.L. NAKATSU, and L.S. WILSON

3. Trigger Waves and Calcium: a New Model for Mitosis
   P. HARRIS

4. Stimulation of Hepatic DNA Synthesis and Cell Proliferation by Ethylene Dibromide
   E. NACHTOMI and E. FARBER

5. The Enthalpy Changes During Cell Cycle in Normal and Malignant Cells
   B.A. NEŚKOVIĆ and D. NIKOLIĆ

6. Control of Mitosis by Cell Size: a Genetical Analysis in Yeast
   P. NURSE and P. THURIAUX

7. Genetic Instability at Mitosis
   J.A. ROPER

8. On the Nature of the Mitotic Clock in *Physarum polycephalum*
   W. SACHSENMAIER, CH. LINORTNER, P. LOIDL, and J.J. TYSON

9. Regulation of Transcription in the Cell Cycle of *Physarum*
   H.W. SAUER and A. HILDEBRANDT

10. On the Nature of the Indeterministic Phase of the Cell Cycle
    S. SVETINA

11. Biochemical Genetics of Mitosis as a New Approach to the Study of an Old Phenomenon
    R.J. WANG

12. Transition Point and Transition Probabilities for Cells Moving into Mitosis
    D.N. WHEATLEY

13. Chromosomal Dynamics and Environments
    B.E. WOLF

14. Surface Changes During Mitosis
    C.A. PASTERNAK and S. KNUTTON

15. Functional Heterogeneity of the Lymphocyte Plasma Membrane Relating to Growth Control in Lymphocytes
    K. RESCH, A. LORACHER, B. MÄHLER, M. STOECK, and H.N. RODE

16. Microfilaments in Dividing *Xenopus laevis* Tadpole Heart Cells
    U. EUTENEUER, J. BEREITER-HAHN, and M. SCHLIWA

34. Chromosomal Organization and Differentiation in *Ascaris*
    K.B. MORITZ and G. ROTH

35. Keratinization and Growth Regulation in a Bladder Tumor Cell Line
    R. TCHAO and J. LEIGHTON

36. Dynamics and Ultrastructure of Monocentric Chromosome Movement
    R. CAMENZIND and TH. FUX

37. Dynamics and Ultrastructure of Unorthodox Chromosome Movements
    R. CAMENZIND, TH. FUX, P. GANDOLFI, and M. ZANAZZI

38. Anaphase Chromosome Movement in the Grasshopper Neuroblast and its Relation to
    Other Similar Studies
    J.G. CARLSON

39. The 1977 Attempt to Formulate an Assembly Hypothesis of Spindle Formation and
    Spindle Function
    R. DIETZ

40. Microcinematographic Studies on Cell Division in Cultures of Spleen Cells
    G. MUNGYER and C. JERUSALEM

41. Birefringence in Mitosis and Meiosis   (16 mm movie)
    H. SATO, Y. OHNUKI, and K. IZUTSU

42. The Cytaster, a Colchicine-Sensitive Migration Organelle of Cleavage Nuclei
    in Insect Eggs
    R. WOLF

43. Ultrastructure of Pathologic Mitosis in Mice
    P. DUSTIN and J.P. HUBERT

44. Two Different Types of Spindle Pole Organelles in the Life Cycle of Monotha-
    lanous foraminifera
    D. SCHWAB

45. RNA and Protein Synthesis During Plant Mitosis
    J.F. LÓPEZ-SÁEZ

46. Effects of Low Temperature on Dividing Cells of Plant Endosperm
    A.M. LAMBERT and A.S. BAJER

47. Chromosomes and Cancer
    R. KORSGAARD

48. Non Tubulin Molecules in Meiotic Cells
    M. HAUSER and B.M. JOKUSCH

1. DNA Synthesis and Differentiation in Imaginal Wing Disks of *Calliphora*

D.J. EGBERTS

Vakgroep celbiologie en morfogenese, Zoölogisch Laboratorium der Rijksuniversiteit, Kaiserstratt 63, Leiden, The Netherlands

Imaginal disks of holometabolous insects have sharply separated periods of proliferation and final differentiation. Analysis of the DNA synthesis in relation to the final differentiation should reveal whether DNA synthesis could play a role in differentiation. As a first approach the DNA synthesis in wing disks of *Calliphora* has been measured during the third larval instar. Two phases of DNA synthesis could be distinguished. The first had a high activity at the beginning of the third instar, decreased according to first-order kinetics, and reached a marginal value around the white prepupa stage. The second phase started when the larve left wet surroundings and entered dry conditions. This phase also ended at the white prepupa stage. There is evidence for a reversible inhibition of the cell cycle between DNA synthesis and mitosis during this second phase. The second phase could be advanced and delayed by manipulating the humidity of the surroundings. Ecdysterone was capable of initiating this second phase of DNA synthesis. This phase is a necessary condition for further development and coincides with the onset of many differentiation processes. [Supported by the foundation for fundamental biological research (BION), which is subsidized by the Netherlands organization for the advancement of pure research (ZWO).]

2. Ultrastructural Probes of DNA Helix Openings During in Vivo Mitoses of Normal and Neoplastic Human Cells

J.H. FRENSTER, S.R. LANDRUM, M.A. MASEK, S.L. NAKATSU, and S. WILSON

Department of Medicine, Stanford University, Stanford, California 94305, USA

RNA synthesis decreases markedly during prophase to almost zero during metaphase, anaphase, and early telophase in normal mitotic cells. Localized openings of the DNA helix are necessary to allow strand-specific gene transcription within living cells [Cancer Res. 36, 3394 (1976)]. These DNA helix openings can be visualized within individual living cells by a high-resolution electron microscopic probe technique [Nature New Biol. 236, 175 (1972)], and their number and size can be determined within each cell [Nature 248, 334 (1974)]. One hundred twenty-six normal and neoplastic mitotic cells within biopsied human bone marrow and lymph nodes were analyzed for the localization, number, and size of DNA helix openings within each cell, and these data were then separately correlated with the ultrastructural stage of mitosis in each cell. All DNA helix openings were confined to the extended chromatin portion of the cell nucleus. The number of DNA helix openings was decreased in early and late prophase, and became zero within normal cells during metaphase, anaphase, and early telophase, but was not zero within neoplastic cells. The number of DNA helix openings increased in late telophase at a time when stable nuclear RNA species return to postmitotic daughter nuclei following their release to the cytoplasm earlier in mitosis [Exptl. Cell Res. 89, 421 (1974)].

These quantitative single-cell data suggest that small stable nuclear RNA species may be important ligands in effecting openings of DNA helices and inducing locus- and strand-specific gene transcription within these mitotic human cells.

224

[Supported in part by Research Grants CA-10174 and CA-13524 from the National Cancer Institute, by Research Grant IC-45 from the American Cancer Society, and by a Research Scholar Award from the Leukemia Society.]

3. Trigger Waves and Calcium Regulation: A New Model for Mitosis

P. HARRIS

Department of Biology, University of Oregon, Eugene, Oregon 97403, USA

The mitotic apparatus of living sea urchin eggs appears as a clear region from which larger formed elements such as yolk lipid granules are excluded. Electron microscopy has revealed this clear region to consist of a massive system of membrane-bounded vesicles or smooth endoplasmic reticulum. Recent evidence of calcium sequestering vesicles in various contractile or motile nonmuscle cells, and the discovery of a Ca-dependent ATPase in the sea urchin mitotic apparatus, strongly suggest that this vesicular system in the eggs may also regulate cytoplasmic calcium and thus affect microtubule polymerization as well as contractile proteins.

Fine-structure studies of normal sea urchin mitosis indicate that the spindle microtubules begin to break down first within the centrosphere of the asters at a time when the asters are still growing in size. If calcium is the regulator, this suggests that both sequestering and release of calcium may be occurring at the same time in different parts of the mitotic apparatus. Assuming a calcium sequestering function for the vesicles, a new model for mitosis is proposed that would account for this situation. Briefly, it supposes that calcium release is a self-recovering reaction, moving outward as a membrane-mediated trigger wave from the cell center followed by a wave of calcium sequestering. Entry of a G2 cell into mitosis begins with a triggered release of calcium, which breaks down the interphase aster or cytoplasmic microtubules, centrioles separate, and a new wave of calcium sequestering is begun, now with two organizing centers instead of one. As the asters grow at their periphery, calcium is again released at the aster centers. Deformation and disorientation of microtubules seen in the centrosphere may be the result of a direct effect of calcium on the microtubules, or the effect of their being enmeshed in a nonmicrotubule contractile system. The result in either case is to pull the aster and spindle microtubules, along with the chromosomes, toward the aster center.

4. Stimulation of Hepatic DNA Synthesis and Cell Proliferation by Ethylene Dibromide

E. NACHTOMI[1] and E. FARBER[2]

[1]Department of Animal Science, Agricultural Research Organization, The Volcani Center, Bet Dagan (50 200), Israel
[2]Department of Pathology, University of Toronto, Canada M5G 7L5

The effect of ethylene dibromide (EDB) on DNA synthesis and the morphology of liver cells were studied in Wistar rats. EDB (10 mg/100 g body weight), administered by stomach intubation, produced enhanced DNA synthesis as measured by thymidine-methyl-^{3}H incorporation. The peak of DNA synthesis was observed after 24 h. Mitotic cells were observed without necrosis in the liver sections. The mitotic waves measured with the aid of colchicine were observed at 24-30 h and 48-54 h after EDB treatment. A maximum of 16 % of the liver cells entered mitosis. Autoradiography of liver confirmed the DNA synthesis in the S phase by localization of the labelling in the nucleus of hepatic cells. The mitotic phenomenon induced by EDB in liver cells without apparent necrosis, indicates that some other cell injury may initiate liver cell proliferation.

5. Enthalpy Changes During Cell Cycle in Normal and Malignant Cells

B.A. NEŠKOVIĆ and D. NIKOLIĆ

Laboratory of Experimental Oncology, Medical Faculty, University of Belgrade, Jugoslavia

We measured the enthalpy changes in "normal" and "malignant" cells during their growth, development, and division in monolayer culture in vitro, i.e., we measured the quantity of heat released to the surroundings under constant pressure conditions. From this enthalpy change, the respective amounts of energy that were used for DNA synthesis, RNA synthesis, protein, and lipid synthesis can be calculated when we know that their quantity is doubled in the cell cycle. From these calculations and the measured value of enthalpy change, the entropic member, which is included in Gibb's equation, could also be calculated.

We have measured enthalpy change during growth, development, and division in the following cells: (1) normal lung cells of human embryo (second passage), (2) L-strain cells 929, and (3) HeLa cells. All these cells grew in monolayer on the bottom of a glass vessel in nutritive media 199 supplemented with 10 % calf serum.

6. Control of Mitosis by Cell Size: A Genetic Analysis in Yeast

P. NURSE[1] and P. THURIAUX[2]

[1]Department of Zoology, West Mains Road, Edinburgh EH9 3JT, U.K.
[2]Institut für Allgemeine Mikrobiologie, Altenbergrain 21, 3013 Bern Switzerland

Cell division is regulated by a control that maintains a constant size at division and therefore coordinates division to cellular growth. Mutants altered in that regulation have been isolated (Nurse, 1975; Thuriaux, Nurse and Carter, unpublished). Their growth rate is comparable to that of the wild type, but they divide at about half the wild-type cell size. They define two unlinked genes *wee* 1 (22 mutants) and *wee* 2 (1 mutant). The properties of temperature-sensitive alleles of *wee* 1 and *wee* 2 upon a temperature shift indicate that they are altered in a control-regulating mitosis.

Wee 1 mutants are semi-dominant, whereas the *wee* 2 allele is almost dominant over the wild type. *Wee* 2 is very close and probably allelic to recessive *cdc* 2 mutants that are temperature-sensitive for mitosis and have been previously isolated by Nurse et al. (1976). Some *cdc* 2 mutants, when grown at the permissive temperature, have a small cell size at division (*wee* phenotype, allele *cdc* 2-56) whereas others (*cdc* 2-35 and *cdc* 2-63) divide at a larger size than the wild type. A tentative model for size control of mitosis will be presented.

226

7. Genetic Instability at Mitosis

J.A. ROPER

Department of Genetics, The University, Sheffield, U.K.

In the fungus *Aspergillus nidulans*, strains with one or more chromosome segments in excess of the balanced genome (either haploid or diploid) are unstable at mitosis. Colonies of such strains produce frequent variant sectors and, from analyses of parents and their variants, the primary cause of instability and certain of its consequences have been established.

Comparisons of balanced and unbalanced strains show that imbalance provokes instability and that most, perhaps all, initial instability is confined to the segments represented in excess. The most frequent sectors result from deletions in one or other "excess" segment; other instability processes occur but interpretation of some of these is tentative. Different extra segments confer different degrees of instability and each segment has a unique distribution of deletion breakpoints. Instability is modified by various factors; the effects of genes involved in DNA repair and of certain chemicals provide indirect information on the instability process.

The findings are interpreted tentatively in terms of the control of chromosome replication and of genome organization.

8. On the Nature of the Mitotic Clock in *Physarum Polycephalum*

W. SACHSENMAIER, CH. LINORTNER, P. LOIDL, and J.J. TYSON

Institut für Biochemie und Experimentelle Krebsforschung der Universität Innsbruck, Austria

Mitoses in multinuclear plasmodia of the myxomycete *Physarum polycephalum* are naturally synchronous. Mixed plasmodia containing different sets of nuclei in a single mass of cytoplasm may be prepared by fusion of two plasmodia at different stages of the mitotic cycle. These nuclear populations become rapidly synchronized, suggesting that the onset of mitosis is triggered by a cytoplasmic factor. Synchronization of heterophasic plasmodia by fusion always occurs at an intermediate point on that part of the phase circle that does not include mitosis. This means that the mitotic clock is reset by a molecular event closely correlated with mitosis (or with the onset of DNA replication, which commences immediately after telophase in *Physarum*). A model is discussed proposing a stoichiometric interaction of the cytoplasmic initiator with a given number of nuclear receptor sites. Mitosis is triggered as soon as all nuclear sites are "titrated." The cytoplasmic initiator is supposed to be synthesized parallel to the overall increase of the cell mass. This agrees with experimental findings indicating a tendency of growing plasmodia to regulate the length of the intermitotic period according to their protein/DNA ratio.

9. Regulation of Transcription in the Cell Cycle of *Physarum*

H.W. SAUER and A. HILDEBRANDT

Zoologisches Institut Lehrstuhl (I) der Universität, Röntgenring 10, 8700 Würzburg, Federal Republic of Germany

1. Transcription is necessary for the progression of the cell cycle of *Physarum*.

2. Transcription of DNA varies in the cell cycle in that (a) a larger proportion of poly A + RNA or rRNA is transcribed in S phase or G2 phase, respectively, and (b) practically no transcription occurs in metaphase.

3. There is no correlation of these variations in RNA synthesis with the levels of salt-extractable RNA polymerases A and B. However, salt-resistant *endogenous* RNA polymerase B activity is absent in metaphase, high in S phase, and low in G2 phase.

4. Maximum level of bound RNA polymerase B is observed at the beginning of S phase (there is no G1 phase in *Physarum*), but only if DNA synthesis takes place.

We propose that replicating chromatin of *Physarum* is charged in from a pool of abundant RNA polymerase B molecules and postulate that (a) some of these bound molecules transcribe DNA immediately, which may explain the observed "replication-transcription coupling" in *Physarum*, (b) other bound molecules are in a "preinitiated" state that can be transferred into an "engaged" state in G2-phase, since the capacity to transcribe DNA by endogenous RNA polymerase B can be increased by a brief incubation with detergent.

5. Consequently, regulation of transcription may be achieved not only by sequential binding and activity of RNA polymerase in S phase but also by modifying the state of RNA polymerase molecules already bound.

10. On the Nature of the Indeterministic Phase of the Cell Cycle

S. SVETINA

Institute of Biophysics, Medical Faculty and "J. Stefan" Institute, University of Ljubljana, 61105 Ljubljana, Jugoslavia

The indeterministic phase of the cell cycle has been characterized as that part of the cell cycle from which cells proceed in a stochastic manner. It was previously assumed that the stochastic nature of the transition from this phase is a consequence of a stochastic nature of a molecular process. However, such a process has not yet been elucidated. It is suggested here that the indeterministic phase of the cell cycle consists of a large number of biochemically distinct states. The cell can be in any of these states and can randomly move from state to state in any direction. The transition to the deterministic part of the cell cycle occurs only from one of these states and is unidirectional. The proposed model predicts that the number of cells in the indeterministic phase decreases exponentially with time. The time lag observed in real systems before the exponential behavior is attained may correspond to the time needed for the cells that have entered the indeterministic phase at one end to become distributed among all possible states stationarily.

11. Biochemical Genetics of Mitosis as a New Approach to the Study of an Old Phenomenon

R.J. WANG

University of Missouri-Columbia, John M. Dalton Research Center, Research Park, Columbia, Mo. 65201, USA

To achieve a better understanding of mitosis, we initiated a program utilizing a genetic approach, coupled with biochemical and cytologic procedures, for dissecting mitosis into its participating components and more clearly defined sequential events. Our initial steps involve isolation and study of a collection of temperature-sensitive mammalian cell mutants, each of which exhibits a defective mitotic step. The first mutant entered mitosis at the nonpermissive temperature and initiated metaphase, but subsequent mitotic events failed to occur normally. The chromosomes, instead of moving to opposite poles, were scattered and fused into chromatin clumps. Microtubules and associated proteins were normal. A second mutant showed defective prophase progession. Chromatin material in the mutant cells condensed, but not into chromosome. A different mutant with defects in postmetaphase chromosome movement was found recently. In these cells the metaphase plate also formed at the nonpermissive temperature. Subsequently, chromosomes did move, but in an abnormal fashion. Defective cytokinesis frequently resulted as a consequence. A fourth mutant was defective in premetaphase events. In these cells the chromosomes condensed but metaphase plate was not initiated. Nuclear membrane re-formed around the chromosomes, and the cells completed the next round of cell-cycle events. The second mitosis was again defective, and the chain of events was repeated.

12. Transition Points and Transition Probabilities for Cells Moving into Mitosis

D.N. WHEATLEY[1]

Department of Pathology, University of Aberdeen, Foresthill, Aberdeen AB9 2ZD, Scotland, U.K.

Before a cell can divide, it must satisfy certain minimum conditions, e.g., replication of its genomic DNA. In a proliferating population, each cell moves through integrated sequences of biochemical processes to satisfy the minimum essential requirements for division - familiar to us as the cell cycle.

A cell's behavior thereafter may be seen as having two distinctly different possibilities: (1) it will automatically enter division as these requirements become finally satisfied (deterministic model); or (2) it retains the option to divide, i.e., it has only a certain probability per unit time of actually moving into mitosis (stochastic model). In simpler terms we may say that the former model represents a commitment to divide whereas the latter model results in a cell having the freedom to divide.

Analysis of the cell cycle, particularly events occurring close to the onset of division, has relied heavily on the effects of agents that disturb the biochemical preparations of cells for division. From these studies, the concept of the transition point has arisen. This will be discussed in terms of the concept described above. In addition, recent studies suggest that transition probabilities also exist, which may be considered as changes in the rate at which cells free to divide take up their option of entering mitosis. An attempt will be made to clarify these concepts and discuss their meaningfulness, using evidence from recent studies with HeLa cells and *Tetrahymena*.

[[1]The work to be presented acknowledges the collaboration of Professor Erik Zeuthen and Dr. Leif Rasmussen of the Carlsberg Biological Institute, Copenhagen, and the financial assistance of the Cancer Research Campaign.]

13. Chromosomal Dynamics and Environments

B.E. WOLF

Labor für Cytogenetik, Ortlerweg 18, 1000 Berlin 45

Numerous results from recent and past investigations have shown that the chromosomes in somatic as well as in germ cells of various organisms are subjected to environmental influences. In the following, comparative observations on mitotic, in some cases on meiotic chromosomes of some insects (*Phryne cincta*, *Drosophila melanogaster*, *Cloeon dipterum*, *Trichocera maculipennis*) and of a plant (*Allium cepa*) are presented. *Phryne* and *Drosophila* additionally permit observations on salivary gland chromosomes. Differences were made between the influences from the outer environment (i.e., temperature, dryness, overpopulation, lack of food) and the inner (extranuclear and intranuclear) environment of chromosomes (i.e., physiologic changes within tissues caused by aging or chemicals, and changes caused by genetic factors such as inversions or additional quantities of α-heterochromatin inserted in one of the chromosomes). As proved experimentally, mitotic and meiotic chromosomes in their contraction and pairing behavior (a) and in their crossover frequencies (b), and polytene chromosomes in their structural modificability (c) strongly depend upon environmental conditions. The fluctuations can be interpreted in terms of a chromosome imminent control mechanism leading to functional adaptations within the genetic material to its environment. Also extreme-reaction mutations and/or lethal effects may arise (reducing the number of individuals in a population). The comparative investigations still leave open the question of fundamental significance: Which forces cause chromosomes to change in structure, arrangement, and replication mode? Some of the results presented here favor a DNA component subjected to alternations within the chromosomes, giving rise either to pairing or repulsion of homologous segments.

14. Surface Changes During Mitosis

C.A. PASTERNAK and S. KNUTTON

St. George's Hospital Medical School, University of London, Blackshaw Road, Tooting, London SW 17 0QT, U.K.

Formation of the cytokinetic furrow during mitosis increases the surface area of a spherical cell by 40 %. What is the origin of this amount of surface membrane? Our experiments indicate that it is derived not by synthesis of new membrane, but by a redistribution of existing surface membrane. The form in which the membrane is accommodated prior to cytokinesis is as microvilli, the biogenesis of which occurs during interphase. During cytokinesis the microvilli unfold and thereby generate the required increases in surface area.

15. Functional Heterogeneity of the Lymphocyte Plasma Membrane Relating to Growth Control in Lymphocytes

K. RESCH, A. LORACHER, B. MÄHLER, M. STOECK, and H.N. RODE

Institut für Immunologie der Universität, Im Neuenheimer Feld 305, 6900 Heidelberg, Federal Republic of Germany

The activation of lymphocytes represents a well-established example of the regulative role of the plasma membrane in the cell proliferation of mammalian cells. Lymphocytes are triggered to grow and divide when only a distinct proportion of plasma membrane receptors interacts with a mitogen such as the lectin concanavalin A (con A). This suggests an intrinsic heterogeneity between mitogenic receptors and bulk binding sites, where mitogenic receptors exhibit higher affinities for the mitogen. To test whether this reflects differences in the molecular architecture of the plasma membrane, small homogenous plasma membrane vesicles isolated from calf and rabbit thymocytes (pure T cells) were fractionated by affinity chromatography on con A-sepharose. Two major membrane fractions were obtained: fraction 1 (MF1), which eluted freely from the affinity adsorbens, and fraction 2 (MF2), which was retained on con A-sepharose, and was eluted by mechanical dissociation. Rechromatography revealed 80-90 % homogeneity of both fractions. Both membrane subfractions exhibited the original orientation; i.e., outside out. Successful separation of plasma membranes from homogeneous tumor cell lines suggested that MF1 and MF2 represented different areas of an individual cell. Both membrane fractions showed the same number of binding sites for con A; the affinity in MF2 was higher than in MF1. The fractions were distinguishable by their content of membrane-bound enzymes: Adenylcyclase showed no preferential location. Mg^{2+}-ATPase was concentrated in MF1. In MF2 (containing the high affinity receptors) the specific activities of K^+, Na-ATPase, p-nitro-phenyl phosphatase, and acyl coA:lysolecithin acyltransferase were higher that in unfractionated membranes, whereas MF1 exhibited lower specific activities for these enzymes.

Evidence has been accumulating that mitosis in lymphocytes is initiated by the interaction of a ligand, i.e., con A with high-affinity receptors. Membrane areas that carry the high-affinity receptors (for con A) are distinguishable from the bulk membrane, suggesting a mosaicism of the supramolecular organization of the plasma membrane in the vicinity of the mitogenic receptors.

16. Microfilaments in Dividing *Xenopus laevis* Tadpole Heart Cells

U. EUTENEUER, J. BEREITER-HAHN, and M. SCHLIWA

Arbeitsgruppe Kinematische Zellforschung, Universität Frankfurt/Main, Federal Republic of Germany

After standard fixation and embedding procedure for electron microscopy, two types of fibrous components can be detected within the mitotic apparatus of *Xenopus laevis* tadpole heart cells in primary tissue culture: 24 nm microtubules and a smaller number of 6 to 8 nm filaments. Microfilaments appear in all stages of cell division from prophase to telophase. They are scattered singly among all "kinds" of microtubules (kinetochore microtubules, aster microtubules, and other nonkinetochore microtubules). They normally can be observed even among the microtubules of the midboy. Since there are only a few reports that describe filaments, be they actin-like or not, in mitotic cells not incubated with heavy meromyosin, it seems necessary to stress that they are present during the whole course of mitosis. It is conceivable that they are a normal constituent of the spindle in *Xenopus laevis* tadpole heart cells. We suggest that they are involved in the function of the mitotic apparatus.

17. Telomeric Associations of Human Male Lymphocyte Chromosomes

L. HENS, M. KIRSCH-VOLDERS, and CH. SUSANNE

Instituut voor Morfologie, Anatomisch en Embryologisch Laboratorium, Eversstraat 2, 1000 Brussels, Belgium

Nonrandom distribution of human chromosomes in metaphase spreads is a well-known phenomenon. Association of the acrocentric chromosomes is especially extensively studied. In these studies the position of the centromers is considered as a relevant parameter for the localization of the chromosome in the metaphase plane. It may, however, be questioned whether telomeric associations exist in human metaphases. These telomeric associations were indeed observed in different plant cells, where they reflect the interphase chromosomal organization.

In a first approach we analyzed telomeric associations (distances between 2 p ends or distances between 2 q ends) in 99 lymphocyte metaphase spreads obtained from 10 normal human males. The relative position of the trypsin-banded chromosomes was calculated in a statistically reduced metaphase plane ("generalized distance" technique). Two reference distributions were calculated: the first one represents the histogram of all possible combinations of pp generalized distances (G.D.) in the 99 metaphase planes; the second one, all possible combinations of qq G.D. in all metaphase spreads. For pp analysis as well as for the qq analysis 276 histograms were constructed representing each possible combination of two pairs of homologous chromosomes and the 23 intrahomologous combinations. Moreover, we calculated 44 histograms concerning each combination of X or Y with the 22 other chromosomes. These 640 histograms were compared with their corresponding reference distribution using a χ^2 test.

Most of acrocentric chromosome combinations and the combinations with chromosome numbers 1-21, 18-21, 19-22, 20-21, and 20-20 lie with their p as well as with their q arms nearer to each other than would be expected by the reference distribution (association). Chromosomes 20, 14-15, 13-17, 13-18, and 17-21 showed associations only with their p arms. Q arms associations were observed only for chromosomes 3, 15, 9-15, 15-16, 19-20, 19-21, and 21-24.

18. Ultrastructural Analysis of the Spindle of *Cryptomonas*

B.R. OAKLEY

York University, Faculty of Science, 4700 Keele Street, Downsview, Ontario M3J 1P3, Canada

Serial sections of mitotic spindles of the marine cryptophycean alga, *Cryptomonas*, were analyzed to determine what types of microtubules were present and which of these came close enough to each other (50 nm or less) for cross-bridging to occur. Interpolar microtubules were rare (< 1 %) but from prometaphase through anaphase there was a substantial interpolar framework of free and polar microtubules that came close enough to one another to cross-bridge and generate anaphase spindle elongation by intermicrotubule sliding. However, simple sliding was not sufficient to account for the extent of spindle elongation, and some additional polymerization of microtubules during elongation would be necessary. Cross-bridging between chromosomal and interpolar framework microtubules is unlikely to function in chromosome-to-pole movement because only about 12 % of the chromosomal microtubules came within bridging distance of interpolar framework microtubules. An analysis of microtubule lengths shows that some polar microtubules lengthen during anaphase while chromosomal microtubules shorten. The spindles contained 5nm-diameter microfilaments associated with chromosomal microtubules and a specific model for the possible involvement of these filaments in mitosis is presented.

19. Ultrastructure of a "Nuclear Mitotic Cycle" in Unfertilized Sea Urchin Eggs After Activation with Ammonia

N. PAWELETZ and D. MAZIA

Institut für Zellforschung, Deutsches Krebsforschungszentrum Heidelberg, Federal Republic of Germany
Department of Zoology, University of California, Berkeley, USA

As shown by Mazia, treatment of unfertilized eggs with ammonia activates a partial mitotic cycle in which the complete chromosome cycle proceeds without the formation of a polarized mitotic apparatus and the separation of the chromosomes. The present study is an electron microscopic examination of this "nuclear mitotic cycle"; essentially, we are examining those features of the mitotic apparatus that are organized in the absence of centrioles.

The chromosomes condense, the nuclear envelope breaks down, and nucleoli disappear. As the nuclear membrane breaks down, the condensing chromosomes are contained in a relatively large clear zone. This zone is seen as a dense coherent mass of vesicles of the endoplasmic reticulum. After the breakdown of the nuclear envelope, microtubules are observed in the clear zone, although centrioles are never seen. Some microtubules are seen in close connection with chromosomes. However, the sites of connection cannot clearly be identified as kinetochores. Microtubules also are observed in the clear zone without any relation to chromosomes. They do, however, seem to be related to clusters of osmiophilic material such as are normally seen in the neighborhood of centrioles in the asters of fertilized eggs. The chromosomes split but the sister chromatids are not moved apart. At the end of the cycle, the chromosomes decondense and a nucleus is reconstituted.

In this nuclear mitotic cycle we observe the formation of all the recognized components of a mitotic apparatus and yet none of the ordering characteristics of a normal spindle that moves chromosomes to poles. The mass of vesicles that make up the bulk of the normal mitotic apparatus of the sea urchin egg is formed. Microtubules are formed. Parts of chromosomes seem capable of organizing microtubules, but the microtubules are not ordered as spindle fibers; whether they originate from discrete kinetochores is still uncertain. Microtubule-organizing centers exist in the clear zone, as osmiophilic structures surrounded by many microtubules. So far as we can see, the only components of a normal mitotic apparatus that are lacking in the eggs activated with ammonia are the centrioles, to which must be assigned the commanding role in the structuring of a mitotic spindle in which chromosomes can be deployed in these cells.

20. New Approaches to the Study of Mitosis in Cellular Slime Molds

U.-P. ROOS

Institut für Pflanzenbiologie, Cytologie, University of Zurich, Switzerland

Recent ultrastructural observations of mitosis in cellular slime molds have revealed a spindle apparatus particularly suited for investigations of the role which microtubules play during chromosome movement and spindle elongation. However, these observations depend on random sections of fixed amebae, whereas it would be desirable to combine observations on living cells with an ultrastructural analysis of spindles sectioned in a predetermined plane. With this goal in mind I have successfully applied techniques that have proved very useful for the study of mitosis in higher eukaryotes.

Amebae of *Dictyostelium discoideum* and *Polysphondylium violaceum* were grown to log phase in liquid medium. Chambers fashioned from microscope slides and coverslips were inoculated with a drop from such a culture and sealed. Amebae remained alive and capable of dividing for up to 24 h. Dividing amebae remained flat and could be easily observed and photographed with Normanski interference contrast or phase contrast. Light microscopic observations were also carried out on amebae fixed after attachment of a microscope slide or coverslip, or pelleted in agar after fixation for electron microscopy. Nuclei in anaphase or later stages of division could easily be distinguished by their characteristic shape, but earlier stages of mitosis were not always clearly identifiable.

For electron microscopy, amebae were seeded onto coverslips coated with carbon or treated with polylysine and left to attach firmly. Following standard fixation and embedding in a silicone-rubber mold the epoxy wafers were scanned with the light microscope for mitotic cells. Dividing amebae, being flatter than amebae in interphase, were easily recognized. Selected cells were photographed with phase contrast and prepared for thin sectioning. Serial sections were cut parallel to the surface of the blocks, but since the spindle axis of mitotic cells is most often approximately parallel to the substrate, it is possible to choose the plane of sectioning at will and to perform a detailed analysis of the mitotic apparatus of different mitotic stages.

[Supported by the Swiss National Science Foundation.]

21. Structural Design Features of the Mitotic Apparatus from Sea Urchin Eggs

TH.E. SCHROEDER

Friday Harbor Laboratories, Friday Harbor, Washington, 98250, USA

A sequence of Normaski photomicrographs illustrates the changing appearance of the mitotic apparatus (MA) from eggs of *Strongylocentrotus droebachiensis*, a sea urchin. Eggs were lysed in hexylene glycol at selected stages during the normal development and decline of the MA of first cell division. Six topographically and structurally distinct zones of the MA are seen: a) chromosomes, b) half-spindles (truncated cones of microtubules and vesicles), c) aster "shells" (hollow radiating arrays of microtubules and many vesicles), centrospheres (cavities within aster "shells" virtually devoid of microtubules and vesicles), e) centrosomes (granular masses surrounding centrioles), and f) the interzone (appearing at anaphase and nearly devoid of microtubules and vesicles).

These images serve as reminders of sometimes-overlooked details of organization in the sea urchin MA: (1) The poleward margins of half-spindles are *planes*, not points. (2) Asters expand greatly from late metaphase onward, apparently as a function of expansion of centrospheres rather than of thickness of aster shells. (3) Telophase aster shells are very fragile. (4) Chromosomes move only about halfway to the poleward margins of half-spindles during anaphase.

(5) Further chromosome separation correlates with progressive cell cleavage. (6) The geometric center of an aster shell (a) is not occupied by the centrosome (which is more equatorial) and (b) does not correspond to the focal point of the half-spindle (which is more poleward). (7) The interzone is a fragile zone but can be isolated at late telophase when surrounded by the cortex of the cleavage furrow. (8) The fibrous material of half-spindles declines rapidly during telophase, as nuclei gradually reform.

Proposals about the function or operation of the MA must take into account the changing patterns in its organization and considerable heterogeneity.

22. Ultrastructural Changes of the Mitotic Apparatus of HeLa Cells Treated with a Lipolytic Drug

N. SENNINGER and N. PAWELETZ

Institut für Zellforschung, Deutsches Krebsforschungszentrum, Heidelberg, Federal Republic of Germany

Earlier investigations on chicken fibroblasts in vitro after treatment with the lipolytic drug SP 54 demonstrated an inhibition of mitosis. SP 54, a heparinoid, normally is used in human medicine to treat, in particular, fatty embolies and hyperlipemias. To study the effects of this drug on human tumor cells, monolayer cultures of HeLa cells have been used. The range of dosage was 12.5 to 100 mg/ml at different intervals from 5 min to 8 h.

The mitotic index doubles during the first hour of treatment. The distribution of the mitotic stages shows an increase of metaphases from 30 % in controls up to 70 % after 8 h treatment. The electron micrographs shown depict some of the most characteristic alterations of the chromatin and kinetochore region as well as of the centriole and the pericentriolar area: All elements of the mitotic apparatus are present in all mitotic stages, but they exhibit marked changes in their ultrastructure and disarrangements in their relation to each other.

The chromatin is heavily condensed. The kinetochores have lost their clear three-laminar structure. Microtubules, although abundantly present, have no contact to the kinetochores. A direct connection between the tubules of the centrioles and the spindle microtubules can often be found. The centrioles themselves have lost their clear-cut appearance and the pericentriolar osmiophilic material is densely accumulated around the centrioles.

From these results we can draw the following conclusions: The main effect of mitotic inhibition seems to be at the site of the chromatin and, primarily or secondarily, of the kinetochores. The lack of contact between microtubules and kinetochores and the increased condensation of chromatin are responsible for the arrest at metaphase. Assuming that the most drastic changes occur at the most active sites, the results offer additional hints that the kinetochore region is an organizer of the mitotic spindle.

23. Normal and Inhibited Mitosis in Cleaving Eggs of Urodela

P. SENTEIN and Y. ATES

Laboratoire d'Histologie et d'Embryologie, Faculté de Medicine,
34060 Montpellier Cedex, France

I. Normal Segmentation Mitosis - Light Microscopy. The mitotic cycle is divided in two parts: a first period (formation of the spindle, peripheral regression of the asters), named *internal fibrillogenesis*, and a second period (expansion of the asters, equatorial regression of the spindle), named predominantly *external fibrillogenesis*.

The regression of the asters is accompanied by formation of a peripheral network, interpreted as a new organization of microtubular material. The equatorial regression of the spindle is followed by the formation of a clear zone or diastema, which plays a role in the determinism of furrowing. During the second period the astral fibers go through the vitelline platelets up to the cortex.

II. Normal Segmentation Mitosis. Electron Microscopy of Centrospheres. The material that surrounds the centrioles changes during the cell cycle. The *dense bodies* appear at teloprophase and disappear at late prophase. The *striated bodies* disappear later. The diameter of the centrosphere is maximal at metaphase and minimal at telophase. The centrospheres are more dissociated, and vesicles become larger and more numerous at anaphase and telophase.

III. Normal Segmentation Mitosis. Electron Microscopy of Kinetochores. Two kinetochores, to which bundles of microtubules are attached, are oriented toward each pole at metaphase and the two chromatids separate at the onset of anaphase.

IV. Segmentation Mitosis After Treatment by Spindle Inhibitors. Light and Electron Microscopy.
1) Short treatment by some spindle inhibitors or long treatment by a weak inhibitor produces a separation of a truncated spindle from the two asters and centrospheres (chloralhydrate 0.05 M 2 h ; ethylaminodesacetylcolchicid, 0.5×10^{-3}M, 1 h ; amphetamine sulfate 1/500, 8 h). The conclusions are: a) the inhibition of microtubule assembly begins in the polar regions and does not involve the equatorial portion of the microtubules, b) this equatorial portion may be rapidly polymerized soon after the breakdown of the nuclear envelope.

2) A stronger and longer treatment with chloralhydrate produces star metaphases, which are not monopolar mitoses (*Pleurodeles waltlii* Michah.). The star configuration of the chromosomes is determined by the presence of a *spindle remnant* or *"common mass"*, which embeds the kinetochores and contains unoriented microtubules not organized in bundles. The chromosomes form numerous lateral protuberances. The centrospheres are surrounded by cisternae from endoplasmic reticulum or they contain vesicles and/or concentric membranes. They are not related to a spindle remnant ("inactivated" centrospheres).

3) A strong treatment with quinoline (0.46 M, 15-30 min) produces a "d-mitosis," in which centrospheres are "blocked." An accumulation of *dense bodies* of increased volume characterizes these blocked centrospheres. These dense bodies persist during the cell cycle. They can fuse into dense masses. There are no membranes around or within the centrospheres. Chromosomes have no protuberances, but are sticky.

4) Colchicine 2×10^{-5}M (16 h) inhibits the formation of a bipolar achromatic apparatus, but it does not stop the cell cycle at metaphase, since anaphasic and telophasic transformations can occur

(eggs of *Triturus helveticus* Raz.). The spindle is dissociated, but its fibers can be partially preserved and bundles of microtubules can persist. Apolar mitoses and nuclei can be seen at all the phases of the cell cycle. In contrast to cycloheximide, colchicine is not phase dependent.

5) Chloralhydrate 0.048 M (8 h) does not completely dissociate the achromatic apparatus, nor does it produce many star metaphases, centered by a spindle remnant. The star metaphase can be transformed into a rosette nucleus. With 0.065 M (11 h 30 min) the chromosomes are dispersed and sometimes altered (eggs of *Triturus helveticus* Raz.).

6) Phenylurethane (saturated solution) rapidly and partially inhibits the achromatic apparatus. Its rapidly reversible action results in a pluripolar system at all phases of the cell cycle. Continuous and short treatment produces star metaphases, which are truly monopolar metaphases with centrioles in the middle.

24. Unusual Component of the Closed Mitosis of the Plasmodial Nuclei of *Physarum polycephalum* (Myxomycetes)

M. WRIGHT, L. MIR, and A. MOISAND

Centre National de la Recherche Scientifique, Laboratoire de Pharmacologie et de Toxicologie Fondamentales, 205, Route de Narbonne, 31078 Toulouse Cedex, France

Physarum plasmodial mitosis is partially intranuclear, allowing a clear separation of cytoplasmic and nuclear components. Although nuclei are small (5 µ) it is possible during mitosis to distinguish by electron microscopy: microtubules, chromosomes, kinetochores, and nucleolar remnants. Another element was found. Its biochemical nature is still unknown. It appears as a darker zone than the surrounding nucleoplasm, but clearer than chromatin. Invisible or almost inapparent during metaphase, around the metaphase plate, it becomes clearly visible during proanaphase in the depth of the furrow that appears between the two chromosomal masses moving away from one another. In late anaphase, this zone is easily visible between the two sets of chromosomes. This material, not clearly visible during metaphase, seems to condense during proanaphase, giving rise to the core that remains in the middle of the nucleus in late anaphase. The condensation of this material could be involved in the mechanism leading to the early anaphase chromosomal movement. In this hypothesis, the only role played by microtubules would be distribution of chromosomes in two identical sets at this stage of mitosis.

25. Immunocytochemical Visualization of Microtubules and Tubulin at the Light- and Electron-Microscopic Levels

M. DeBRABANDER[1], J. DeMEY[2], M. JONIAU[3], and S. GEUENS[1]

[1]Laboratory of Oncology, Janssen Pharmaceutica Research Laboratories, B-2340 Beerse, Belgium
[2]Laboratory of Biochemical Pathology, Free University of Brussels, (V.U.B.), B-1640 St. Genesius Rode, Belgium
[3]Interdisciplinary Research Centre, Catholic University of Louvain, B-8500 Kortrijk, Belgium

An antibody against rat brain tubulin was purified by affinity chromatography and used for the immunocytochemical staining of tubulin-containing structures in tissue-cultured cells at the light- and electron-microscopic levels. For this purpose the unlabeled antibody enzyme method (PAP method) was used. Tissue-cultured cells were fixed with glutaraldehyde or formaldehyde postfixed with acetone and stained with the PAP procedure. With the light microscope, a dense network of darkly stained fibers was visible in nondividing cells. The fibers converged toward a point in the nuclear vicinity and radiated throughout the cytoplasm toward the cell periphery. This cytoplasmic microtubule complex could be demonstrated both in nontransformed and in transformed cells. Mitotic spindles were also stained heavily in all stages. The cytoplasm of nondividing and dividing cells contained a variable amount of diffuse staining. The nucleus was unstained. The network could be destroyed by treating the cell with microtubule inhibitors. An intense diffuse staining was visible in the cells treated with colchicine or nocodazole. The bundles of 10 nm filaments were unstained. The paracrystalline tubuline precipitates that appeared in cells treated with vinblastine were heavily stained.

Preliminary ultrastructural observations were done on sections through cells that were fixed with glutaraldehyde, stained with the PAP procedure, and embedded in Epon. The microtubules were seen to be covered with stained PAP complexes. The cortical microfilaments, the 10 nm filaments, and the plasma membrane were unstained as were the contents of the nuclei, the mitochondria, the endoplasmic reticulum, the Golgi elements, lysosomes, and vacuoles. A variable amount of diffuse staining was visible in the cytoplasm and the membranes of several organelles often showed PAP deposits. This staining was presumed to localize tubulin in nonmicrotubular form. It is indeed probable that at least part of this staining was immunologically specific since no PAP complexes were seen in the cells when the antitubulin antibody was omitted from the procedure or when it was replaced by an unrelated antibody solution.

In cells treated with microtubule inhibitors only diffuse cytoplasmic staining was left both in interphase and mitotic cells. The vinblastine-induced tubulin crystals were heavily stained. Cytoplasmic staining was more intense and could be demonstrated with higher antibody dilutions in cells treated with cholchicine or nocodazole than in cells treated with vinblastine. The bundles of intermediate filaments were unstained. The following structures, which are probably composed of tubulin, were not stained with this method: the centrioles, the microtubules in the midbody, and the macrotubules induced by vinblastine. This is supposedly due to the close association of nontubulin molecules with these structures hampering the access of the antitubulin antibody.

26. Reversible in Vitro Polymerization of Tubulin from Ehrlich Ascites Tumor Cells

K.H. DOENGES[1], B.W. NAGLE[2], A. UHLMANN[1], and J. BRYAN[2]

[1]Institute for Cell Research, German Cancer Research Center, 6900 Heidelberg, Federal Republic of Germany
[2]Department of Biology, University of Pennsylvania, Philadelphia, Pennsylvania 19174, USA

Microtubules from nonneural cells (Ehrlich ascites tumor cells) could be assembled in vitro by two cycles of polymerization and depolymerization. The formation of microtubules is favored by 4 M glycerol. The tubules are morphologically identical to neurotubules and are temperature sensitive. In contrast to cycled tubulin from brain, neither high-molecular-weight (HMW) components nor 36/30 or 20S species (rings) have been observed. Gel permeation chromatography of the depolymerized material yielded a fraction with an $S^{o}_{20, w}$ value of > 200 and another fraction of 6S. Sodium dodecyl sulfate-polyacrylamide gel electrophoresis of the protein in the gel fraction with the higher s value showed that it consists almost entirely of tubulin, the rest being distributed in 20-25 minor bands. The other fraction is composed of tubulin and a larger amount of nontubulin proteins. On electron-microscopic examination the first fraction contains large aggregates consisting of dense clusters of filaments. These aggregates had no tendency to form microtubules whereas the 6S fraction could be easily polymerized into microtubules. The results demonstrate that HMW proteins are not essential for the in vitro formation of microtubules from ascites tumor cells and that rings are neither intermediates nor required nucleation centers in tubule assembly.

27. Separation, N-Terminal Sequences, and C-Terminal Characterization of Subunits from Pig Brain Tubulin

M. LITTLE, H. PONSTINGL, and E. KRAUHS

Institut für Zellforschung, Deutsches Krebsforschungszentrum, 6900 Heidelberg, Federal Republic of Germany

Tubulin subunits were obtained in μmolar quantities by chromatography on hydroxylapatite in 0.1 % SDS using a linear gradient from 0.20 to 0.40 M sodium phosphate pH 6.4. Sequences were determined in an automated sequencer using 0.1 M quadrol as a buffer. The N-terminal regions of both subunits are homologous and appear to be strongly conserved during evolution. α and β are present in equimolar amounts according to the yields of phenyl-thiohydantoins on automated degradation of native tubulin.

By hydrazinolysis, Glu was found to be the C-terminal residue in α tubulin. This is relevant to reports of an enzyme in brain supernatant that adds tyrosine to the C-terminal residue of α tubulin in the presence of ATP or GTP. Since our isolation buffer does not contain nucleotides, the terminal Glu should reflect the original state of tubulin.

28. Immunofluorescence Microscopy of the Anastral Mitotic Apparatus of *Leucojum* Endosperm Cells after Reaction with Antibodies to Tubulin from Porcine Brain

W. HERTH[1], E. SEIB, W.W. FRANKE, M. OSBORN, and K. WEBER

[1]Lehrstuhl für Zellenlehre der Universität Heidelberg, 6900 Heidelberg, Federal Republic of Germany

Structures binding an antibody against tubulin from porcine brain were localized in the giant anastral mitotic apparatus of endosperm cells of the monocotyledonous plant, *Leucojum saetivum*, by indirect immunofluorescence microscopy. Both continuous and chromosomal spindle fibers were strongly stained. Positive fluorescence was also noted in polar cap regions and, in prometaphase stages, to some extent at the fragmented nuclear envelope. Intermingling and branching of subfiber elements was frequently noted. These studies demonstrate an antigenic relation between tubulin from porcine brain and the tubulin in the spindle apparatus of a higher plant.

29. Tubulin Immunofluorescence in Mitotic Spindle and Unpolymerized Spindle Subunits

H. SATO[1], Y. OHNUKI[2], and K. FUJIWARA[3]

[1]University of Pennsylvania, Dept. of Biology, Philadelphia, Pa. 19174, USA
[2]Pasadena Foundation for Medical Research, Pasadena, California 91109, USA
[3]Dept. of Anatomy, Harvard Medical School, Boston, Massachusetts 02115,USA

Direct and indirect immunofluorescence of spindle microtubules and unpolymerized microtubule subunits was attempted in various dividing cells with antibody prepared against mitotic tubulin paracrystals (vinblastine crystals) isolated from sea urchin gametes. The fluorescent images provide clear differentiation between tubulin organized into spindle fibers and astral rays and amorphous material in the nonbirefringent area around the spindle.

In the treated tissue culture cells of salamander lung epithelium and rat kangaroo Ptk2 intense fluorescence surrounds the nucleus reflecting the accumulation of tubulin in the perinuclear "clear zone" in prophase. In prometaphase to metaphase, the fluorescent image of the spindle is similar to the in vivo polarized light image of the birefringent spindle. Chromosomes show no affinity for either antibody or anti-IgG. In anaphase, asters and chromosomal spindle fibers stain more intensely than the interzonal spindle region. Amorphous material, which presumably reflects the amount of unstructured tubulin, increases in prophase, decreases during prometaphase, reaches minimum at metaphase, starts increasing during anaphase, and finally becomes maximum at late ana- to telophase.

Immunofluorescent reaction is specific; however, polarized light and electron-microscopic examination revealed dissociation and artificial clumping of "riddled microtubules" during the procedure. As much as 50 % of unstructured tubulin may also be leached out during this period. Plant spindles, such as dividing endosperm cells of *Haemanthus katherinae* or pollen mother cells of *Lillium* sp., show no affinity to the present antibody, suggesting an interesting phylogenic problem. We believe this technique provides important information on mitosis, complementing data obtained by polarized light. (Supported by grants NIH 9 RO1 GM 23475-11 and NSF BMS-7500473.)

30. Effects of Cold Treatment on Cell Functions Permitted by Microtubules in Cultured Mouse Fibroblasts

L. SCHIMMELPFENG and J.J. PETERS

Institut für Genetik, Weyertal 121, 5000 Köln 41, Federal Republic of Germany

Assembly of microtubules in vitro is inhibited by colchicine or low temperature. We tested the effectiveness of these procedures in vivo. Cold treatment preferentially affected cytoplasmic microtubules; A9 mouse fibroblasts rounded up and resembled morphologically mitotic cells. Cold treatment could be applied for up to 48 h without impairing cell viability. When cells recovered after shifting the temperature to 37°C, they showed a rapid spreading, bleb formation, and ruffling. The cells regained their typical fibroblastoid shape within 1 h. Monolayer cells kept for 24 h at 4°C were protected against rounding up by addition to the medium of 30-50 % glycerol, an agent known to stabilize microtubules in vitro. When incubated at 14°C, freshly trypsinized cells rounded up totally. Moreover, during cold treatment for 30 h at 14°C, attached monolayer cells showed mitotic activity of 25 %.

Rapid restoration of the typical cell shape after the shift to 37°C was unaffected by colchicine (1-80 µg/ml) but was arrested by cytochalasin B.

These results indicate a preferential action of cold treatment on the cytoplasmic microtubules, whereas colchicine exhibits a much greater effect on spindle formation in vivo. In addition, the cytoplasmic events connected with cold treatment and subsequent shape recovery are comparable to those taking place during mitosis.

31. Polymerization Mechanism of Tubulin Purified from Rat Glial-C_6-Cells

G. WICHE, L.S. HONIG, R.D. COLE

Department of Biochemistry and Department of Molecular Biology and Virus Laboratory, University of California, Berkeley, USA

The formation of polymers from purified rat glial-C_6-cell tubulin was studied in the absence of glycerol by electron microscopy. It was found that the assembly of microtubules proceeded via intermediate structures in the form of rings, spirals, and ribbons (planar and twisted), that were very similar to the ones observed during the formation process of hog brain microtubules. The polymer structures observed in the presence of colchicine or vinblastine also closely resembled their hog brain counterparts. The ability of depolymerized microtubule preparations to repolymerize declined rapidly with time; measurements of the half-life by quantitative electron microscopy gave values of approximately 30 min only. Preparations of depolymerized rat glial cell microtubules, which had completely lost their ability to repolymerize on their own, could be "revitalized" by the addition of purified hog brain microtubule-associated protein factors (MAPs). Tubulin, purified to homogeneity by phosphocellulose chromatography neither formed polymers on its own nor after reconstitution with endogenous MAPs. However, polymerization was induced by the addition of heterologous MAPs. Therefore, it was concluded that although the mechanisms of in vitro microtubule assembly of glial cell and hog brain tubulin are similar in their basic features, there apparently are distinct differences in the properties of microtubule-associated protein factors.

32. Translocational Head-to-Tail Polymerization of Actin

A. WEGNER

Department of Biophysical Chemistry, Biocenter of the University of Basel, Switzerland

The irreversible ATP hydrolysis connected with each association step of actin monomers to actin filaments and the polar structure of filaments are prerequisites for a polymerization mechanism in which actin filaments lengthen at one end and simultaneously shorten at the other. This leads to very fast, directed exchange of protomers and to a translocation of the filaments. Without the irreversible step in the association reaction, subunits could exchange only in a slow diffusion-like way at the two ends of a polymer.

We distinguished between the diffusion-like and the translocational type of polymerization on the basis of the large difference in the velocity of subunit exchange. Measurements of the subunit exchange with the aid of radioactively labeled actin molecules were consistent with the translocational type of growth and excluded a diffusion-like polymerization.

33. Altered Controls of Cell Division Following Interrupted Cytokinesis in Cultured Erythroid Cells

P. MALPOIX

Department of Molecular Biology, U.L.B., 67, rue des Chevaux, 1640 Rhode St Genese, Belgium

Normal commitment of hemopoietic cells to differentiate along the red cell pathway implies that the cells are programmed for the synthesis of a specific protein, hemoglobin, and also for a finite number of cell divisions. A homogeneous population of precursor cells isolated from 12-day-old mouse fetal liver purified by immune lysis can be induced by erythropoietin to differentiate in the presence of cytochalasin B at concentrations that completely block cytokinesis and lead to bi- and tetranucleate cells. Removal of the drug, maintenance of the cells in culture for several days, and subsequent renewal of the treatment to inhibit cytokinesis permits the survival of multinucleate cells with enlarged nucleoli. Further growth in normal culture medium lacking cytochalasin permits cell division to proceed in several ways: 1) by extrusion of nuclei surrounded by sufficient cytoplasm to give rise to viable cells, 2) by cell fragmentation ("clasmatosis"), and, 3) by normal and multipolar mitoses. Extruded karyoplasts surrounded by a too-thin shell of cytoplasm fail to survive. Further culture under favorable conditions leads to a population of permanently dividing uninucleate cells resembling proerythroblasts, but which no longer require the presence of erythropoietin for the maintenance of their proliferative state. Nor do they respond to the presence of the hormone by synthesizing hemoglobin. Differentiation can be induced by polar solvents like dimethylsulfoxide, but remains incomplete.

The full experimental procedure from primary to established permanent cultures that can be cloned takes only one month and therefore offers an interesting model for the study of the change in cellular controls of the proliferative state. The requirement for passage through a multinucleate state invites numerous hypotheses, including the possibility of: accumulation of mitotic inducers in the cytoplasm of multinucleate cells, disorganization of membrane sites and cytoskeleton resulting in modified genetic expression, permanent derepression of genomic control sequences that might be related or equivalent to endogenous viral sequences, karyotypic modifications, or specific mutations. The model described is convenient for the analysis of such hypotheses.

34. Chromosomal Organization and Differentiation in *Ascaris*

K.B. MORITZ and G. ROTH

Zoologisches Institut der Universität, Luisenstraße 14, 8000 München 2, Federal Republic of Germany

Under the influence of cytoplasmic factors of the *Ascaris* egg, the chromosomes of the presoma cells break up at specific sites into many segments. Only the intercalary elements separate into chromatids and move to opposite spindle poles where they constitute nuclear envelopes. The akinetic hetrochromatin is resorbed in the cytoplasm. In the germ line all chromatin is preserved, retaining its holokinetic property and continuity.

In *Ascaris megalocephala univalens* the germ line genome is integrated into one *Sammelchromosom* with distally located blocks of germ line-limited chromatin; in *Ascaris meg. bivalens* there are two plurivalent chromosomes haploid. In *Ascaris suis* germ line cells the heterochromatin is scattered over the chromosomes. After diminution heterochromatin is absent from all somatic chromosomes.

The analytical value of the germ line genome size corresponds the kinetic value, determined by DNA reassociation studies using *E. coli* DNA as a standard. Therefore, the germ line genome constitutes the unit genome. Proteinase K incubation of isolated germ line chromosomes does not cause their fragmentation, but endonuclease breaks them. Therefore, the germ line chromosomes contain one single DNA molecule.

The amount of germ line-limited DNA equals the fraction of fast-reassociating satellite DNA. Cot analysis has shown that in the germ line genome there is no appreciable amount of germ line-limited single copy DNA.

In the soma genome, highly repeated sequences are missing.

T_M-, Cot-, buoyant density-, and restriction analyses indicate that the basic nucleotide sequence (s) of the germ line satellites are highly conserved. They may differ in the two species studied. There is evidence for a rectification mechanism acting between chromosomes, which results sometimes in new chromosomes differing in their sat–DNA content.

Diminution is probably not a DNA modification restriction mechanism, because 5MC could not be detected in germ line DNA. In preparing diminution the ultraviolet absorption spectrum of the chromosomes is altered. The protein composition may be modified before chromosomal fragmentation.

35. Keratinization and Growth Regulation in a Bladder Tumor Cell Line

R. TCHAO and J. LEIGHTON

Department of Pathology, The Medical College of Pennsylvania and Hospital, 3300 Henry Avenue, Philadelphia, PA. 19129, USA

We have observed that a rat bladder cancer cell line NBT II spontaneously keratinized in monolayer cultures as well as in suspension cultures or aggregates, in collagen-coated cellulose sponge cultures, and in meniscus gradient cultures. Keratinization only occurred when cell cultures reached high density. Vitamin A, at 5-10 I.U., effectively prevented keratinization and also stimulated mitotic activity of a dense monolayer culture. NBT II cells formed aggregates by the Moscona techniques and showed distinct keratinization after 3 days. Vitamin A also inhibited this process. Cells were grown for over 30 passages in the presence of vitamin A. Yet upon aggregation, in the absence of vitamin A, these cells readily keratinized. We have also shown that BudR inhibited keratinization only if the drug was added to cultures 24 h before aggregation. These results show that 1) cell contact and interaction are key factors in the differentiation of these tumor cells, and 2) the inhibitory effect of BudR and vitamin A on keratinization are different. When NBT II cells divide, the daughter cells always have two options, viz., to divide or to differentiate, depending on the environment presented to the cells. Supported by NIH grant # CA 14137.

36. Dynamics and Ultrastructure of Monocentric Chromosome Movement

R. CAMENZIND and TH. FUX

Department of Zoology, Swiss Federal Institute of Technology, Universitätstraße 2, CH-8006 Zurich, Switzerland

In all gall midges (Cecidomyiidae, Diptera) spermatogenesis is unorthodox. In the first meiotic division in males of *Mycophila speyeri* three chromosomes (sperm chromosomes) segregate from the residual chromosomes by monocentric chromosome movement. In prometaphase a morphologically bipolar spindle is formed. It contains only the three sperm chromosomes. They look like normal metaphase chromosomes with two kinetochores pointing to the two spindle poles. The chromosomal fibers are long individual bundles. In addition to microtubules they contain endoplasmic reticulum-like membranes that are connected to the microtubules by bridges. During anaphase the sperm chromosomes do not divide. Their segregation from the residual chromosomes involves three components:
1. The sperm chromosomes move within the spindle toward the pole nearer the cell surface.
2. The spindle elongates.
3. The spindle as a whole shifts away from the cytoplasmic region containing the residual chromosomes.
Cytokinesis separates the small spermatocyte with the three chromosomes from the large residual cell. The second meiotic division occurs only in the spermatocyte.

37. Dynamics and Ultrastructure of Unorthodox Chromosome Movements

R. CAMENZIND, TH. FUX, P. GANDOLFI, and M. ZANAZZI

Department of Zoology, Swiss Federal Institute of Technology, Universitätstraße 2, CH-8006 Zurich, Switzerland

I. Chromosome Elimination. Chromosome elimination in the 3rd cleavage division of the gall midge *Heteropeza pygmaea* is described. The chromosomes that move all the way to the poles (S chromosomes) are included in the presumptive somatic nuclei, while the lagging chromosomes are eliminated (E chromosomes). Until mid anaphase, the E and S chromosomes cannot be distinguished from each other either morphologically or topologically, and they all behave like chromosomes in a normal cleavage division. After variable amounts of anaphase movement, the E chromosomes return toward the equator with their kinetochores still oriented toward the poles and chromosomal microtubules still present.

II. Restitutive Fertilization. Male-determined eggs of *Heteropeza pygmaea* develop parthenogenetically. After meiosis the metaphase plate of the first cleavage division is formed by the chromosomes of the egg nucleus and of two additional nuclei of maternal origin. There are several ways in which the metaphase configuration can be achieved. In one case the egg nucleus forms a spindle while the two additional nuclei are in close lateral contact with it. Their chromosomes move toward the equator of the spindle. The kinetochores of these moving chromosomes are associated with fewer, but more divergent microtubules than the chromosomes already present in the spindle.

38. Anaphase Chromosome Movement in the Grasshopper Neuroblast and its Relation to Other Similar Studies

J.G. CARLSON

Department of Zoology, The University of Tennessee, Knoxville, Tennessee, USA

The positions of the two sets of chromosome kinetochores, the spindle poles, cell membranes adjacent to the poles, and cleavage furrow (in half the cell) of grasshopper neuroblasts in culture at 38°C were marked at short-time intervals, using a camera lucida. The mean positions at half-second intervals from the end of metaphase to the end of anaphase (or for a maximum of 8 min) were determined. The percent of motion due to poleward movement and spindle elongation, which coincide in time, were calculated for each minute, the former falling from 61 % in the first minute to 15 % in the seventh minute and increasing to 86 % in the final minute, probably as a result of pressure and bending of the spindle. The maximum velocity of a set of kinetochores was 3.41 µm/min and the mean velocity 1.86 µm/s (one-half the rate of separation).

Certain general conclusions based on many related studies can be drawn with respect to poleward movement and spindle elongation as components of anaphase movement and with respect to anaphase velocity in relation to temperature, size of chromosomes, kinds of cells and the distance the chromosomes move.

39. The 1977 Attempt to Formulate an Assembly Hypothesis of Spindle Formation and Spindle Function

R. DIETZ

Max-Planck-Institut für Zellbiologie, Melanchthonstraße 36, 7400 Tübingen, Federal Replic of Germany

The hypothesis follows the possibility that assembly and disassembly of the spindle represent the immediate cause of mitotic transport. Since there are divisions in which the kinetochores are the only spindle-organizing centers, and since kinetochores are firmly anchored within the spindle through their chromosomal spindle fibers, the following assumptions are made: In producing a diffusable substance, mitotically active kinetochores surround themselves with a field in which assembly-disassembly rates change with distance; assembly being maximum next to kinetochores, disassembly being maximum farthest away from kinetochores. Kinetochores expose a number of sites that readily bind tubulin and act as nucleation sites. Since chromosomal spindle fibers react upon moderate mechanical strain with an almost elastic response, it is assumed that subsequent microtubule ends are interconnected through islands of a reticular polymer and that transition between the two phases is not only concentration-dependent but modified according to the principle of Le Châtelier. If, however, the strain exceeds a critical value, microtubules are thought to collapse cooperatively. The resulting free subunits and oligomers shift the equilibrium locally toward polymerization. In other words, tension-induced collapse of spindle microtubules causes the stabilization or even the growth of a few adjacent microtubules. In making these assumptions, a large body of mitotic phenomena becomes deducible, including chromosome movement, chromosome orientation, akinetochoric transport, and phase-dependent changes of spindle architecture.

40. Microcinematographic Studies on Cell Division in Cultures of Spleen Cells

G. MUNGYER and C. JERUSALEM

Laboratorium für Cyto- und Histologie, Universität Nijmegen, Geert Grooteplein Noord 21, Nijmegen/The Netherlands

Time-lapse photographs of spleen cells from mice on the 7th day of a malaria (*Plasmodium berghei*) infection, grown in tissue culture, were taken.

The cell population consisted of large, often multinucleated fixed cells and rapidly moving smaller ones with very active ruffling membranes.

A mitotic division was recorded inside a vacuole surrounded by the cytoplasm of a large multinucleated cell. The two daughter cells showed an intense motility but remained in close contact with the large fixed cell for about 140 min.

Several further cell divisions occurred among the actively moving cells around the multinucleated fixed cell, whereby the dividing cells lay mostly in close contact with the large cell.

From our time-lapse records we were unable to determine whether the cell in the first mitosis described was completely surrounded by the cytoplasm of the multinucleated cell; but the regular observation that dividing cells lie in close contact with the larger, less motile cell points to a possible metabolic cooperation, e.g., interactions between (re-) circulation and fixed immunocompetent cells, preceding and secondary to sensibilization by antigens, where such interactions appear to be a crucial step.

41. Birefringence in Mitosis and Meiosis   (16 mm movie)

H. SATO[1], Y. OHNUKI[2], and K. IZUTSU[3]

[1]University of Pennsylvania, Department of Biology, Philadelphia, Pa. 19174, USA; [2]Pasadena Foundation for Medical Research, Pasadena, California 91109, USA; [3]Mie University, School of Medicine

Mitosis in salamander (*Taricha granulosa*) lung cell cultures, and meiosis in grasshopper (*Chrysochraon japonicus*) spermatocytes were recorded by time-lapse cinematography using rectified polarized light optics. Images of phase contrast microscopy and differential interference microscopy were inserted for comparison. In both cases we observed 1) the prometaphase spindle, which is composed mainly of continuous fibers and astral rays, elongates past the nuclear envelope to the interior of the nucleus and rapidly gains in birefringence as it lengthens. The advancing pole of the spindle occasionally undergoes saltation, tilts, and swings. 2) The prometaphase spindle is not a stable structure. It twists, jerks, swims, and sometimes migrates within the cell with the chromosome attached. 3) Occasionally centrosomes of both poles in the weakly birefringent, over-elongated metaphase spindles, appear to split and then form three tandem spindles. 4) However, the spindle becomes stable at full metaphase and the poles show little oscillatory movement during the anaphase. 5) Anaphase chromosomes move at constant velocities of 1.2-2 μm/min at 24°C. This rate is not affected by the amount of birefringence of spindle fibers to which the chromosomes are attached. 6) Lateral oscillation of continuous spindle fibers is seen in the interzonal region. This phenomenon is comparable to the observation of dividing endosperm cell of *Haemanthus katherinae*. 7) Occasionally, leftover centrosomes from previous division become active, migrate toward the newly developing spindle, and so organize a tri- or tetrapolar spindle with or without chromosomes. (Supported by grants from NIH 9 RO1 GM 23475-11 and NSF BMS-7500473).

42. The Cytaster, a Colchicine-Sensitive Nuclear Migration Organelle in Insect Eggs

R. WOLF

Zoological Institute I, University of Würzburg, Röntgenring 10, 8700 Würzburg, Federal Republic of Germany

Time-lapse analyses of nuclear multiplication in the eggs of the gall midge *Wachtliella persicariae* L., documented in an educational film, furnished evidence for an autonomous nuclear migration organelle ("migration cytaster"), which originates from a polar cytaster of the mitotic apparatus. The daughter nuclei are pulled actively through the ooplasm by short-term peripheric insertions of their large cytaster, which are combined with tractive forces occurring along the polar rays.

Electron microscopic studies demonstrate that the polar rays are composed of single microtubules. After treatment with colchicine, the nuclei are only passively shifted over short distances by means of rhythmic ooplasmic flows. The resulting inhibition of active nuclear migration gives further evidence that microtubules play an essential part in it. Conversely, under the influence of cytochalasin B, normal active nuclear migration continues, although ooplasmic flow is inhibited. Thus, it can be concluded that the active and the passive modes of nuclear migration proceed independently of each other.

On the assumption that one and the same fundamental mechanism is responsible for generating tractive forces along the astral rays as well as along the spindle fibers, our results are compatible with the assembly concept of spindle function (DIETZ, 1972; INOUE, 1976), whereas the microtubular sliding model (BAJER, 1973; McINTOSH et al., 1969) does not seem to be applicable in the case of single microtubules.

43. Ultrastructure of Pathologic Mitosis in Mice

P. DUSTIN and J.P. HUBERT

Department of Pathology, Université libre de Bruxelles, Brussels, Belgium

1. Metaphase with Polar Chromosomes. This metaphase abnormality, with the smallest chromosomes remaining located at the poles of the spindle, is frequently observed in some human neoplasms (e.g., preinvasive epidermoid carcinomas of the cervix). In mice, the injection of 0.2 mg/g of hydroquinone reproduces this type of metaphase in various tissues. Preliminary ultrastructural studies of hydroquinone mitoses in the Lieberkühn glands of the intestine demonstrate the integrity of microtubules, of the kinetochores of the polar chromosomes, and of the centrioles.

2. Action of cacodylate on mitosis. This arsenical was one of the first chemicals to be demonstrated to arrest mammalian mitosis. Ultrastructural studies of the intestinal mitoses of the mouse indicate that after 1 or 2 h following injection of 2 mg/g of sodium cacodylate all mitoses are arrested, with centrally located centrioles and no spindle microtubules. These results, to be compared with those recently published after diamide poisoning, confirm the role of -SH groups in spindle function and microtubule assembly.

44. Two Different Types of Spindle Pole Organelles in the Life Cycle of Monothalamous Foraminifera

D. SCHWAB

Fachbereich 3, Fachrichtung Anatomie, Universität des Saarlandes, 6650 Homburg/Saar, Federal Republic of Germany

In the monothalamous foraminifer *Myxotheca arenilega* Schaudinn (Rhizopoda, Protozoa) centrioles appear at the poles of the mitotically dividing nuclei during gametogenesis, which leads to the formation of biflagellate gametes. During meiosis, instead of centrioles, centrosomal bodies can be demonstrated. They are spherical, about 4µm in diameter, and are composed of fibrillar and granular material. They are surrounded by a membrane and are localized within the perinuclear space. *M. arenilega* thus exhibits two types of spindle pole organelles that alternatively appear depending on the division stage within the life cycle.

In *Allogromia laticollaris* Arnold during mitosis in agar mounts, centrosomal bodies with the same fine structure and localization as are seen in *M. arenilega* can be observed. During gametogenesis centrioles are present at the dividing nuclei although this foraminifer develops nonflagellate gametes.

During mitosis in *Kibisidytes* sp. centrosomal plaques about 2 µm in length are visible outside the nucleus at the persisting nuclear envelope.

45. RNA and Protein Synthesis During Plant Mitosis

J.F. LÓPEZ-SÁEZ

Instituto de Biologia Celular. Dpto. Citologia C.S.I.C., Velázquez, 144 Madrid-6, Spain

Using synchronous cell populations, the concurrent biosynthesis of certain macromolecules has been demonstrated to be an essential requirement for mitosis in onion root-tip meristems.

Protein synthesis is required by early and middle prophase, and its inhibition induces a morphologic return to nuclear interphase.

RNA synthesis at late prophase appears to be a requirement for nuclear membrane breakdown and subsequent mitosis development.

Nucleolar telophase reorganization appears to be dependent on parallel RNA synthesis, while it is surprisingly accelerated by protein synthesis inhibition.

46. Effects of Low Temperature on Dividing Cells of Plant Endosperm

A.M. LAMBERT[1] and A.S. BAJER

[1] Institut de Botanique, 28, Rue Goethe, 67083 Strasbourgh Cedex, France

Lowering the temperature of dividing endosperm of *Haemanthus* from 24°C to + 3.5°C or below results in reversible arrest of chromosome movements within 2-4 min in all stages of mitosis. These movements can be reinitiated within 2-5 min by returning the cells to room temperature, even if temperature shock is extended for 2 h. Over 95 % of the reversed cells survive, and complete division. The same cells observed in the light microscope during experimental treatment were fixed under the same precisely monitored temperature, and then studied in the electron microscope. This technique permits a precise record of the relation between chromosome behavior and ultrastructure.

Low temperature affects both the number and the arrangement of micro-tubules (MTs). After 10 min at +3.5°C, the total number of spindle MTs is reduced to 1/3 - 1/4 that of anaphase controls. Continuous MTs are no longer present, but some kinetochore MTs persist even after 2 h at + 3.5°C. The arrangement of these persisting kinetochore MTs changes during temperature shock from a divergent configuration (cf. controls) to a parallel one (experimental). These "parallel kineto-chore fibers" of cells arrested at early anaphase contain more MTs than fibers in late anaphase controls, when chromosomes are still moving.

Time-lapse analysis of living cells show that, during temperature shock, kinetochores are either motionless or moving a few μm back-ward, i.e., during MT disassembly and rearrangement.

After returning low temperature-shocked cells to room temperature, the number of MTs increases, and their arrangement reverses from parallel to divergent configuration, and chromosome movement is re-initiated.

Detailed discussion: see Lambert-Bajer. Cytobiologie, 15, 1-23, 1977

47. Chromosomes and Cancer

R. KORSGAARD

Department of Tumor Cytogenetics, the Wallenberg Laboratory, Univer-sity of Lund, S-220 07 Lund and Department of Lung Medicine, Univer-sity Hospital, S-221 85 Lund, Sweden

The presence of chromosome abnormalities in human cancer cells was demonstrated by the German pathologist von Hansemann in 1890. Karyo-type analysis, however, was not applied in research or routine diag-nostics until 1956, when the exact chromosome number of man was es-tablished by TJIO and LEVAN in Lund.

The advantages of the new chromosome-banding techniques, including G-, Q-, N-, and C-banding, have markedly increased the applicability of cytogenetic analysis as a research tool in cell kinetics, growth control, cancer, etc., as well as a tool in clinical diagnostics.

At the departments of Lung Medicine and Tumor Cytogenetics in Lund we have analyzed cytogenetically cells from 1298 pleural, pericardiac, and ascitic effusions during the period 1973 to 1977. Exudates from our own hospital have been examined in vivo and, following cloning and establishment as cell cultures, in vitro. Specimens from outside hospitals have been sent to our laboratory by usual postal service - as cytolysis normally does not affect karyotype analysis within the first week - and analyzed in vivo.

The application of stemline aberrations as a criterion of malignancy has proved a valuable complement to standard cytologic examination of effusions and has even shown itself to be superior to these examin-ations in the case of mesotheliomas and tumors of a low degree of dif-ferentiation. The chromosome-banding techniques have also rendered possible studies of marker chromosomes and comparative cytogenetic studies of various tumors.

The posters demonstrate preparation of chromosome slides, staining techniques, various chromosome-banding techniques, karyotypes from malignant human tumors in vivo and in vitro. Scanning electron microscopy of cells from malignant exudates, and computer displays of a marker chromosome after photometric scanning densitometry.

48. Non Tubulin Molecules in Meiotic Cells

M. HAUSER and B.M. JOKUSCH

University of Bochum and University of Basel

We studied the distribution of tubulin, actin and alpha actinin in meiotic cells of the grasshopper *Locusta migratoria*. Antibodies against brain tubulin, smooth muscle actin and skeletal muscle alpha actinin were elicited in rabbits. Monospecific gamma globulins were purified on sepharose affinity columns and used in indirect immunofluorescence (IIF) with FITC-coupled goat ant-rabbit gamma globulins. Examples for the specificity of the antibodies are given.

Male meiotic cells were prepared from grasshoppers by disrupting the testal tubes in insect Ringer solution and collected by centrifugation onto a glass slide.

For IIF, live meiotic cells were incubated for 7 min in a "lysis buffer" after J.R. McINTOSH, containing 0.03% Triton x 100 and tubulin (1 mg/ml). These cells and formaldehyde fixed cells were then incubated with the first and the second antibody according to the procedure of LAZARIDES and WEBER.

For electronmicroscopy, cells were fixed with a modified aldehyde fixation (M. HAUSER, Manuscript in prep.) and processed for thin sectionning in conventional manners.

Our preliminary results demonstrate the following:
1. Meiotic cells, after proper fixation, reveal a large number of actinlike filaments surrounding meiotic microtubules. In many instances, direct contact between such filaments and microtubules can be seen.

2. Meiotic cells contain actin in their spindles, as seen in IIF. The distribution is similar to that found by J.W. SANGER and J.R. McINTOSH in mitotic cells: the spindle poles show bright fluorescence, the area of which is clearly exceeding the centriales. In contrast to mitotic cells, however, meiotic cells in anaphase show also bright fluorescence between the separating chromosome populations. This reflects probably the formation of the contractil ring, since in meiotic cells cytokinesis starts already in late anaphase. These results were found with formaldehyde fixed as well as with Triton treated live cells.

3. Meiotic cells contain also alpha actinin in their spindles, as revealed by IIF. During prophase, the fluorescence pattern changes considerably, possibly reflecting a redistribution of alpha actinin molecules during some stages of meiotic prophase. Early anaphase cells showed a bright fluorescence throughout the cytoplasm.

Subject Index

ELECTRONIC COMPUTERIZED/CALCULATER.

Radiation and Cellular Control Processes

Editor: J. Kiefer

176 figures. XIV, 321 pages. 1976
(Proceedings in Life Sciences)
ISBN 3-540-07878-9

The book contains the complete and carefully edited papers of a conference held in October 1975 at the Strahlenzentrum, Gießen (Fed. Rep. of Germany). The main emphasis is placed on eucaryotes. The central theme being – though not exclusivley – "Simple systems" (yeasts, algae, slime mould). The material was divided into the following sections: biochemical key processes, phenomena of repair and recovery, cell division and progression. Each section contains an explanatory introduction (Authors: J. M. Boyle, A. R. Lehmann, W. Sachsenmaier), and a further general paper by T. Alper deals with modern trends in radiobiology. As the participants of the conference deliberately choose not to be limited solely to radiobiology, the range of material covers not only radiation – induced processes, but attempts to elucidate general principles of intercellular regulatory processes by means of a study of radiation effects, of both UV and ion irradiation.

R. Rieger, A. Michaelis, M. M. Green

Glossary of Genetics and Cytogenetics

Classical and Molecular

Springer Study Edition

4th completely revised edition
100 figures, 8 tables. 647 pages. 1976
ISBN 3-540-07668-9

The fourth edition of the well-proven Glossary is now complete, and has been completely reworked and enlarged since the third edition. Special care has been given to including the important new terms which have meanwhile come to be used in genetics. The style of the Glossary has been preserved, so that only a definition of scientific expressions is given when that suffices, but when a term can only be understood by more detailed explanation, it is described in a short essay, including scientific data. The inclusion of a short history of the subject, together with the crossreferences and quotations from the literature, make this Glossary a handbook for scientists and a valuable textbook for students. The lower price of this edition as compared to the last should encourage a wide distribution of this new work.

Springer-Verlag Berlin Heidelberg NewYork

Chromosoma

Title No. 412

Editorial Board: H. Bauer (Managing Editor), Erlangen; W. Beermann, Tübingen; J. G. Gall, New Haven; B. John, Canberra; H. C. MacGregor, Leicester; R. B. Nicklas, Durham; Ir. J. Sybenga, Wageningen; J. H. Taylor, Tallahassee; D. von Wettstein, Copenhagen

Advisory Board: T. Caspersson, Stockholm; R. Dietz, Tübingen; J. E. Edström, Stockholm; W. Hennig, Tübingen; T. C. Hsu, Houston; H.-G. Keyl, Bochum; C. D. Laird, Seattle; G. F. Meyer, Tübingen; H. Stern, La Jolla; Ch. A. Thomas, Jr., Boston.

Chromosoma, founded in 1939, publishes original contributions concerning all aspects of nuclear and chromosome research. Current studies in this field range from those on protozoan chromosomes to research on the nuclei of higher organisms and frequently apply techniques and material from fields as diverse as molecular and population genetics. Pertinent biochemical and biophysical approaches to cytological problems are often included.

Differentiation

Official Organ of the International Society of Differentiation

Title No. 258

Editor: Dimitri Viza, Paris

Editorial Board: E. J. Ambrose, London; R. Auerbach, Madison, WI; J. T. Bonner, Princeton, NJ; J. Brachet, Rhode-St. Genese; G. Brand, Minneapolis, MI; A. Braun, New York, NY; F. Chapeville, Paris; D. A. L. Davies, High Wycombe, Bucks; B. Ephrussi, Gif-sur-Yvette; M. Feldman, Rehovot; W. Franke, Heidelberg; H. Holtzer, Philadelphia, PA; N. K. Jerne, Basle; C. Kafiani, Moscow; T. J. King, Bethesda, MD; L. G. Lajtha, Manchester; C. L. Markert, New Haven, CT; R. G. McKinnel, Minneapolis, MI; B. Mintz, Philadelphia, PA; A. A. Moscona, Chicago, IL; C. Pavan, Austin, Sao Paulo; B. Pierce, Denver, CO; A. Ruthmann, Bochum; S. Spiegelman, New York, NY; M. Sussmann, Pittsburgh, PA; S. Toivonen, Helsinki; T. Yamada, Lausanne.

Differentiation, an international journal devoted to the problem of biological diversification, focuses on the differentiation of eukaryotes at the molecular level and is directed toward the coverage of biological differentiation and evolution from the subcellular level to species differentiation. It publishes original contributions in the following areas: embryonic differentiation, normal cell growth and division, carcinogenesis and the cancer problem as an aspect of cell differentiation, inter-tissue reactions in vivo and in vitro, genetic mosaicism, nucleocytoplasmic interactions, cell hybridization, membrane control of the cell, plant evolution and differentiation, immunological events relevant to differentiation. The journal is the official organ of the International Society of Differentiation.

Springer-Verlag
Berlin
Heidelberg
New York